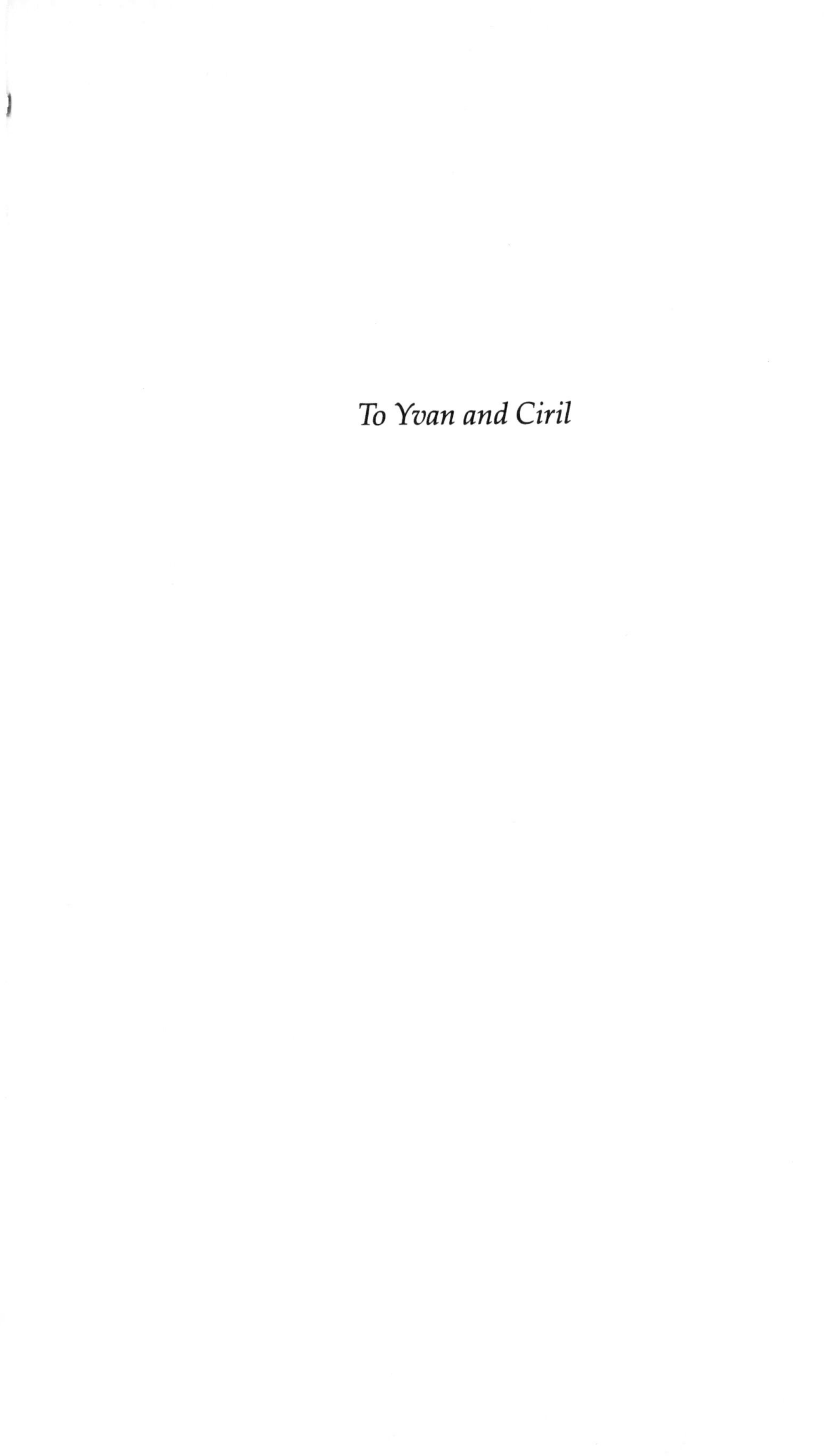

To Yvan and Ciril

Marie-Françoise Schulz-Aellen

Aging and Human Longevity

Birkhäuser
Boston • Basel • Berlin

Marie-Françoise Schulz-Aellen
Institutions Universitaires de Gériatrie
et de Psychatrie de Genéve
Domaine Belle-Idée
Chêne-Bourg, Genéve
CH-1225 Switzerland

Library of Congress Cataloging In-Publication Data

Schulz-Aellen, Marie-Françoise, 1947-
Aging and human longevity / Marie-Françoise Schulz-Aellen.
p. cm.
Includes bibliographical references and index.
ISBN 0-8176-3875-X (hard : alk. paper). - ISBN 0-8176-3964-0 (soft : alk. paper)
1. Aging. 2. Longevity. I. Title.
QP86.S356 1997
612.6'7-DC20 97-17985
CIP

Printed on acid-free paper
 Birkhäuser 

ISBN 0-8176-3875-X (Hardcover) ISBN 0-8176-3964-0 (Softcover)
ISBN 3-7643-3875-X (Hardcover) ISBN 3-7643-3964-0 (Softcover)
Typset by Best-Set Typesetter Ltd., Hong Kong
Cover art "Metamorphoses" by Geneviève Delaunay
Printed and bound by Maple Vail Press, York, PA
Printed in the U.S.A.

9 8 7 6 5 4 3 2 1

Acknowledgments

In this book on aging, I had the privilege of reviewing the remarkable accomplishments of many women and men who patiently strive to understand the complexity of the phenomenon of aging at all levels, from the broad societal and psychological to the intricate intracellular. Information in this book comes from many scientific articles and books as well as from enjoyable discussions with scientists and physicians. I express my deepest gratitude to all these people for their support. The problem of aging also triggered fascinating discussions with friends, women and men in their thirties and forties, who are concerned with aging since they will be the elderly people of tomorrow. Their opinions on aging and on elderly people were a source of inspiration. I thank Laurent Roux, Victor Bartanusz, Jose Barral, Ruth Marti, Sandy Baehni, Françoise Mauron, Lore Aellen and Anne Schulz for their helpful suggestions. Claude Rossel played an important role in the implementation of this monograph. This book benefited from the useful editing of Antony Fincham and excellent comments of George Adelman and the skillful drawings of Magali Husler-Leeman. I thank them very much. I am grateful to my husband, Pierre Schulz for his suggestions on the successive versions of the manuscript, and for his constant support. I thank my two sons, Yvan and Ciril, for their patience and understanding throughout the progression of the writing. Their close relationship with their grandparents is a vivid illustration of the type of mutual comfort that different generations bring to one another.

This work was supported by grant No 3234 038856 from the Fonds National Suisse de la Recherche. Additional funding was provided by the Clinique Bon-Port, in Montreux Switzerland.

Dr. Marie-Françoise Schulz-Aellen received her degree in medical biology and her doctorate in molecular biology at the University of Geneva. She did her postdoctoral research at the University of Stanford, Palo Alto, California. She then worked in molecular biology and biotechnology at the pharmaceutical companies Biogen, Ciba-Geigy, and Glaxo, as well as at the University of Geneva. Her current work is on biological mechanisms of stress at the Institutions Universitaires de Gériatrie et de Psychiatrie de Genève, Switzerland. She is the recipient of a Marie Heim-Vögtlin grant from the Fonds National Suisse de la Recherche.

Contents

PART I: THE MECHANISMS OF AGING

INTRODUCTION 3

1 **Variability of life span between species** **7**
Methods to quantify life span 7
Life span of humans 10
Life spans of invertebrates 13
Life span of cells in culture 17
Are some plants or animal species immortal? 23

2 **Factors that influence human longevity** **25**
Genetic influence on longevity 25
Influence of personality, social class and social interactions 27
The role of happiness and marriage 29
Sexuality 30
The role of stress 32
Diseases of premature aging 40
A note on diseases and longevity 44

3 **Structural and physiological changes occurring with age** **51**
Major functional systems 51
Connective tissue 54
Nervous system 55
Endocrine system 60
Biological clocks and circadian rhythmicity 62
Immune system 66
Aging of intracellular functions 70
Homeostasis and multisystem regulation 72

4 **Theories of aging** **77**
Stochastic theories 77
Programmed theories 80
Overview of aging theories 83

5 **Mechanisms of cellular aging** **85**
Intrinsic mechanisms 86
The influence of genes and chromosomes on health and longevity 86

Human genetic diseases due to somatic mutations 95
DNA methylation 110
DNA and protein glycation 111
Garbage molecules 112
Extrinsic mechanisms 112
Free radicals 113
Effects of radiation 119

6 Apoptosis, disease, and senescence **121**

CONCLUSION OF PART I **124**

PART II: THE PROLONGATION OF LIFE

INTRODUCTION **129**

1 Life extension **131**
Life extension in nonhuman species 131
Food restriction in nonhuman species 132
Life extension in humans 136
Human studies on food restriction 140

2 The role of diet in longevity **142**
Recommendations 143
Perspectives 145
Vitamin supplements 145
Mineral supplements 151
Unsaturated fats, omega-3 and omega-6 fatty acids 155

3 Antioxidants and the chemoprevention of cancers and cardiovascular diseases **158**
Vitamins, minerals, coenzyme Q10 158
Aspirin 166

4 Life styles **168**
Social network 168
Sexuality 171
Physical exercise 172
Smoking 177
Alcohol consumption 183
Controlling stress 187

5 Substitution of endogenous compounds and hormones **190**
Melatonin 190
Thymus hormones 193

Sex hormones 193
Dihydroepiandrosterone 196
Human growth hormone 198

6 Medications and alternative medicine 201
Anti-aging medications 201
Alternative medicine 204

7 Surgical strategies 207
Plastic surgery 207
Replacement of body parts 207

8 Perspectives in the detection and treatment of diseases 210
Biotechnology contributions in medicine and gene therapy 210
Contribution of genetic engineering techniques to medicine 214
Application of gene therapy to neurological diseases 218
New targets for drug therapy 225

CONCLUSION **228**

Glossary 233
Relevant books on aging 241
References 247
Index 277

Preface

The proportion of elderly people continues to increase in the western world—nearly a quarter of the population will be over 65 years by the year 2050. Since aging is accompanied by an increase in diseases and by a deterioration in well-being, finding solutions to these social, medical and psychological problems is necessarily a major goal for society. Scientists and medical practitioners are therefore faced with the urgent task of increasing basic knowledge of the biological processes that cause aging. More resources must be put into this research in order to achieve better understanding of the cellular mechanisms that underlie the differences in life span between species and to answer the difficult questions of why some individuals age more quickly than others, and why some develop liver problems, some have heart problems, and others brain problems. The results of such a wide program of research will provide important information about the causes of many life-threatening and/or debilitating diseases of old age; it will help find ways to prevent some of the ailments that result from aging, and it may well lead to discoveries enabling the prolongation of human life.

In this monograph, we discuss the mechanisms of aging at the cellular level, as well as some of the consequences of tissue or system aging in humans. Our goal is to provide recent data on the phenomenon of aging and to give a view of the variety as well as the complexity of the aging processes. We also review interventions and preventive measures, whether they are medical, psychological or social, that will postpone, and perhaps even eliminate, various of the handicaps and diseases associated with old age. As the saying goes, the goal is not to add years to life, but rather to add life to years.

Note to the readers: Words in boldface type are defined in the glossary.

Geneva, June, 1996
Marie-Françoise Schulz-Aellen, Ph.D.

PART I

THE MECHANISMS OF AGING

Introduction

For the first time in the history of mankind, average life span has increased to more than 70 years for men and almost 80 years for women. This dramatic increase in life expectancy is recent and has occurred in developed countries during the course of this century. People often think with nostalgia of the good old days—no traffic-jams, no smog, no noise—but forgotten are the threats of high infant mortality and early adult death in these earlier times in which prenatal and postnatal medical care, drugs against infectious diseases, and improvements in nutrition and personal or public hygiene were unheard of. Indeed, who, a century ago, would have anticipated that smallpox, poliomyelitis, tuberculosis, and many other infectious diseases would have practically disappeared from many countries or have become, to all intents, curable? Despite the impressive increase in the mean longevity of the population, many scientists and medical doctors consider that we still die too soon. They estimate that our potential life span is around 120 years. So, why do we fall short by some 45 years, since we die at an overall average age of 75 years? If we want to live longer, we need to understand the mechanisms of the currently incurable diseases that shorten life span. Even more importantly, we must better characterize the process of normal senescence (senescence and aging are used interchangeably throughout the book).

Aging is accepted as a normal biological process that affects all complex living organisms; it is presented sometimes as the consequence of processes that occur because of the passage of time, and sometimes as the cause that modifies normal physiology and functions. Arking (1991) selected five criteria for aging and named them with the somewhat contrived acronym of CUPID: aging is cumulative (C), universal (U), progressive (P), intrinsic (I), and deleterious (D). Gerontologists agree that aging can be characterized as follows:

- Aging is accompanied by documented changes in the chemical composition and the macroscopic structures of the body.
- These changes affect the ability of the organism to respond adaptively to the environment. This is the case at all levels, from the synthesis of molecules to cognitive abilities (Adelman, 1980).
- In elderly individuals, the recognized, but poorly understood, increased vulnerability to diseases and to environmental changes and demands leads to a higher risk of dying (Shock, 1985).

An important part of this book is devoted to the description of the causes and the mechanisms that cause aging. For a long time, aging was assumed to be an eroding process, a sort of end-product of development, and, as such, it was considered an uninteresting area of research. This lack of interest by scientists to study the mechanisms of aging may be explained in part by the difficulty in tackling scientific issues concerned with aging. Aging is a universal process, with no definite starting time. Until recently, therefore, scientists have shown more interest in deciphering the mechanisms of embryonic development. Embryologists, in contrast to gerontologists, are able to describe the successive steps involved in the transformation of an egg into an adult organism. They know that the complex biological phenomena they observe begin at the moment of fertilization, even though they do not always have insight into the exact mechanisms regulating and controlling each phase of the development of an embryo. During the last decades, it became clear that a scientific approach designed specifically to study the processes of aging was necessary, and biologists are now faced with the challenge of exploring the various trajectories of the aging processes in different living organisms.

The study of aging processes is now part of modern biology; it is accepted as the science called gerontology. It is not a unique scientific discipline, as are, for example, biochemistry or genetics; rather, it is multidisciplinary, bringing together knowledge and techniques from several disciplines. Gerontology attracts the interest of a large number of scientists who come from different fields of research and approach the study of aging with their own professional orientations or biases (which may explain some of the disagreement on the basic causes of aging).

Scientists interested in the phenomenon of aging are faced with a vast array of questions ranging from the deciphering of the mechanisms of aging (for example, finding the reasons why most unicellular organisms escape aging, or studying the causes and mechanisms of aging in species that show particularly short or long life spans), to understanding the role of the different extrinsic and intrinsic causes of senescence.

The relevance of these and other questions is obvious, and finding answers to these questions is a challenging task. It is the goal of a subspeciality of gerontology called biological gerontology.

Living organisms are complex, hierarchical, and interactive systems, and, according to Murphy's law, the more complex a system is and the more functions it can display, the more "things can go wrong". This implies that there might be an almost infinite number of causes of aging in living organisms. Any biological change could affect virtually all components of the system and be a possible cause for the process of aging. Aging is a multifaceted phenomenon that affects different systems in specific ways; it is also a mosaic process, in that different organs and tissues do not age at the same rate within the individual. Thus, it appears aging does not have a single cause, any more than embryological development has a single mechanism. Yet, to this day, no clear picture of the multiple and combined aspects of aging has emerged.

Senescence is a complex and diverse process of deterioration. It is therefore necessary to measure the manifestations of senescence at different organizational levels: at the population level (the life span of individuals), at the individual organism level (the changes in physiological-biochemical functions), at the cellular level (structural and biochemical properties), and at the subcellular level (changes in molecules, e.g., in the activity of various enzymatic and repair systems operating in cells). All organisms can serve as model systems to analyze the phenomena of aging, and Nature has provided an extraordinary diversity of life forms for study. The study of the aging processes in each one provides valuable information for the overall understanding of aging. All biological systems, from the so-called simple (e.g., invertebrates) to highly complex human beings, display the effects of aging in different ways. The use of invertebrate species has the advantage that both genetic and biochemical studies can be performed easily, and they represent good model systems to test and refine the various theories of aging (see Chapter 4).

However, the choice of a model or of an experimental organism is, of course, dictated by the goals that one wants to achieve. If the goal is to understand human aging, it seems that the study of organisms that are biologically similar to humans should intuitively be of more value than organisms that are evolutionarily distant. It is only when scientists have a better understanding of aging phenomena at all levels and in many different organisms that they will be able to judge whether these phenomena can be unified into a single aging process, one that governs the life history of most organisms, or whether the multifaceted model, alluded to above, should remain the prevailing one.

In the first part of this monograph, we present a description of the life span of humans and of other species to illustrate the dual influences of heredity and environment on longevity. We also address the problem of aging at the cellular level and report on the way cells age when they are isolated from different tissues and are cultured in vitro. The study of cultured cells is an invaluable tool in identifying the genes that determine aging.

Cells or organisms that do not age and, at the other end of the spectrum, organisms that age quickly, are useful models to understand the physiology and the genetics of aging. For example, bacteria, yeasts, and the clones of many plants do not exhibit signs of aging. They propagate vegetatively, i.e., through division of a mother cell into two **totipotent** daughter cells. In humans, a few people live to be 120 years old and thus reach their theoretical maximum life span; others suffer from diseases that make them age prematurely. Genetic diseases that accelerate senescence, called **progerias**, serve to identify and characterize specific genes and factors that influence life span in humans.

We then focus on aging in humans and discuss the role played by factors such as personality, stress, and disease. This leads us to a description of the changes observed with aging in important physiological systems: the neurological, endocrinological, and immunological systems. The way these pivotal systems age brings us naturally to some of the aging mechanisms that act at the lowest component levels of living organisms, that is, at the cellular and subcellular levels.

Aging is regulated by highly conserved and species-specific intrinsic factors, taken here in the broad sense to include genes and their products. Aging processes are also influenced by extrinsic factors, comprising influences and traumas from the external environment as well as **epigenetic** factors. These are cellular events that play a role in the functional differentiation of cells and in regulation of gene expression but do not involve the genetic code per se. The nature of these intrinsic and extrinsic factors and their influences on aging are described.

Chapter 1

Variability of Life Span Between Species

Methods to Quantify Life Span

Nature has provided millions of plant and animal species that display a wide range of life spans[1] and different scenarios of aging. There is no rule on how long plant and animal species can live, and the time between birth and death (total life span) may be almost any length. Flies and nematodes may live for only a few days to a month, whereas bamboo plants may live for a hundred years, and certain species of trees survive a thousand years or more. Some organisms undergo gradual senescence during the course of their life while others show no signs of degenerative changes (or aging) for most of their life and then pass through a period of rapid senescence followed by sudden death: this occurs in organisms with a wide range of life spans. For example, many plants display long vegetative growth phases that last for decades; they then go through a period of fruiting after which they suddenly age and die within a few months. The same applies to insect queens: they die suddenly after a long adult life span. All these scenarios of life spans and aging processes merit detailed study, and knowledge gained from different organisms will contribute to a better understanding of what causes aging.

In scientific publications, and for reasons of simplification, most scientists provide data on mean life span, but several other parameters are used to characterize the life span of living organisms. The two simplest parameters are the maximum and mean life span. *Maximum life span* refers to the age of the oldest individual of a given species. *Mean life span* of a population is the average figure of the maximum life span of all individuals born alive. It is commonly referred to as the *life expectancy* in human populations. Caution should be used when comparing mean life span values between studies, because factors such as high infant mortality have a significant influence on this parameter. The maximum and mean life span values provide useful measures of the longevity of individuals or species, but they do not necessarily reflect the *rate of senescence*. Several methods

---

[1] In this book, we use the terms life span and longevity interchangeably. Both terms delimit the period of time between birth and death.

have been devised to calculate this rate. For example, in human and animal populations, an exponential increase in mortality (from cancers and degenerative diseases) is observed as a function of age (after completion of the developmental period of individuals). This change in mortality rates in a population is a major determinant of maximum life span. The *average mortality rate* or AMR is an accurate estimate of the rate of senescence. However, even this method of analyzing senescence may not be optimum in cases where mortality is high because of malnutrition, infectious diseases, or social conflicts. These conditions prevent individuals from surviving to advanced ages, and thus no exponential increase in mortality is observed. In fact, in wild-life conditions, few adult animals survive to ages when senescence becomes manifest. Two other parameters describe the rate of senescence at the population level. The *mortality rate doubling time* (MRDT), is held to be a fundamental measure of senescence; it gives an estimate of the acceleration of mortality rate in relation to age. The *initial mortality rate* (IMR), is a measure of the risk of dying independently of age; it gives the limits of a theoretical life span that may prevent individuals from achieving ages when senescence could be manifested.

In theory, both the MRDT and the IMR must be known in order to conclude whether spontaneous or experimentally induced changes in life span result from differences in the rate of senescence rather than from environmental factors. Readers interested in more technical details on the quantification of longevity and senescence are advised to consult the first chapter of the review by Finch (1990).

Variability of Life Spans

Living creatures show a remarkable variability in life span. Table 1 lists the extremes of life spans encountered among plants and various animal species. Variability among plants is extreme, ranging from annual plants that live one season, to perennials such as sequoias that live over 5000 years. Although not as spectacular, the variability of life spans in the animal kingdom, and among vertebrates in particular, is far from negligible. The life span of mice is 2 years whereas that of tortoises can reach 150 years. Interestingly, the variability of life spans is often high within a given group (phylum).

To illustrate further the variability of life spans and aging processes and to emphasize further the relative role of intrinsic and extrinsic mechanisms of aging and their respective influences on longevity, we have selected a few examples.

The first concerns the differences in longevity among humans. Of particular interest are life span studies on **monozygotic twins**, which provide a model to evaluate the importance of genetic versus environmental influ-

Table 1. Variability of life spans for various plants and animals

Group	Species	Approximate maximum life span (in years)
	PLANTS	
Trees and shrubs	Sequoia	5000 and more
	Rhododendron	50
	ANIMALS	
Coelenterates	Anemone	50–90
Annelids	Common earthworm	>6
Mollusks	Octopus	3–4
Spiders	*Tarantula*	15
Insects	*Drosophila*	70 days
	flour beetle	2–3
Vertebrates		
Amphibians	frog, toad, etc.	1–60
Chordata	lake sturgeon	>150
	river perch	25
Reptiles	Alligator	50–60
Birds	Goose	30–40
Rodents	Rat	4–5
	Rabbit	10–13
Carnivores	Cat	20–30
	Bears	40
Proboscidea	African elephant	>70
Primates	Rhesus monkey	>30
	Chimpanzee	45
	Human	>110

These examples were compiled from several sources.

ences. The second example concerns relevant aspects of aging in insects. It shows that, within a given species, the characteristics of senescence can vary widely, depending on the pathways chosen during early development of the individual. For example, the different developmental pathways and life spans of the various castes of honeybees argue strongly that senescence depends on external influences. In the case of the fruit-fly, *Drosophila*, longevity studies point to a link between aging and reproduction and emphasize the plasticity of these organisms in adapting to their environment. The third example points to the relevance of studying cells in culture, as scientists made the interesting observations that the in vitro potential of cells to proliferate differs depending on whether they are taken from young or old individuals. Finally, the question of whether immortality exists among plant or animal species is addressed.

Life Span of Humans

Richard Cutler (1984) noted that longevity is related to several factors such as rate of development, length of the reproductive period, maximum consumption of calories, and size of the brain. He concluded that the potential maximum life span for man, estimated from these variables, should be around 115–120 years. His estimate agrees with an earlier observation by George Buffon, a biologist of the 18th century, who claimed that animals tended to live six times the period needed to complete their growth. As humans reach their skeletal maturity at approximately 20 years, the maximum projected life span would be about 120 years. Interestingly, the maximum life span reported corresponds to this value and does not vary much among different human populations. There are spectacular reports on the supracentenarians from the Caucasus, the Andes, and the "Sacred Valley" south of the Equator. Unfortunately, these claims are frequently not documented with bona fide records of birth dates. Despite the lack of formal proof, some scientists believe that a few human beings have survived beyond 130 years. Jeanne Calment, a French woman who celebrated her 121st birthday in 1996, is the oldest person whose age has been confirmed.

In 1693, the astronomer Edmund Halley published the results of a study on longevity in Breslau, Germany, a city where births and deaths had been carefully registered. Of 100 newborn children, only 51% were still alive at the age of ten; 43% reached their thirties; 28% their fifties; and only 11% attained the age of 70 years. Life expectancy was no better two centuries later, in England, where about 50% of all deaths occurred among children below the age of 14 years. The life span of many great artists was dismally short. Wolfgang Amadeus Mozart died in 1791 of a presumed fever at the age of 35, Frédéric chopin died in 1849 of tuberculosis at the age of 39, Robert Schumann died in 1856 of syphilis at the age of 46, Paul Gauguin died in 1903 of syphilis at the age of 55, and the list could be extended indefinitely. These early deaths are a loss for humanity: consider how many masterpieces have not been created because of diseases that today could have been cured or prevented. Of course, one may argue that, in past centuries, a number of artists died at what was then considered very old age. For example, Leonardo De Vinci died at 67, Galileo at 78, and Newton at 67. While this is true, their risk of dying prematurely was infinitely greater than ours.

From 1830 to 1980, a significant increase in mean longevity has been observed in Europe and other developed countries, as illustrated in Figure 1. The dramatic 30-year increase in longevity has occurred mainly during this century.

In the developing nations of the third world, life expectancy is 15 to 20 years lower than it is in Europe. High child mortality is the primary reason for this. It is interesting to note that the curves in the figure representing increases in life expectancy are sharper in eastern Europe

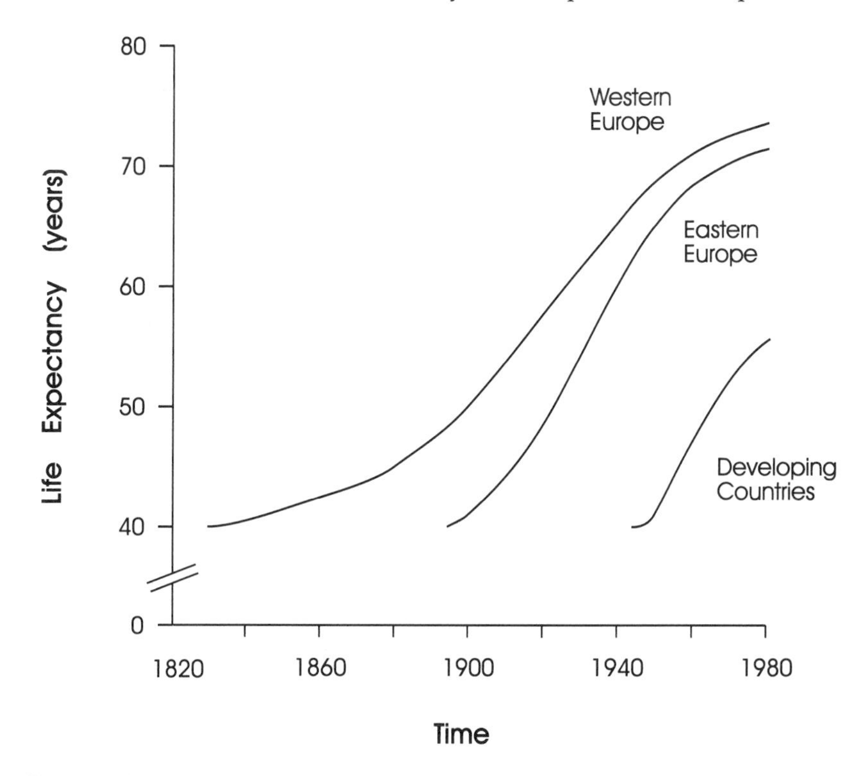

Figure 1. Mean life expectancy (years) in western and eastern Europe, and in developing countries, from 1830 to 1980.
The mean life spans represent an estimated average figure on a national scale. (Redrawn from Gwatkin and Brandel, 1982.)

and in developing countries than in western Europe. Apparently, the introduction of better hygiene and the use of antibiotics in developing countries have increased the mean life expectancy in a shorter time than have further improvements achieved in hygiene, food, and medicine in western Europe.

Life Expectancy Increases Throughout the World

In the USA, life expectancy has increased at a regular rate since the beginning of this century, by about three years every decade. A tangible way to appreciate this increase is to record the number of centenarians. In the United States, in 1950, there were about 4500 persons 100 years of age and older. In 1990, this number was 54,000. The sharpest increase in life expectancy in the 20th century has been observed in Japan (Figure 2). Interestingly, it has been accompanied by a mean increase in young Japanese body height, now equivalent to that of most Europeans and North Americans (Anderson, 1982). These changes in mean life span result mainly from better quality of the diet and improved public health. They are also observed in most third world nations, despite the widespread poverty.

Notwithstanding the rare cases of men and women living to 120, the mean life span for men and women does not exceed 75 years, a figure much lower than the theoretical life span. We have to conclude, therefore, that, at the present time, the vast majority of human beings still die several decades earlier than their theoretical biological limit. A number of factors influence the mean longevity in humans. Table 2 displays a list of factors for which there is documented proof that they influence longevity.

Several of the factors listed in Table 2 are interrelated: for example, heredity determines some personality characteristics, and these can predispose to healthy or unhealthy behavior patterns. Such patterns also

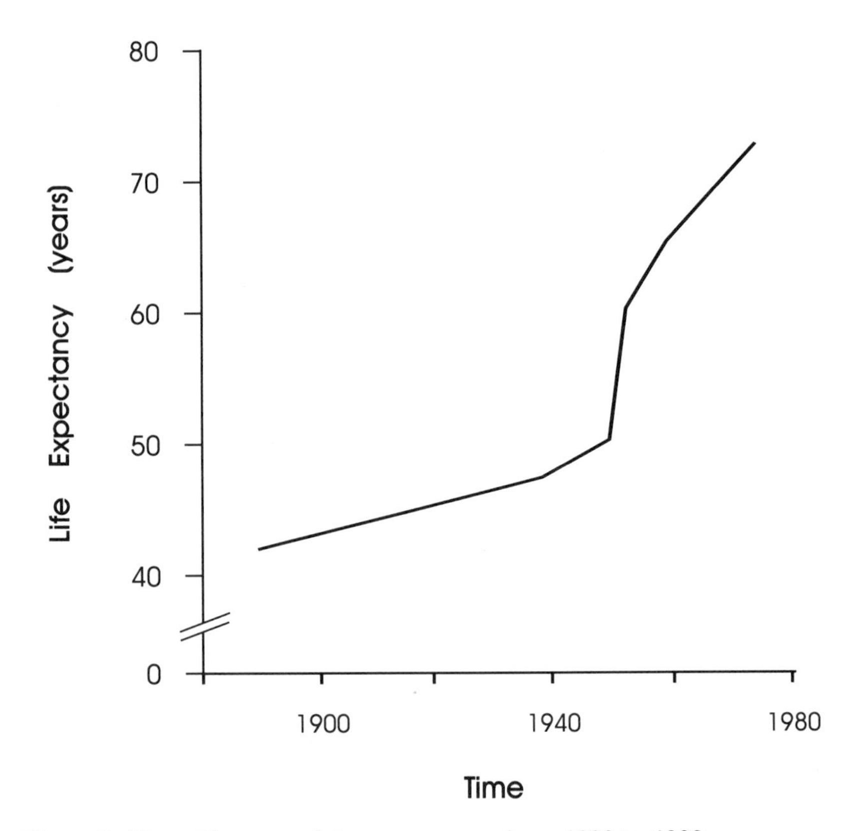

Figure 2. Mean life span of Japanese men, from 1890 to 1980.
Note the sharp increase in mean life span since about 1950. (Redrawn from Anderson, 1982.)

Table 2. Factors that have an influence on human longevity

Factor	Influence
Socio-economic status	Higher status has positive influence on longevity
Geographical location	Positive or negative
Life styles	Improved diet, positive Tobacco, alcoholism, negative
Improved public health	Positive
Sex	Men die earlier than women in all human races and societies. This is also observed in many animal species.
Heredity	Hereditary predisposition to diseases is well recognized in humans; longevity and death genes have been discovered and play a role in animal life span.
Personality	The personality factor includes a wide range of factors such as sensitivity to stress, optimism and pessimism, involvement in social interactions, etc. (positive or negative).
Environmental toxins	Negative

depend on cultural backgrounds: North Americans are more inclined to perform regular physical exercise than Europeans. Furthermore, other factors such as the level of growth **hormone** or the level of thymic and sexual hormones are claimed to influence longevity. It is not an easy task to judge the importance of each of these factors on longevity, and their definitive role continues to be debated. At present, most scientists and medical doctors agree that genes, life styles, personality, and stress have the strongest influence on life span. Their role, as well as the role played by diseases, are discussed in Chapter 2.

Life Spans of Invertebrates

Honeybees

Honeybees are referred to as social insects, because they have a complex social organization. They provide an example of how the genome may be programmed during development to yield short- and long-lived adults within the same species through the interactions of genes with environmental and epigenetic factors.

The differing life spans of the various castes are shown in Figure 3.

Honeybee queens live at least 5 years, which is up to 50 times longer than workers; yet, both castes develop from the same batch of fertilized

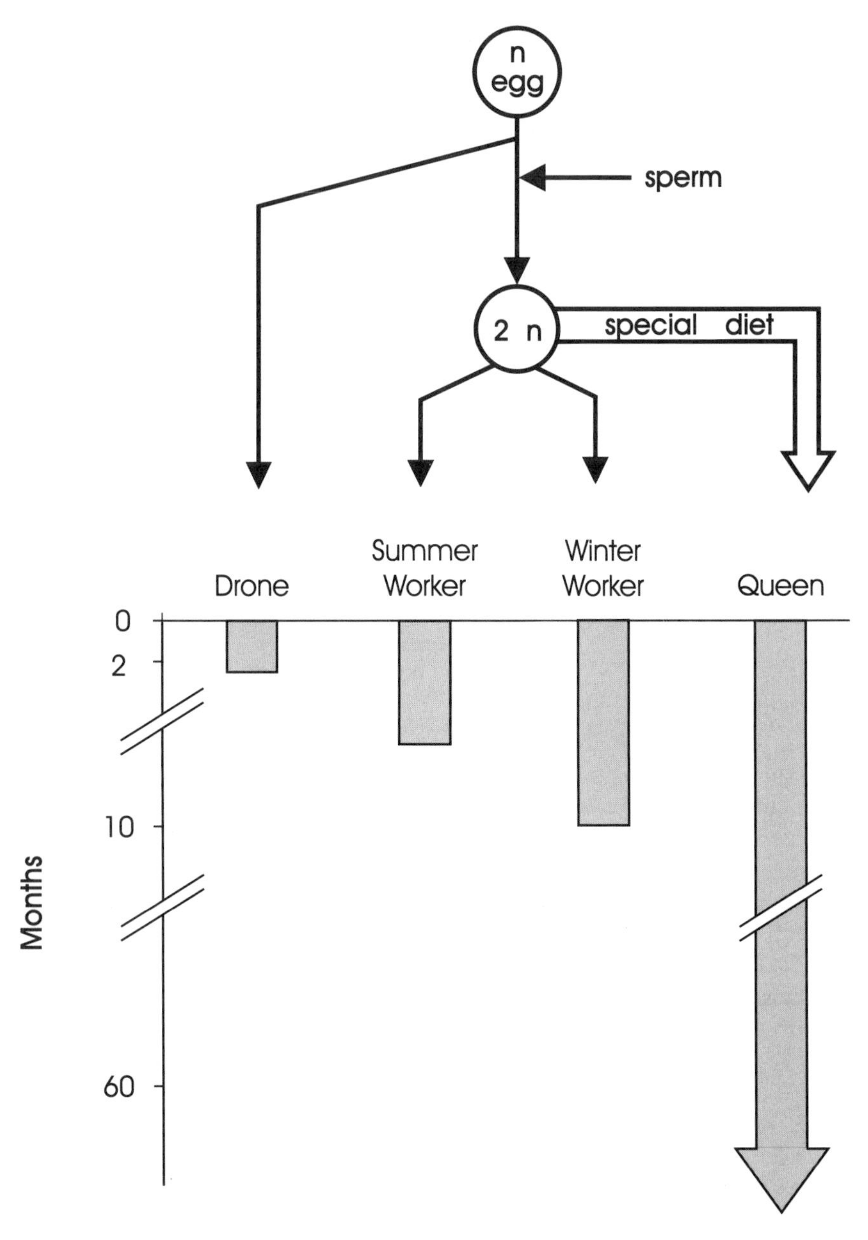

Figure 3. Life spans of the different castes of honeybees.
Fertile males or drones arise **parthenogenetically** from unfertilized eggs and live one to two months. Female workers and queens arise from fertilized eggs. When born in the summer (summer workers), they have a shorter life span than when born in the autumn (winter workers). Queens can live more than 5 years. When their sperm stores are exhausted, queens are killed by workers. (Redrawn from Finch, 1990.)

eggs. These different developmental pathways and life spans are regulated by complex interactions of neuroendocrine hormones and nutrition, i.e., a special feeding provided to the queens at a specific larval stage. Queen-destined larvae are fed ten times more often by the nurse bees, with a diet that contains three times more fructose and glucose. This type of diet influences the secretion of juvenile hormone (JH) by a neuroendocrine gland, the *corpora allata* (Asencot and Lensky, 1984).

Hormones and diet play a determining role in the fate of worker bees as well. The task of young hive worker bees, in the first three weeks after hatching, is to attend the queen. During this time, they feed mostly on protein-rich pollen. They then become field worker bees and forage for food, at which point they feed mainly on carbohydrates. Major physiological changes occur during this transition, including increases of JH and major decreases of **vitellogenin** levels in the **haemolymph**. Important differences in life spans are observed between summer and winter worker bees. Summer workers have life spans of one to two months; they show major increases of JH during their transition from hive to field bees. The short life span of summer bees is associated with foraging flight, and death is partly caused by this intense flight. Winter workers, on the other hand, survive through the winter and are active for six to eight months, feeding the queen, the males, and the larvae; during that time, they maintain low JH levels. Interestingly, death may be induced prematurely by injecting them with JH (Robinson, 1987). Thus, senescence and death are not determined by age alone but seem to be linked to the onset of foraging, and senescence caused by strenuous physical efforts can be mimicked by the injection of hormones. The level of JH therefore determines a branching point leading to alternate pathways of senescence in workers. It is interesting to speculate whether postponing the onset of foraging even longer would extend the life span of worker bees further. Another interesting observation is the relation between reproduction and senescence. The fertilizing males or drones live no longer than one to two months. They arise by **parthenogenesis**, from unfertilized (**haploid**) eggs. Their role is to inseminate young queen bees by placing sperm in the queen's spermatheca. Following the physical contact with the female, they leave part of their endophallus in the female genitalia and die (Winston, 1987). This injury causes what could be called an acute lethal mechanical senescence. Incidently, the 5 million spermatozoids delivered to the queen will suffice for her entire production of eggs. The long survival of honeybee sperm in the female genitalia is as astonishing as are the life span and fertility of the queen. Eggs are produced from the queen's ovaries, which are enormous compared with those of the worker bees. Each ovary is made up of about 150 ovarioles, which can produce an almost unlimited number of eggs (about 200,000 eggs per year). Workers, in comparison, have only 2 to 12 ovarioles. The eggs pass down the oviduct where a pump and valve, leading from the spermatheca to the duct, open to release minute amounts of sperm. The controlled release of only a few spermatozoids at a time is important for the queen, since she will be superseded and killed by the colony when her store of sperm is used up. Thus, the life span of the queen appears to depend largely on the amount of sperm held in her spermatheca.

Drosophila

The tiny fruit fly *Drosophila melanogaster* has served as another model in the study of the influence of genetics on life span. The genetics of this insect are well known, and a number of inbred strains have been obtained by appropriate genetic crossings. Many studies on longevity have shown unambiguously the existence of a genetic determinant (or genetic deter-

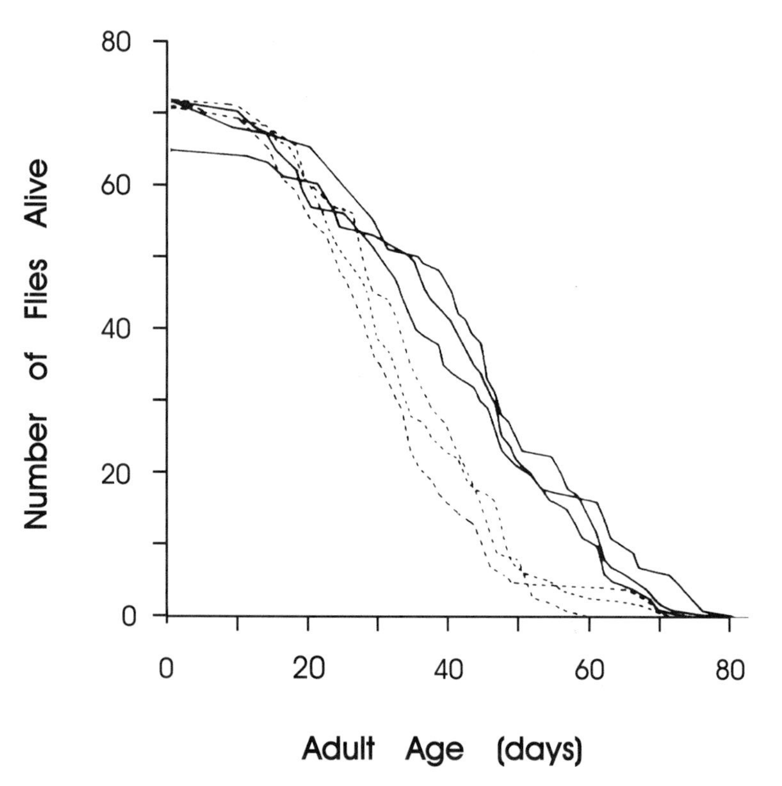

Figure 4. Survival of *Drosophila melanogaster* females born from females of different ages.
The results of three different experiments are shown. Dotted lines represent three control populations (stocks born from younger females). Solid lines represent three populations selected for late reproduction (stocks born from older females). (Redrawn from Rose, 1984.)

minants) in the life span of *Drosophila* (see Chapter 5, pages 88–90). However, some scientists have raised objections to these studies: an artificial selection for increased life span, as applied to inbred laboratory animals, could yield potentially different results from those obtained in natural populations of species with **polymorphic** forms. The same objections are raised to studies with inbred strains of rodents.

With this in mind, Rose (1984) analyzed the genetic influences on life span in genetically heterogeneous populations of *Drosophila*. His hypothesis was that populations selected for late reproduction would also show delayed senescence. His study was based on selecting *Drosophila* females that reproduced early versus females that reproduced late. In the late-reproducing population, *Drosophila* cultures were maintained using adult females that had survived a number of weeks before being allowed to contribute eggs to the next generation. The age at which females were used to produce eggs was then progressively increased from generation to generation. Control lines were females exposed to males within a few days from adult eclosion, i.e., flies reproducing early. The results of this study were quite spectacular: after fifteen generations, the mean and maximum life spans of females belonging to the late-reproducing population had increased by about 30%, while male life spans increased by 15%, compared with the early-reproducing populations (Figure 4). Postponing reproduction artificially resulted not only in an increased life span in females from late-reproduced populations, but also in the extension of the fertility period (Luckinbill et al, 1984). These results indicate that aging can be postponed when a strong environmental selection is applied, such as postponing reproduction to a later age. They also support the evolutionary theory of aging, which stipulates that the probability of survival drops after reproduction has occurred (see Chapter 4).

Life Span of Cells in Culture

Because cells are the cornerstone of biological processes and express the identity of each organ, it is logical to conduct research on the senescence of cells isolated from tissues and cultured in vitro in the absence of extrinsic regulatory mechanisms. In 1961, Hayflick and Moorhead were among the first scientists to propose that cultivated human diploid cells could serve as a model to study aging. Their idea was based on the observation that many changes that occurred during in vitro senescence were reminiscent of changes that occurred in vivo during aging. They observed that the timing of cellular and physiological processes was important: fetal cells displayed a consistently greater growth potential (as assessed by total number of cellular divisions, referred to as the in vitro cell population proliferative doubling limit, or PDL), than cells isolated from adult tissues. This finding was not readily accepted, until a number of other laboratories confirmed Hayflick's results. Martin and his collaborators (1970) and later Schneider and Mitsui (1976) showed that the replicative potential of human fibroblast-like cells in culture is inversely related to the age of the donor (Figure 5). Other data supporting Hayflick's initial observation are that embryonic stem cells put into culture 4 days after fertilization do not undergo the same clonal senescence as cells from later stages of differentiation: they reached 250 cumulative doublings, in comparison to 10 to 30 cumulative doublings for fibroblasts from adult mice (Suda et al, 1987). The observation that culture cells from progeria individuals with Werner's syndrome have an attenuated growth potential (PDL) emphasized further the importance of host genetic influences (Figure 5). Factors such as age-related diseases also play an important role on cellular replicative potential: fibroblasts from diabetics, patients with cystic fibrosis, or patients with cancer of the colon have a

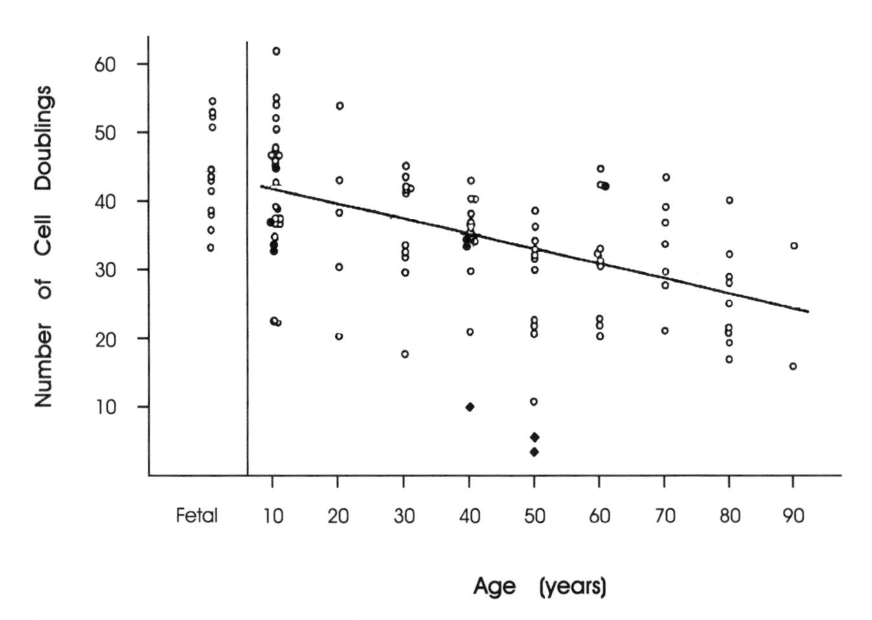

Figure 5. Replicative potential of cells relative to donors' age.
Skin fibroblasts were isolated from human donors of different ages, placed in individual cultures, and replaced in a new culture dish after each cellular division. The number of total cell divisions recorded until cells died is plotted against the age of the donors. Analysis of these data indicates that about 30% of the capacity to proliferate is lost during the life span of an individual. Note that some of the cells from very old donors, however, retained a capacity to proliferate similar to that of much younger donors.

Open circles: fibroblast cultures from individuals of different ages (control cultures); closed circles: patients suffering from age-related diseases; closed diamonds: Werner's syndrome homozygotes (individuals affected by progeria). The linear regression line (solid line) for the control groups is drawn between the first and ninth decades. (Redrawn from Martin et al, 1970.)

reduced proliferative capacity in vitro (Martin et al, 1970). This age-dependent decline of replicative potential is referred to as the Hayflick limit (see below).

These studies have provided the most compelling support for the use of fibroblast-like and other cell types as a model for the study of cellular aging. Studies have since been extended to other cell types, such as lens epithelial cells (Tassin et al, 1979), endothelial cells, arterial smooth muscle cells (Bierman, 1978), glial cells, lymphocytes, and other types of cells, with similar results. This decline is programmed, and individual cells exhibit a form of memory of the number of divisions (doublings) they have undergone. This is indicated by the fact that, if cells are frozen after a given number of cell divisions and then thawed and allowed to divide again, they will resume the expected number of remaining cell divisions.

Additional arguments in favor of genetically determined cellular senescence are provided by the positive correlation between the potential of fibroblast cultures to proliferate in vitro and the mean life span of the species from which the fibroblasts were taken (Figure 6).

The decrease of the in vitro capacity to proliferate is commonly referred to as cell senescence or aging in vitro. It is more accurate to use the term **clonal senescence**, because the relation between the life span of cells in culture and normal cellular functions during senescence of the whole organism is unknown. Clonal senescence is still poorly understood. It is unlikely to be an artifact of cell culture, such as a lack of some **essential nutrients** in the culture media, since similar results are obtained with cells grown in various media containing different growth factors and nutrients.

A closer analysis of possible intracellular mechanisms underlying cellular senescence reveals differences between young and old cells in the successive cell cycle events. During the course of serial subculturing, many of the early cell cycle events initiated by the addition of growth factors to culture media occur in a similar way in both young and old cells. However, old cells do not complete the successive phases of the cell cycle as do young cells (Figure 7). The percentage of senescent cells involved in DNA replication (or S phase) is found to decrease and their average cell cycle time to increase.

The perturbation in the cell cycle is primarily due to an increase in the G_1 phase; eventually, cells become blocked in G_1, just before the DNA synthesis phase (Cristofalo and Stanulis-Praeger, 1982). Scientists had provided evidence earlier that the switching from G_1 to S phase requires the presence of specific growth factors. Whereas young cells stimulated with serum or a mixture of growth factors trigger the events that lead to DNA replication and then replicate DNA, old cells display many of the characteristic activities of the early G_1 phase, but they become blocked in late G_1, just prior to synthesizing DNA. The important question is to

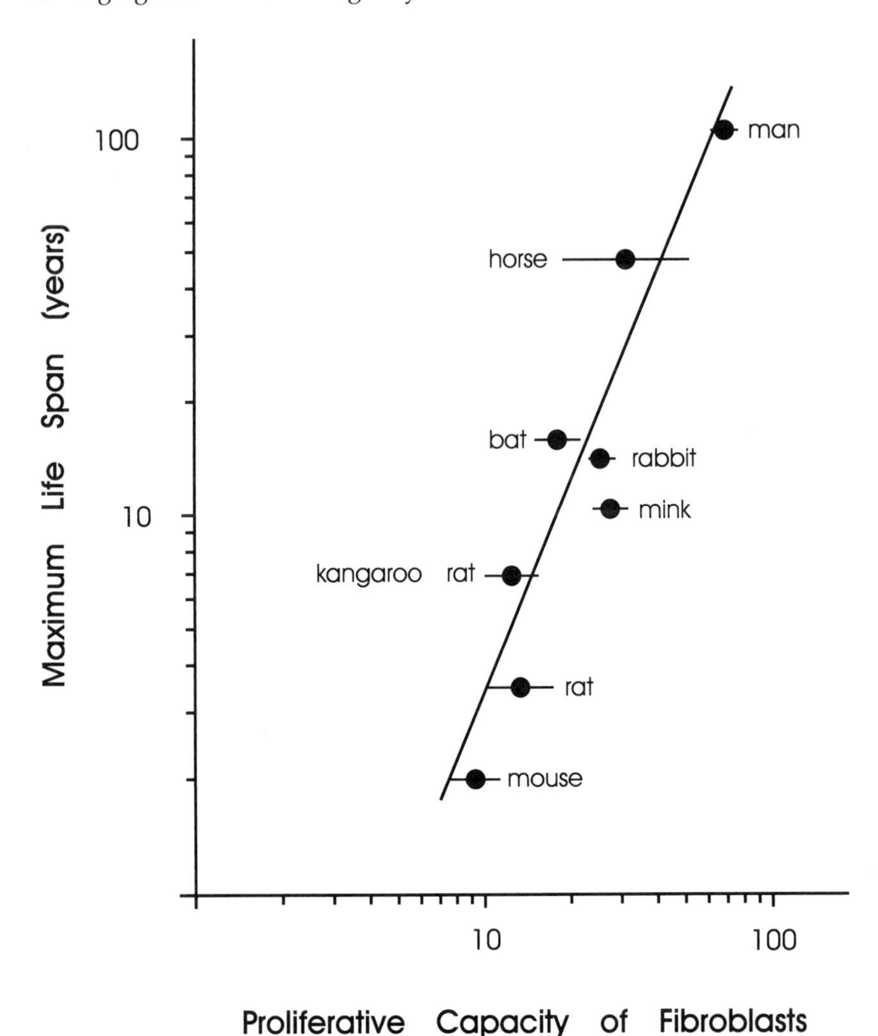

Figure 6. Relation between the potential of fibroblasts to proliferate in vitro and the maximum life span of mammalian species.
Proliferative potential of fibroblasts was estimated by counting the total number of cellular divisions in culture. Note that the life span of mammalian species differs by more than thirtyfold. (Redrawn from Röhme, 1981.)

elucidate the exact mechanisms of regulation of cell proliferation by DNA and, in particular, to understand why and how the triggering actions of specific growth factors to synthesize DNA fail in senescence (Cristofalo et al, 1989).

Several studies have provided evidence of a genetically controlled negative regulation of growth: (1) results of studies on DNA isolated from fibroblast-like cells suggest that negative growth regulation could be mediated by **interspersed repeated sequences** (Howard et al, 1988); (2) Smith and Lumpkin (1980) suggested that the gradual loss of repeated DNA sequences was the trigger for the activation of inhibitors of DNA synthesis (or negative growth factors). They referred specifically to the ends of the replication points of DNA molecules (**telomeric regions**) postulated to be involved in the maintenance of gene expression. In support of this hypothesis, a loss of repeated sequences has been observed with aging in vitro as well as in vivo (Shmookler-Reis and Goldstein, 1980); (3) DNA **demethylation** has been proposed as yet another possible mechanism of cell cycle regulation. These topics are discussed in more detail, in Chapter 5 (pages 110–111).

Finally, the conclusions of recent studies support the view that the genetic mechanisms that suppress neoplastic transformation could be involved in the negative regulation of growth during senescence as well. Lee and his collaborators (1995) showed recently that fibroblasts that manifest a decreased proliferative capacity, in vitro, express increased

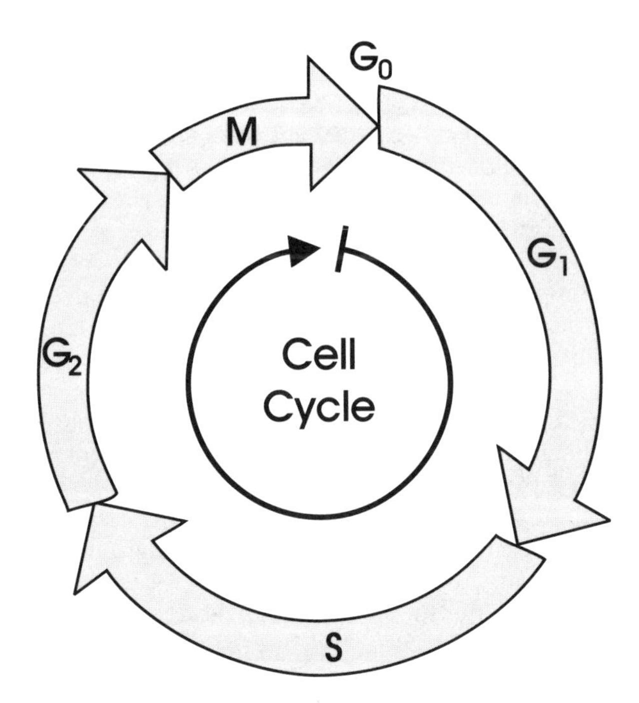

Figure 7. Phases of the eukaryotic cell cycle.
Schematic representation of the cell cycle. G_0 = beginning of cell cycle, G_1 = period before DNA synthesis, S = DNA synthesis, G_2 = period between DNA synthesis and **mitosis**, M = mitosis.

levels of one of the nuclear receptors for retinoic acid (RAR) in response to retinoic acid (RA). This is interesting because RA and its derivatives were previously known as effective negative regulators of growth and inhibitors of tumor growth. In the presence of ligand, RAR binds to DNA and represses gene expression. Moreover, mac25, a member of the family of insulin-like growth factor binding proteins, also has been found to have tumor suppressive activity. Whereas its expression is down-regulated in mammary carcinoma cell lines, it is stimulated by retinoic acid and has been found to accumulate in senescent cells (Swisshelm et al, 1995). Thus, the tumor suppressive role of these two proteins may involve a senescent pathway.

All the observations made so far in different systems demonstrate the limited capacity of cells from older donors to proliferate in vitro and suggest that cells contain biological clocks that ultimately limit their life span. The cellular senescence, in all cells studied so far, seems to be driven specifically by the number of cellular replications and by the activation of precise genetic mechanisms responsible for the negative growth regulation (and not by the passage of time itself).

Further discussions have centered on the interpretations of the Hayflick limit. Investigators who observed a continual decrease in the fraction of cells synthesizing DNA during aging or an exponential increase in the fraction of nondividing cells, favor the interpretation that, at each cell population doubling, an all-or-none event occurs. Cells that lose their replicative potential are called senescent cells, and, by extension, senescence is described as a prolonged period of arrest of cellular division preceding the death of cells. Other investigators favor the idea that specific intracellular events occur at each doubling and lead to an increased heterogeneity of the population, with a spectrum between the two extremes, i.e., complete inhibition and abnormal division cycle. Macieira-Coelho (1995) supports the idea of a functional evolution or a drift created by cell division, because it is more compatible with the process of aging observed in vivo: ". . . it is not significant for aging to reach an end point, but rather that cells are modified through proliferation so that they become more heterogeneous in their response to growth stimuli, and progress differently through the division cycle". This implies new cell interactions and regulations. Proponents of this idea point out that there is no evidence that terminal **postmitotic cells** play any role in the physiology of aging of the organism. Replicative senescence is not synonymous with cell death, and the mechanisms that lead to senescence are not necessarily the same as those that trigger programmed cell death. In fact, when cultured cells stop dividing (i.e., become terminal cells), they remain viable and metabolically active for long periods of time. Moreover, postnatal tissues contain a fraction of cells that have acquired the structural characteristics of terminal cells. Macieira-Coelho (1995) proposed that postmitotic cells, having reached their terminal differentiation process, might play a role in pathological conditions associated with aging; the presence of an excess of these cells might disturb tissue homeostasis and lead to pathological states (see Chapter 3, pages 72–74).

Are Some Plants or Animal Species Immortal?

As early as 1891, August Weismann had an explanation for the so-called immortality of unicellular organisms (Weisman, 1891): "Normal death could not take place among unicellular organisms, because the individual and the reproductive cell are one and the same: on the other hand, normal death is possible, and as we see, has made its appearance, among multicellular organisms in which the somatic and reproductive cells are distinct." Bacteria never have the opportunity to age because they divide before aging processes produce measurable changes. Some multicellular organisms, such as sea anemones (belonging to the phylum Coelenterata) seem also not to age. Over periods of observation greater than 50 years, they show no evidence of decline in their physiological functions or in their capacity to reproduce. The absence of senescence is attributed to the fact that bacteria, anemones, and several other organisms retain the ability to replace their cells or parts thereof: molecules and subcellular **organelles** do deteriorate, but they are continuously replaced. Unravelling the mechanisms by which these immortal organisms escape deteriorative processes is one of the important issues. Another interesting issue is to find out to what extent bacteria and other immortal organisms would age if they were prevented from dividing; it is likely that the very conditions that inhibit cellular division would induce degenerative changes. With the exception of these immortal species, organisms that reproduce exclusively by sexual means all undergo senescent changes. In some plants, these changes can be very slow and life spans very long; the famous redwood conifers undergo only a negligible senescence during thousands of years (Noodén, 1988). Scientists have postulated that repair mechanisms in plants are analogous with those of sea anemones and that their slow aging could be due to the presence of particular cell populations that degenerate and turn over regularly, without damaging the whole organism.

A few species from the animal kingdom, in particular tortoises and turtles, are reported to live up to a few hundred years. They are probably the longest-lived animals on earth (the ages of some specimens living in the Galapagos were estimated, using carbon-14 dating methods, to be between 170 to 250 years). Turtles have long life spans, but like the vast majority of animal species, they do undergo senescence. It is suggested that animals with thick shells tend to live longer than animals without armour. This is also true among invertebrate species: the thick-shelled bivalves have a mean longevity of about 15 years, while other molluscs with their generally thinner or nonexistent shells live four years or less (Comfort, 1979).

Chapter 2

Factors That Influence Human Longevity

As mentioned earlier, heredity has a strong impact on longevity but is not the only factor. The quality of the environment and the geographical location have strong influences on our life span as well. Longevity also depends on other factors that directly concern each individual: personality, the ability to control one's environment, and the level of social integration are variables that have a predictive value and perhaps a causal relationship with health and longevity (Rowe and Kahn, 1987; Costa et al, 1983; House et al, 1988). These factors are discussed below.

Genetic Influence on Longevity

Are there genes that enhance or decrease longevity in humans? Studies on the longevity of family members, adopted children, and twins provide useful information on the effects of hereditary factors and corroborate that longevity is strongly influenced by genetic determinants. Familial trends have been found, in spite of the difficulties of interpreting data that do not always take into account accidental deaths or the impact of environmental variables. Among 7000 adults, longer-lived parents had longer-lived children. Children with parents who lived to at least 81 years of age lived six years longer than children whose parents died in their 60s (Abbott et al, 1978). To distinguish between the influence of nurture versus nature, scientists analyzed the incidence of diseases and death rates of adopted children. The results of a study by Sørenson and his collaborators (1986) showed that adopted children whose biological parents had died early of nonaccidental causes were twice as likely to die early from nonaccidental causes themselves. If one of the parents had died early of heart disease, the child was four times as likely to be struck by an early heart attack. If one of the biological parents had died prematurely from an infectious disease, the adopted child was five times as likely to die early from the same cause. The exception to this observed trend was cancer: while there was no relation between cancer deaths among adopted children and their biological parents (an unexpected finding!), adopted children whose adoptive parents had died of early cancer were five times more likely to die of early cancer themselves. This favors the idea that stress, diet, or environmental toxins play a greater role overall than heredity in the occurrence of cancer. This observation does not include all types of cancers, and it is now well established that many cancers are genetically determined (see Chapter 2).

Another observation is that males have a 10% to 20% shorter statistical life span compared with females in many mammalian species, including humans. (Incidentally, the difference in longevity between men and women is diminishing for reasons that are not clear.) The shorter life span of males has been attributed to the **Y chromosome**. Many male secondary sex characteristics are associated with testicular hormones determined by the Y chromosome (Hamilton et al, 1969). The likelihood that male sexual hormones could reduce life span was supported by the observation that eunuchs live slightly longer than normal men (Hamilton and Mestler, 1969). Similarly, postnatal castration of male cats and other animal species increases life span to values close to those of females. The shorter life span of males could be due also to the absence of a second **X chromosome**. Indeed, an abnormal gene on the X chromosome in females has less detrimental consequences than in males, because the second X chromosome may carry a corresponding normal **allele**, thus masking the deficient gene. Longevity studies conducted on **monozygotic** and **dizygotic twins** are useful in that they minimize the issue of environmental influences and add weight to the existence of a genetic basis for longevity. Monozygotic twins die within three years (mean value) of each other, whereas the life span of dizygotic twins differs by more than six years (Jarvik et al, 1960, 1980). The small difference in life span between monozygotic twins can be explained in part by their physical similarities, for example, in weight or blood pressure (Schieken et al, 1992). These similarities are obvious early in life. Twins inherit specific genes that may set limits on longevity, for example, through predisposition to obesity or **atherosclerosis**. Psychological resemblance can also be the cause of similar longevity among monozygotic twins, at least during their childhood, adolescence, and early adulthood (Vogel, 1986). This similar longevity could occur through mechanisms involving sensitivity to stress, a tendency to indulge in toxicomania, and other psychological and behavioral characteristics. Another observation comes from studies on twins' fibroblast cultures. There is a smaller difference in the proliferative potential (or proliferation doubling limit, PDL, see pages 17–23) between fibroblasts isolated from monozygotic versus dizygotic twins (Ryan et al, 1981).

It must be emphasized that, despite their remarkable resemblance, even monozygotic twins are not identical. Adult monozygotic twins show significant psychological differences (Vogel, 1986) as well as considerable differences in the surface areas of brain cortical gyri, analyzed by a two-dimensional mapping procedure (Jouandet et al, 1989). These differences are presumably the consequence of distinct developmental pathways and/or environmental impact.

Influence of Personality, Social Class and Social Interactions

Personality

During the last two decades, several researchers have presented evidence that personality characteristics such as tolerance to stress, ability to cope, and sensation-seeking influence health. (Only a few references are selected, but an extensive literature covers this issue: Eysenck, 1984, 1988; Booth-Kewley and Friedman, 1987; Eysenck and Grossarth-Maticek, 1991.) Statistical analyses carried out on the relation between social habits and longevity have emphasized the important role played by personality variables in the incidence of diseases. These studies have provided valuable clues to the link between personality, the state of the immune and endocrine systems, and susceptibility to cancer (see Chapter 3).

In the often-cited Grant Study, several hundred men of the Harvard classes of 1939–1942 were followed from their college years to the year 1977. During the 35 years of this study, material success, social success, mental and physical health were recorded on a yearly basis to determine how psychological variables could predict future life course (Vaillant, 1977). Conclusions of this and later studies (Seligman and Elder, 1986) were that pessimism in early adulthood was a risk factor for poor health in middle and late adulthood.

Various personality types were identified by psychological tendencies. These different personality traits were found to have a profound influence on the way people reacted to stressful events or to social environment (Argyle, 1987). For example, personalities described as type A personalities tend to have a hard-driving and aggressive behavior, they are highly involved in their profession, and they have a feeling of time urgency, the feeling that time is passing too fast with respect to what has been achieved and has to be achieved. Type A personalities try to control aspects of their environment that could hurt them psychologically or physically; they tend to have frequent and strong emotions of anger toward others and toward themselves. Perhaps as a result of the constant stress they impose on themselves, they are twice as likely to have heart attacks than Type B subjects, who have a more relaxed attitude towards life. Type B personalities exhibit the converse of the behavior pattern of Type A people. In particular, they exert less control over their environment and exhibit less **adrenocortical** reactivity than Type A subjects.

A Type C personality was identified by several authors. People belonging to this group seem unable to express anger; they are compliant, conforming, unassertive; they become highly sensitive to stress and are less able to discharge their inner tensions (Greer and Watson, 1985). They are often depressed with a feeling of helplessness. Type C subjects are more likely to develop cancer, and this could be due to the combined effects on their immune systems, of stress and the lack of social support. Stuart-

Hamilton (1991) notes that some personality types may be better adapted to one given period of life than to another. It is possible that Type A personalities are best suited to the early phase of life, when competition is high, whereas the reverse would be true for Type B personalities. On the other hand, Type B personalities would be better suited to later years of life. However, it appears that in their old age, people display the same personality traits as when they were younger. As personality appears to be fixed long before old age, all that elderly people can do is to adjust their personalities; they cannot alter them radically.

A different definition of personality types has been assigned to elderly people: Stuart-Hamilton describes personalities as "integrated", "armoured-defensive", "passive-dependent" or "disorganized". People belonging to the first category are able to reorganize their activities according to their physical and mental abilities, and they are able to focus on activities that are rewarding. People in the second category keep a high level of activity and refuse to accept the limitations of age or to remind themselves about what they have lost as a result of aging. In the third and fourth group, people are dependent and rely on others to help them. These personality traits determine the way different people age.

Social Class

Social class also has been found to influence life expectancy. Unskilled workmen live on average 10 years less than people with a university education. This may be due to two main reasons: first, people with higher levels of education are more inclined to adopt behaviors that promote health. Second, the highly educated have higher incomes and higher socio-economic status. While money does not buy happiness, it is unmistakenly associated with a longer life span. In a study supported by the American government, the Mortality Study of One Million Persons, Rogot and his colleagues (1988) showed that up to age 65, death rates in the poorest families (those earning $8000 a year or less) were 2.5 times higher than in families with an income of $65,000, which was about twice the typical American family income at the time of the study.

The choice of a partner and his or her level of education can also be a decisive factor on obesity and this, by extension, will influence longevity. Women who married men with a high school education only had a tendency to be up to 12 pounds heavier than women who married college graduates.

Social Interactions

Social ties play an important role in health and may be predictive of longevity. Berkman and Syme (1979) analyzed the impact of social interactions on mortality in a population of 6928 adults between the ages of 30 and 69. Social contacts with friends and/or relatives were assessed by asking questions such as "How many close friends do you have?" "How many relatives do you have that you feel close to?" "How often do you see these people each month?" The types of social interactions were, for example, marriage, contacts with close friends, church membership, and informal and formal group associations. This nine-year follow-up study showed that people who maintained social connections and who remained active in social affairs, had lower mortality rates than people lacking social and community ties or people who retired from active personal involvement. In his study on 2700 older people, House confirmed this observation: "Regular volunteer work, more that any other activity, dramatically increased life expectancy" (House et al, 1988).

The mechanisms by which social networks influence health and mortality rates are not clear. At the onset of their study, Berkman and Syme (1979) checked the influence of factors such as health status and health practices (smoking and obesity, alcohol consumption and physical activity), measures of income and educational level, that potentially could have biased the conclusions of their study. But they found that these factors were independent of the association between the social network index and mortality. A likely explanation is that lack of social ties and resources may influence host resistance, cause an imbalance of neuronal, hormonal, and immunological control systems, and increase vulnerability to diseases (Cox and MacKay, 1982).

However, social contacts are closely linked to the subjective level of stress, and they may not always be beneficial. For some people, social contacts after a tedious and stressful day at work are an unbearable burden. Others, in identically stressful situations, may find relaxation in socializing. The types of social contacts that involve mostly friendship, moral support, confidence, and love are the ones capable of conferring the necessary relaxation and feelings of relief.

It is difficult to assign good health or longer life span specifically to mental attitude, health, or social interactions. These three factors are interdependent, and positive aspects in one of these areas could act as positive reinforcement on the other two. Whether people with a positive mental attitude in life are more likely to establish social ties (and would be more apt to fight diseases) seems a reasonable assumption, although still open for discussion. Personality, education, and social contacts determine the way people live, including the way they eat, whether they exercise, to what extent they are health conscious, how much alcohol they drink, or whether they smoke tobacco. The influence of these factors on health is now established and is further discussed in the second part of this book (see Chapter 4).

The Role of Happiness and Marriage

Feelings of happiness and satisfaction might lead to a longer life. However, this is a complicated question to which there are only partial answers. Happiness is a complex combination of emotions and cognitions that includes having an overall feeling of well-being, having a positive attitude toward life events, having friends, and tending to be satisfied with one's life. If positive feelings may protect against disease, negative psychological factors, on the other hand, can be injurious to health. More and more evidence shows that negative emotions play a causal role in the incidence of many diseases. For instance, depression and stress can provide a defective biological environment leading to the emergence of diseases. A study was conducted in a cohort of women physicians to analyze the factors that were likely to influence health at midlife (Cartwright et al, 1995). These women were interviewed and their psychological condition tested, first, while they were still in medical school and then at midlife. Women who were in good health were characterized by certain personality characteristics. In particular, they showed more empathy and relatively less hostility than their peers and had optimistic and trusting relationships with people. Other factors appeared to determine good health. These included a higher degree of intellectual efficiency, being better able to use intellectual resources, and having better educated parents.

On the other hand, health can have a strong influence on feelings of well-being: healthier people report greater happiness (Kozma and Stones, 1983). Positive but weak associations have been found between happiness and education, wealth and physical attractiveness. The aptitude to lead a happy and satisfactory life, or, on the contrary, to feel constantly overwhelmed by negative emotions, is likely to determine the way a person ages.

Interestingly, happiness has been found to depend on marital status. Married persons are happier than those who are single; this is particularly true for men (Argyle, 1987). An analysis of the influence of marriage on longevity was carried out in the United States, western Europe, and Asian countries, including Japan, Taiwan, and several others. In all these countries, over the last 35 years, the average death rate of single men and women was higher by a factor of 2 and 1.5, respectively, than that of married people. The situation of divorced and widowed people seemed especially woeful: six widowed men for every married man of the same age died of tuberculosis. The difference in mortality for married versus single persons was greatest between the ages of 25 and 44 years (Lynch, 1977). These puzzling observations are open to different interpretations: physically and mentally healthier people might be more likely to marry; alternatively, having a partner with whom to share the ups and downs of life makes a person better able to cope with stress and diseases.

Sexuality

Contrary to popular belief, the loss of sexuality is not an inevitable aspect of aging, and the majority of healthy people remain sexually active on a regular basis until advanced age. Brecher (1984) analyzed the sexual behaviors of 4200 individuals ranging from 50 to 93 years. A very high percentage of men and women (65–98%) in this age group were sexually active; active includes sexual intercourse and/or masturbation. In his early seventies, Brecher wrote: "Having successfully pretended for decades that we are nonsexual, my generation is now having second thoughts. We are increasingly realizing that denying our sexuality means denying an essential aspect of our common humanity. It cuts us off from communication with our children, our grandchildren, and our peers on a subject of great interest to us all, sexuality. The rejection of the aging and aged by some younger people has many roots, but surely the belief that we are no longer sexual beings, and therefore no longer fully human, is one of the roots of that rejection." Marsiglio and Donnelly (1991) reported that 53% of people 60 years of age and older, and 24% of those 76 years and older, had had sexual relations at least once within the past month and for many of these men and women, about four times during that month. They pointed out that a person's sense of self-worth and competence and his/her partner's health status were two factors positively related to the frequency of sexual relations. While results on studies vary in the percentage of sexually active men and women, they all show that elderly people have sexual desires.

Sexual interest and behavior in healthy elderly people does not seem to be influenced by gender or race; it is of universal concern. Bretschneider and McCoy (1988) showed that the frequency of sexual encounters was fairly consistent in different parts of the USA and in different countries. On the other hand, there are differences in sexual habits among countries. For instance, the results of a large survey of the sexual behavior of Japanese men (2228 men, aged from 42 to 94 years of age) indicated that the age of first sexual intercourse was more than seven years after that of American men. The frequency of sexual intercourse before 40 years of age was higher than that of American men, but after 40 years of age was less. The age at which sexual intercourse was no longer practiced was almost the same in the two countries (Nakagawa et al, 1990).

Mulligan and Moss (1991) reported in their study that intercourse frequency was diminished from a mean of once per week in 30- to 39-year-olds to once per year in 90- to 99-year-olds. However, even if there is a significant diminution of sexual desire, arousal, and activity and an increasing prevalence of sexual dysfunction with age, it is important to know that there is no age difference in sexual enjoyment and satisfaction (Schiavi et al, 1990).

The decline in sexual function with age has been attributed to changes in physiological functions in both men and women. In men, the frequency, rigidity, and duration of erections are less in aged compared to younger men. Sex steroid hormones clearly decline with age in both sexes, and there is some evidence that **testosterone** levels correlate with libido in both sexes (Morley, 1991). Schiavi and his collaborators (1992) proposed an explanation for men's decline in sexual function. They showed that there is a decrease in gonadal function during sleep in healthy aging men and suggest that the presence of sleep disorders and/or changes in sleep architecture are associated with alterations in gonadal activity in these older individuals. However, these physiological changes play a smaller role in the decline in sexual activity than is generally acknowledged (Morley, 1991), indicating that sexuality in humans is less dependent on hormonal status than in other species. Roughan and her colleagues (1993) emphasized that the physiological changes that accompany aging in women have not been shown to impair enjoyment of physical intimacy or sexual desire, arousal, and orgasm. Bachmann and Leiblum (1991) reported that in their group of healthy, post-**menopausal** women between 60 and 70 years of age, women who were sexually active had higher levels of sexual desire and greater sexual satisfaction. On pelvic examination they were noted to have less genital atrophy than the abstinent group. Interestingly, women who reported increased sexual desire had higher serum levels of free testosterone. A problem of great concern to elderly women is the absence of a sexually functioning partner in their old age. There are approximately four single women for every single man in the age range over 65, and 70% of nursing home residents are women.

The gradual decline in sexual function is also due in great part to the increased incidence of chronic diseases that become more prevalent in old age and render many elderly men impotent. Furthermore, the consumption of some medications can affect sexuality (Weiss and Mellinger, 1990).

The Role of Stress

It was Hans Selye who first drew attention to the pathogenic features of the stress response. In his initial observations, Selye found that a variety of chronic stressors caused peptic ulcers, hypertrophy of the adrenal gland, and immune deficiencies (Selye, 1936). Recently Sapolsky defined stressors and stress responses as follows: "Stress physiology is the study of the imperfections in our world and the attempts of our bodies to muddle through them. A stressor can be defined in a narrow, physiological sense as any perturbation in the outside world that disrupts homeostasis, and the stress response is the set of neural and endocrine adaptations that help re-establish homeostasis" (Sapolsky, 1992).

Exposure to stressors induces numerous biological responses. Stress affects the **hypothalamo-pituitary-adrenal** system (HPA axis, see below), the monoaminergic system, the cholinergic system, the immunological system, the neurological system, and so on. The secretion of the steroid hormone, cortisol, in stress is one of the best publicized, but other hormones and neurotransmitters are secreted during stress as well. The stress response is not only deleterious, it is also a vital and protective reaction of

Table 3. Disorders related to specific aspects of the stress response

Stress response	Stress-related disorders
Mobilization of energy	Myopathy, fatigue, **diabetes**
Increased cardiovascular tone	Stress-induced **hypertension**
Suppression of digestive functions	Digestive ulcers
Suppression of growth	Dwarfism
Suppression of reproduction	Amenorrhea, impotency, loss of libido
Suppression of immune system	Increased risk for several diseases
Neuronal overstimulation	Neuronal death

the body against acute psychological and physical stressors. Individuals who fail to exhibit this response lack appropriate defense mechanisms. The deleterious effects of stress include the difficulty to interrupt stress responses after the stressor has disappeared or the permanent activation of stress responses. Psychological stressors rank high out of the large variety of stressors affecting humans. Sapolsky listed the kind of stress-induced disorders that are related to specific aspects of the stress response (Table 3; Sapolsky, 1992).

Since the discovery by Selye (1936) (for a review, see Selye and Tuchweber, 1976), much evidence has been published supporting links between stress and diseases, although there are still investigators who think that psychological and physical factors are unlikely to have major influences on the incidence of diseases and, more specifically, on the occurrence of cancer. Solomon and Amkraut (1979), for example, conclude their work with the following statement: "Stress-induced changes in the immune system are generally small and determine the course of the disease chiefly by shifting the balance between toxic factors and defense mechanisms in disease processes." Bammer (1982) states that: "There are no known data to support the concept that stress will produce a primary tumor", and Fox (1981) adds that: "the contribution of psychological factors to cancer incidence is probably relatively small . . . and if it exists, is almost certainly specific to certain organ sites and depends on several things."

It is not easy to prove a causal relationship between stress and diseases that have a long-term development; cancers have a time-course of development over many years and are caused by an array of different factors, including radiation, chemicals and carcinogens, viruses, hormonal actions, and tissue damage (Chapter 2). Carcinogenesis is also modulated by genetic influences, by age, and by the state of the immune system.

To find answers to these questions, scientists have studied the effects of stress in animals, particularly rodents. The stressors used are physical constraint, electric shock, and restricted feeding. Even though animal models provide useful answers, they do not always mimic human situations. It is difficult if not impossible to design animal studies that will reproduce the experience of major stressful events such as the loss of a close parent or separation from a spouse. Thus, studies with rodents or nonhuman primates are necessarily restricted in the psychological and emotional factors they can investigate. Even studies on the role of stress and of detrimental psychological factors on the onset of diseases can be misleading in that they focus on factors that are in addition to the general or everyday stresses of living. It is likely that basic living itself can be psychologically stressful enough to contribute to the production of primary cancers. Riley (1981) showed that removing mice from the stresses of normal housing conditions may reduce the likelihood of tumor growth. Another important point is that the limit between good stress and excess of stress varies for each individual. McEwen and Stellar (1993) pointed to the importance of individual differences in the susceptibility to stress. The behavioral responses of people to environmental challenges are specific for each individual and are coupled to specific physiological and pathophysiological responses.

Gerontologists are interested in stress responses, particularly because aging is accompanied by a decreased capacity to respond to stressors due to a less efficient regulation of homeostasis. A well-known example is the longer time required to return to normal body temperature after a cold or hot challenge (Shock, 1977). Also, chronic stress is thought to accelerate neuronal degeneration during aging. Most scientists agree that intermediary factors exist between stress and personality on the one hand and diseases on the other. Such an intermediary factor is believed to be the immune system (Solomon, 1987). The negative effects of stress on the efficiency of the immune system, in particular on the activity of natural **killer cells** (Rodin, 1986; Levy et al, 1987), are well described. Cancer might then occur as a consequence of a deficient immune system, secondary to stress.

The Biology of the Stress Response

Steroid hormones are implicated (together with other endocrine or neurotransmitter systems) in the adaptive and pathogenic aspects of the stress response. One of these hormones, cortisol, a **glucocorticoid** hormone synthesized from **cholesterol**, is secreted by the adrenal glands into the bloodstream in response to a variety of stressors. This secretion is the last step in a series of hormonal responses of the hypothalamo-pituitary-adrenal (HPA) axis, starting in the brain, proceeding to the pituitary gland, and ending in the adrenal glands. A simplified sequence of events is as follows. First, the brain perceives and identifies a stressor, i.e., concludes that there is a significant discrepancy between predicted and measured internal states or external events. This identification depends in part on automatic, unconscious, and rapid analysis and in part on a secondary reflection on the exact meaning of the stressor for the individual. Within seconds, this causes the secretion of **corticotropin-releasing factor** (CRF), **vasopressin** (VP), and other secretagogs including **oxytocin** and **catecholamines** (epinephrine and norepinephrine) into the hypothalamic-pituitary portal circulation. This signal triggers the release of **adrenocorticotropic hormone** (ACTH) from the anterior pituitary, which in turn activates the synthesis and release of glucocorticoids in the adrenal gland. Glucocorticoids increase in the bloodstream within a few minutes after stress and reach their target tissues (McEwen and Brinton, 1987). In normal circumstances, the HPA axis is under a negative feed-back regulation, as are most neuroendocrine systems (Figure 8a). When the feed-back threshold is reached, further secretion of CRF and ACTH is inhibited by cortisol itself. If circulating concentrations of glucocorticoids are too low, because of sustained or excessive stress for example, secretagogs and ACTH continue to be released.

Recently, two classes of intracellular glucocorticoid receptors have been characterized: a high-affinity type I receptor, binding glucocorticosteroids and **aldosterone**, and a low-affinity type II receptor, binding endogenous and synthetic glucocorticoids. Both types of receptors are present in the brain. Whereas type II receptors are ubiquitous in the brain, type I receptors are mostly in the hippocampus and septum. The hippocampus has in fact more corticosteroid receptors of both types than any other brain region and is among the major neural targets for glucocorticoids. The regulation of receptor numbers in the brain is controlled by negative feed-back: adrenalectomy causes a sharp drop in glucocorticoid synthesis and a secondary increase in the number of regulatory brain corticosteroid receptors, while sustained glucocorticoid exposure during stress can decrease these regulatory receptors. For example, three weeks of a regimen of various stressors such as cold, ether, or histamine administration, and immobilization, caused a notable decrease in brain glucocorticoid receptors in rats, without changing the affinity of these receptors. Interestingly, this decrease was not induced in all brain regions but was most prominent in the hippocampus and amygdala, where a 50% decline in the cellular concentrations of receptors was found (Sapolsky et al, 1983).

In recent years, considerable research has focused on the intracellular transport of glucocorticoids and on their molecular mode of action. These hormones pass through the cell membrane and bind to receptors located in the cytoplasm and in the nucleus. Once a hormone-receptor complex is formed, the receptor is in its activated form and passes through the nuclear membrane. The hormone/receptor complex undergoes a conformational change that reveals a site known as the *steroid-hormone-responsive elements,* which binds to specific DNA sequences. One of the main functions of steroid hormones is to change the rate of transcription of new proteins: transcription is increased or decreased. The direction of the positive or negative regulation of transcription depends on the number and nature of defined DNA sequences (or DNA-responsive elements) located upstream with respect to the genes, the spacing between them, and also on the interactions between glucocorticoid receptor complexes and a variety of other transcriptional regulators (Burnstein and Cidlowksi, 1989). To add another degree of complexity, steroid hormones themselves can be positive or negative regulators at the *same* DNA-responsive elements depending on the presence of other binding cofactors (Diamond et al, 1990). One may wonder about the justification for such complex interactions at the genomic level. One reason is that hormones have dose-related effects on target tissues, which probably implies the existence of a hierarchy of steroid-induced responses. A low concentration of hormones activates only a few highly responsive genes, while higher concentrations of the hormones will activate those genes further, and possibly recruit new genes. Furthermore, the interactions with other transcriptional factors generate additional subtleties in the transcriptional regulation of genes.

At the physiological level, glucocorticoids have a wide range of actions besides mobilizing glucose. Low basal concentrations of glucocorticoids exert anabolic effects, while higher levels have inhibitory effects, and, when produced in excess, extremely damaging effects. At high concentrations, glucocorticoids inhibit transport of nutrients to target tissues, promote catabolism of proteins, **glycogen** and **triglycerides**, and stimulate gluconeogenesis. They enhance catecholamines synthesis and potentiate their cardiovascular effects. They also inhibit growth, digestion, inflammation, and immunity through different mechanisms (Rose, 1985). Sapolsky (1992) listed the known inhibitory actions of glucocorticoids (Table 4), based on data by Krieger (1982), Munck et al (1984), and Sapolsky et al (1986b).

Most studies so far have been based on rodents, but the neuroanatomy of many other mammalian species is quite similar. The primate hippocampus, for example, communicates with the hypothalamus by means of neuronal projections that are analogous to those of rodents; its corticosteroid receptors show many of the same functional properties. The primate hippocampus, therefore, is probably as much a glucocorticoid target tissue as is the rodent hippocampus. The inhibitory role of the primate hippocampus on the hypothalamus has been demonstrated by specific hippocampal lesions that lead to hypercortisolism. Incidentally, in patients with Alzheimer's disease, early neuronal damage is seen in the hippocampus, and the more severe the hippocampal atrophy, the stronger the hypersecretion of cortisol into the bloodstream (De Leon et al, 1988).

Table 4. Physiological and biochemical processes inhibited by glucocorticoids

Immune system:
Interferon
Natural killer cell activity
Interleukin-1
Macrophage activating factor
Lymphocyte activating factor
T-cell growth factor
Inflammatory processes:
Prostaglandins synthesis
Leukotrienes synthesis
Bradykinin synthesis
Metabolism:
Insulin secretion
Vasopressin secretion
Growth hormone secretion
Insulin-like growth factor secretion
Amino acid transport and protein synthesis
Triglyceride synthesis
Glucose transport and glycogen synthesis
Calcium transport into bone
Hormone secretion:
CRF secretion
ACTH secretion
β-Endorphin secretion
Thromboxane synthesis
Gonadal responsiveness to **LH**

(CRF: corticotropin-releasing factor; ACTH: adrenocorticotropic hormone; **LH: luteinizing hormone**).

Aging of the Hypothalamo-Pituitary Axis (HPA)

When aged rats were subjected to various stressors, they were perfectly capable of secreting glucocorticoids; but they secreted an excessive amount of these hormones and showed a marked inability to terminate secretion after stress. This hypersecretion points to a failure of the negative feedback inhibition of the HPA axis by glucocorticoids. The origin of the failure is, at least in part, in the brain, possibly at the level of the hippocampus (Sapolsky et al, 1986a, b) where about 50% of cytosolic corticosteroid receptors are lost in aged rats (Sapolsky et al, 1983). While the receptor loss is of considerable significance, it seems to be secondary to the loss of neurons themselves, specifically those neurons that express a high concentration of corticosteroid receptors. Sapolsky hypothesized

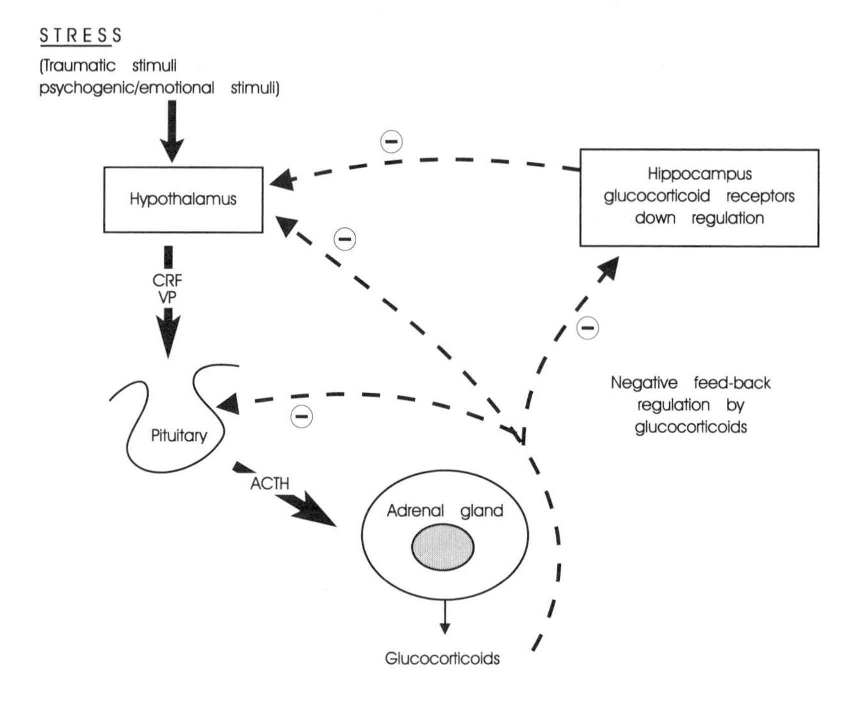

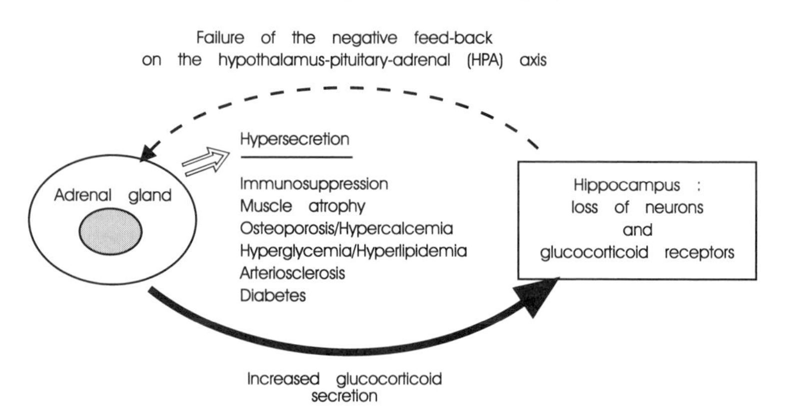

Figure 8. Physiology of the endocrine stress response and its modification with aging.

Normal secretion of hormones and negative feed-back that occurs in stressful conditions are shown in Figure 8a (see text for abbreviations). Figure 8b shows the feed-back insensitivity due to senescent hippocampal neurons that results in glucocorticoid hypersecretion. Well-known consequences and/or diseases caused by glucocorticoid hypersecretion are listed (Sapolsky et al, 1986b).

that the death of hippocampal neurons could explain the loss of sensitivity to glucocorticoid feed-back and might be the actual cause of the glucocorticoid hypersecretion with aging (Figure 8b). The hypothalamus or the pituitary are also implicated in the negative feedback regulation of glucocorticoids. The hippocampus, however, is probably the most important, because it is the one damaged most readily during aging.

Studies in adult rats have shown that the synthetic glucocorticoid dexamethasone normally exerts a negative feedback signal and lowers endogenous glucocorticoid secretion. But, the older the rats, the more resistant they are to the inhibitory feedback effect of dexamethasone (Sapolsky et al, 1986a). A similar regulation of glucocorticoid secretion is observed in elderly people. Those over 80 years of age showed marked basal hypercortisolism and resistance to dexamethasone. Interestingly, the regulation of the HPA axis remains relatively normal in people with an average age of 60 years (Indeed, 60-year-old men and women are not considered old.). Glucocorticoid hypersecretion is also seen in about half of patients with severe depression, in patients with Alzheimer's disease, and in those who absorb alcohol both acutely and chronically (Majumdar et al, 1981; Stein et al, 1991; Young et al, 1991).

Stress, the HPA axis, and Accelerated Senescence

The question of whether chronic stress can damage primate hippocampal neurons sufficiently to accelerate the rate of aging has not been solved yet, but a few studies on wild-born vervet monkeys kept in captivity provide some pointers. The monkeys died spontaneously of what seemed to be a syndrome of sustained social stress. They had ulcers, hyperplastic adrenals, splenic lymphoid depletion, and colitis. Interestingly, their behaviors were analogous to socially subordinate animals. When their brains were examined, they showed pronounced hippocampal degeneration (Uno et al, 1989). Scientists then investigated whether glucocorticoids were mediators of this degeneration in primates. They implanted cortisol-secreting pellets into one hippocampus and cholesterol-secreting pellets into the other hippocampus of four vervet monkeys. Although there are many caveats inherent to this type of study, they observed after one year, a preferential damage in the cortisol-treated hippocampus (Sapolsky et al, 1990). These results, as well as those of other studies, suggest that many features of the glucocorticoid-induced neurodegeneration are shared by rat and human and support the idea that sustained severe stress in primates is associated with hippocampal damage and early death.

The model proposed by Sapolsky of the effects of stress on hippocampal neuronal loss and on the resulting hyperadrenocorticism is attractive and has experimental support, but it is not yet proven. More research needs to be carried out in humans in particular to show that the aging process is accelerated by chronic stress.

Diseases of Premature Aging

More than 150 human syndromes have been identified as having potential relationships to the biology of aging. Patients afflicted with these diseases called progerias have an accelerated rate of senescence and show at an early age many of the **phenotypic** features associated with old age: greying and loss of hair, changes in fat distribution, hypertension, degenerative vascular disease, **osteoporosis**, cataracts, cancers, diabetes mellitus, disturbed lipid metabolism, hypogonadism, **autoimmune** changes, chromosomal abnormalities, fibrosis, **amyloid deposition**, **lipofuscin** deposition, and mitochondrial abnormalities (Martin, 1977).

The substantial phenotypic overlap between progerias and normal aging suggests that they might share some mechanisms. However, none of the progerias corresponds to a true global progeria disease, i.e., these syndromes include several but not all the characteristics of normal aging (Martin, 1982). A few progeria syndromes that display characteristics that best fit the criteria applied to normal aging are listed in Table 5.

The first four progerias from Table 5 are the best studied, and an outline of their features is given below.

Werner Syndrome

Werner Syndrome (WS), also called progeria of the adult, has been reported worldwide. WS typically has its onset in late adolescence. The

Table 5. Representative premature aging diseases (progerias)

Werner syndrome
Hutchinson-Gilford syndrome or classical progeria
Cockayne syndrome
Down syndrome
Turner's syndrome
Louis-Bar (ataxia telangectasia) syndrome
Seip syndrome
cervical lipodysplasia
myotonic distrophy

clinical phenotype of WS is consistent among affected individuals from all populations. WS affects virtually all tissues: greying of hair and baldness, short stature, cataracts, sclerodermatous appearance of the skin, and atrophy of peripheral muscles. WS patients have delayed sexual maturation, somewhat of a paradox for a disease in which senescence is accelerated. They develop glucose intolerance, increased incidence of cancers, soft tissue calcification, atherosclerosis, etc. They very rarely, if ever, develop **dementia**. Death occurs between 30 and 60 years of age, most often as a result of cancers or cardiovascular diseases. The deteriorations observed in WS patients suggested at first that this phenotype was conferred by several **mutations**. Surprisingly, however, the genetics of the disorder argue for a simple, single-gene mutation. Indeed, this rare disease displays a clear **autosomal** recessive mode of inheritance and is highly penetrant. Extensive studies on the clinical and biochemical features of WS have led recently to the exciting discovery of a candidate gene responsible for the disease. First, the locus of the primary defect was identified by Goto and his collaborators by **genetic linkage** studies (Goto et al, 1991). They reported a close linkage of the WS mutation to a group of **markers** on chromosome 8. These genetic markers enabled prenatal and carrier diagnosis in families with WS members and served to identify the defective gene responsible for the disease. The sequence of the WS gene encodes a protein of 1432 amino acids. It closely resembles the sequences of genes that code for helicases, enzymes that unwind DNA strands (Yu et al, 1996). Mutations that cause WS may exert their effects by affecting the unwinding of the genetic material, and thus perturbing different activities such as DNA repair, replication, gene expression, or chromosome recombination. Researchers are now devising experiments to confirm that the WS gene encodes a helicase and to check what effects mutations in this gene have on DNA repair and other aspects of DNA metabolism. These findings may provide important clues to the understanding of aging.

Hutchinson-Gilford Progeria Syndrome

In Hutchinson-Gilford's progeria syndrome (HGS), children age prematurely and have reduced growth between the age of 6 to 12 months. Their skin shows thinning and loss of fat. They have a characteristic face with prominent eyes, a beaked nose, and facial disproportion resulting from a small jaw and a large cranium. Interestingly, they do not show the increased skin lipofuscin pigment, glucose intolerance, cancers, diabetes, dementia, and osteoarthritis that often occur during aging. Their intelligence is usually preserved. They die from severe atherosclerosis with cardiac and cerebral involvement at the time when normal children become adolescent. Sexual maturation often does not occur. This very rare condition (one in a million to one in ten million births) is a sporadic autosomal dominant disease. Many investigations have been conducted to find a genetic marker or a cellular defect that could help define the underlying cause of the disease. For example, one approach consisted in analyzing the life span of HGS progeroid fibroblasts in vitro. Initial studies claimed that progeroid fibroblasts in culture underwent fewer than normal cellular divisions. However, subsequent studies with these cells showed that, once difficulties in establishing a tissue culture were overcome, a normal or only modest reduction in in vitro cellular life span was observed (Goldstein, 1969; Goldstein and Moerman, 1976).

Likewise, the original observations that these fibroblasts were deficient in repairing X-ray-induced single-strand DNA breaks were refuted by subsequent studies. These progeria cells were found to repair X-ray-damaged and ultraviolet-damaged DNA normally (Brown et al, 1980). No enzymatic or metabolic abnormality has yet been found. The etiology of HGS could be related to the excessive production of hyaluronic acid; hyaluronic acid is known to inhibit blood vessels and connective tissue development (Brown et al, 1984). It is possible that this finding might lead to interesting discoveries. Current hypotheses also favor the idea that HGS patients suffer from a defect in normal homeostatic metabolism or in overall gene expression (Brown et al, 1980). Obviously, more research is needed to identify the cause of this disease.

Cockayne Syndrome

Cockayne syndrome (CS) patients have characteristic body features: long arms and legs, enlarged hands, microcephaly, prominent ears, eyes deeply set in their cavities, loss of subcutaneous fat. CS patients develop hypogonadism as well as optic atrophy, deafness, and progressive ataxia. An unusual feature is the occurrence of a progressive intracranial calcification detected by computed tomographic (CT) scan. Unlike patients with HGS, most CS patients show a progressive mental retardation with onset at variable ages. Death is usually secondary to progressive neurodegeneration during late childhood or early adolescence. Cultures of fibroblasts from CS patients show an increased sensitivity to UV irradiation (Schmickel et al, 1977) resulting in a reduced cell growth in culture, as assayed by **colony forming ability**. Despite this hypersensitivity to UV, no defect in excision or DNA repair has yet been found (Cleaver, 1982). CS is one of the rare premature aging diseases with a well-documented abnormality at the cellular level. This abnormal sensitivity to UV can be used for prenatal diagnosis (Sugita et al, 1982), but the genetic abnormality has yet to be identified.

Down Syndrome

Down syndrome (DS) is often considered the progeria displaying the most features in common with normal old people, aside from the typical mongoloid face and mental retardation of the patients. DS patients show early degenerative vascular disease, hypogonadism, cataracts, greying of hair, hair loss, increased tissue lipofuscin, a high incidence of cancers, **amyloidosis**, and increased autoimmune diseases. Life expectancy is reduced to 40 to 50 years. DS is usually due to an extra copy of chromosome 21, i.e., cells are **trisomic** with regard to this chromosome. However, about 4% of DS cases are due to **mosaicism**: in other words, subjects have both normal diploid and abnormal trisomic cells. Another 4% are due to translocations of the whole or part of chromosome 21 to other chromosomes. Thus, unlike what appears to be the case in other progerias, the cause of DS is not a qualitative gene defect, but a disturbance in gene dosage: the disease is due to a quantitative difference in expression of genes located on chromosome 21. The products of these genes could secondarily affect the expression of genes located on other chromosomes. In almost all DS patients older than 40 years of age, neuropathological lesions reminiscent of senile dementia of the Alzheimer's type (SDAT) are present. Interestingly, chromosome 21 is implicated in the pathogenesis of the familial form of SDAT, since the gene encoding the β-peptide amyloid precursor (APP), a primary constituent of the **senile plaques** seen in SDAT patients, is located on chromosome 21. Recent studies sustain the hypothesis that the extra copy of the APP gene results in increased expression of APP which, in turn, may trigger abnormal processing and lead to βA4-amyloidosis, as in SDAT (Iwatsubo et al, 1995) (see pages 49–50).

As seen from the above description, the progerias share many of the characteristics of aging. However, these various phenotypes also reveal significant differences from normal aging. The fact that the progerias result from various genetic defects suggests that senescence is likely to be due to the action of several genes, i.e., to have a polygenetic rather than a monogenetic basis.

Little is known about the mechanisms responsible for the progerias. At present the identification of a few genetic regions localized on chromosomes 8 and 21 and the recent characterization of the WS gene are important steps forward in the search for genes that play a role in the pathophysiology of WS and DS, respectively. Whether these genetic defects are sufficient to explain all signs and symptoms of these two progerias is a key question. A better understanding of WS and DS will undoubtedly help determine whether similar genetic mutations occur in all progeroid syndromes and whether they share the same defective mechanisms (Salk, 1982; Epstein et al, 1966; Salk et al, 1985). How many genes will be involved in the different progerias, and what types of proteins they code for, are still open questions. Research on progeria will have important implications for the understanding of normal aging.

A Note on Diseases and Longevity

As we have indicated, humans theoretically are programmed genetically to live about 120 years. This means that if the slow deterioration of all physiological functions associated with aging proceeds normally, the human body has the potential to live significantly longer than one hundred years. In fact, the human life span is shorter mainly because of infections or noninfectious age-associated diseases. While no one would argue that we die because of diseases, the causal relation between diseases and longevity is not so simple. The increase in the incidence of some diseases with age occurs with a precise timing that is often concomitant with severe deterioration of the homeostatic balance (see Chapter 3, pages 72–74). However, the loss of homeostasis, as observed in normal aging, does not necessarily lead to the onset of diseases. This dilemma is a key issue in gerontology research. The question gerontologists try to solve is whether it is possible to determine the breaking point, at the cellular level, where imbalance in homeostasis triggers a signal that switches from physiological to pathological aging.

The slow processes leading to deterioration of physiological functions and the increased vulnerability to all sorts of stresses are the causes of a number of diseases, but there is evidence that many diseases are due to genetic predispositions, result from given environmental factors, or, as is often the case, are due to a combination of both genetic and environmental factors. In these cases, diseases override the loss of homeostasis and act as determinant factors in diminishing longevity.

Alzheimer's disease or senile dementia of the Alzheimer's type (SDAT) and Parkinson's disease (PD), the most common neurodegenerative diseases in the elderly, are illustrations of the difficulties of pinpointing the causes of diseases. Patients affected by SDAT or PD represent about 10 percent of those aged over 65 years. SDAT leads to dementia, characterized by the loss of the ability to learn new information and to recall and use previously acquired knowledge. It is characterized by a progressive neuronal dysfunction.

The cytoskeleton of neurons deteriorates, and this is accompanied by severe neurocytological lesions of the major filament systems (microfilaments, intermediate filaments, and microtubules). **Neurofibrillary tangles** and **neuritic plaques** are the two major structures indicative of cerebral cortex degeneration in SDAT. They contain high amounts of abnormal proteins. In the last ten years, the primary cause of neuronal degeneration and of SDAT has been attributed to defective neurofilaments and to the small β-amyloid protein (βA4 peptide fragment, about 40 amino acids in length) which is part of the β-amyloid precursor protein (APP). Scientists think that other factors may also be linked to SDAT. They are focusing on genes, toxins, infectious agents, head trauma, stress, and on changes in the immune and endocrine systems, as well as changes in normal metabolic processes. Even though the possibility that SDAT might occur as age-related changes common to everybody, the hypothesis that it is caused by gene defects is gaining more support (see below).

PD is a progressively disabling neurological disorder caused by the degeneration of **dopamine**- and **noradrenaline**-producing nerve cells that originate in areas of the brain called the *substantia nigra* and the *locus ceruleus*. It results in tremor, rigidity, undirected movement, and postural instability. Quite often, these features are accompanied by dementia and behavioral abnormalities. PD, as well as many other neurodegenerative diseases, are likely to be a syndrome, i.e., the final product of a variety of disease processes (Chiu, 1989; Henderson and Finch, 1989).

If many diseases are the result of combined genetic and environmental factors, some of them are caused solely by genetic anomalies or exposure to external injuries. A few examples are given below.

Genetic Predisposition to Diseases

It is estimated that about 20% of age-associated diseases have a genetic basis; for example, diabetes, arthritis, human leukocyte antigen (HLA)-associated diseases, hyperlipidemia, α-1-antitrypsin deficiency, cystic fibrosis, and genetic forms of mental retardation.

The human leukocyte antigen (HLA) system, or major histocompatibility complex (MHC), deserves particular attention because it plays a major role in diseases and senescence. More than other genetically determined diseases, the HLA-associated diseases illustrate the crucial role of specific genetic factors on the incidence of diseases. Several HLA-associated diseases are described below.

For a long time, scientists have stressed the role of genes located in the MHC in the aging process. These genes play a decisive role in immunological defenses and in tissue recognition and rejection. In mice, the MHC genes are on chromosome 17, and in humans they are on the short arm of chromosome 6, in the region called human leukocyte antigen locus (HLA). This genetic region consists of a highly polymorphic series of a dozen or more loci that code for proteins found in the plasma and on cell membranes. Originally, scientists noticed that mice with the longest life span had the slowest rate of decline of their T-cell immune functions. Later, this observation was related to specific MHC genotypes (Popp, 1982), which suggested that T-cell functions are controlled by the MHC region. Proust and his colleagues (1982) and Takata and his colleagues (1987) made a similar observation in humans. The latter showed that the 20 centenarians and 80 nonagenerians included in their study had a statistically significant association with a particular allele, HLA-DR1. The DR1 allele was predominant, while DRw9, the cell marker that has been linked to autoimmune and immune deficiency diseases, such as diabetes, colitis, and liver disease, was present in very few of these individuals. The DR1 allele (and maybe other genes) could thus be viewed as a longevity-enhancing gene.

The high frequency of a given HLA allele or a combination of HLA alleles in patients with a particular age-related disease suggests an association between the HLA locus and the occurrence of that disease.

Ryder and his colleagues (1981) described associations between specific HLA genotypes and diseases of the joints, including rheumatoid arthritis (RA), juvenile arthritis, and ankylosing spondylitis (AS), a degenerative bone disease. RA is a chronic inflammation involving the deposition of immunoglobulin complexes in the connective tissues of joints; the immunoglobulin complexes then stimulate the infiltration of lymphocytes. This creates a vicious cycle of inflammatory responses and results in stiffness (ankylosis) and destruction of the joint. While the incidence of rheumatismal diseases increases steadily with age and becomes extremely common in both sexes by 70 years, the risk of juvenile and adult onset of RA are about fourfold greater in carriers of the alleles, HLA-DR4. Similarly, an individual carrying the HLA-B27 genotype is 87 times more likely to develop AS than an individual who does not carry this genotype.

These diseases are not necessarily life-shortening in some socioeconomic subgroups, but the handicaps that they cause contribute to limiting the possibilities of leading a healthy and fulfilling life. The mechanisms that link given HLA haplotypes and diseases are not always known. In some cases, the consequences of viral infections may be influenced by HLA haplotypes. For example, the Epstein-Barr virus has a coat protein (gp110) with an amino acid sequence that is shared by T-cell epitopes of protein coded by the HLA-DR4 allele. Because 70% of patients with RA are positive for HLA-DR4, it is possible that infections with Epstein-Barr virus could trigger autoimmune reactions in people with this haplotype, leading to RA in a subset of these individuals.

At present, researchers are involved in finding how a given HLA genotype favors a particular disease state and, by extension, how it might influence life span. Walford (1987) suggested that the MHC not only influences immune processes, but also a variety of other systems involved in aging processes, such as free radical scavengers, DNA repair, and neuroendocrine regulation. For example, patients suffering from systemic lupus erythematosus, a disease closely linked to the HLA complex in humans, have a severe defect in their DNA repair ability. Similarly, the response of individuals to corticosteroids is also influenced by genes of the HLA complex. On the basis of the evidence linking the HLA complex to many diseases or physiological functions, Walford (1987) suggested that: "The MHC is an antibiosenescent, homeostatic system, probably derived from evolutionarily old genes concerned with free radical scavenging. Its widespread effects may relate to its regulation of a number of inducers including, particularly, hormones and cAMP. In many cases, it may interact with genes at a distance or even on other chromosomes."

Genetic Prediposition to Cancer

In the last twenty years, a steady increase in mortality due to cancer has been noted and cancer is now the second leading cause of death in developed countries, being responsible for one in five deaths (Fobair and Cordoba, 1982).

The recent interest in hereditary transmission of cancer came to a large extent from the major advances in the field of molecular genetics, which made possible the identification of genes and the characterization of the corresponding proteins. Table 6 lists several cancers in which genetic inheritance has been identified. Some of these cancers are rare, such as retinoblastoma; others are more common, such as breast or colon cancers. With some of these familial cancers, different types of cancers may aggregate in the same family. For example, early-onset breast cancer can occur together with osteosarcomas, brain tumors, or **leukemias** in different members of the same family. In other cases of familial cancers, the same specific type of cancer occurs in several members of a family (usually at a younger age than in the general population). It is interesting to note that some cancers such as breast cancers, for example, are associated with germ line mutations in different genes. Moreover, mutations in some of the genes listed in Table 6 were found to occur in types of cancers other than those with which they were originally linked. For example, carriers of the BRCA1 mutation seem to have an increased risk not only of breast and ovarian cancer, but also of colon and prostate cancer.

Situations in which cancers are clearly associated with a preexisting **germ line** mutation are rare; interest in finding other transmitted cancer genes would have waned if studies on major monogenic predisposition to develop cancer had not led to the identification of genetic sites that are frequently mutated at the beginning or during the evolution of many cancers (see Chapter 5, pages 95–106). The identification of these germ line or **somatic** mutations in given genes makes it now possible to identify patients with a strong predisposition to develop tumors.

Table 6. Examples of monogenic predispositions to develop cancers with dominant inheritance

Cancers	Description	Protein involved
Retinoblastoma	Eye tumor in children. Occurs in a familial and sporadic form. Gene defect on chromosome 13.	Transcriptional factor (RB)
Wilm's tumor	Renal cancer in children. Gene defect on chromosome 11. Other genes may be involved.	Transcriptional factor (WT1)
Malignant melanoma	Skin cancer. Gene defects on chromosomes 9 and 1.	Chromosome 9: **cyclin-**dependent kinase inhibitor (MTS1)
Multiple endocrine neoplasia (MEN 1 and 2)	Cancers in endocrine glands and other tissues. Gene defects on chromosomes 11 (MEN1) and 10 (MEN2).	MEN2: tyrosine kinase receptor (RET)
Neurofibromatosis type 1	Cancer of the nervous system. Gene defect on chromosome 17.	Protein controlling Ras protein (NF1)
Neurofibromatosis type 2	Cancers of the auditory nerves and tissues surrounding the brain. Gene defect on chromosome 22.	Protein interacting with cytoskeleton (NF2)
Familial adenomatous polyposis	Numerous colorectal polyps at adolescence; later colon cancer. Gene defect on chromosome 5.	Protein interacting with cytoskeleton (APC)
Nonpolyposis colorectal cancer	Colon cancer. Gene defect on chromosome 2 (Peltomäki et al, 1993).	Protein involved in DNA replication and microsatellite DNA stability (FCC)
Breast cancer	Several genes are implicated in familial breast cancer; the most important of these, BRCA1 and 2, are located on chromosomes 17 and 13, respectively (Evans et al, 1994; Wooster et al, 1994). The defective gene causing ataxia telangiectasia may be the single largest hereditary cause of breast cancer (Savitsky et al, 1995).	Possible tumor suppressor protein (BRCA1)

(For more information, see review by Thomas, 1995).

Genetic Predisposition to Neurodegenerative Diseases

Alzheimer's Disease

Senile dementia of the Alzheimer's type (SDAT) is a genetically heterogeneous and complex **disease**. In the last ten years, important breakthroughs in the identification of genetic susceptility to SDAT have been achieved. Researchers have identified genetic and biological markers for SDAT that should help clinicians make an early, accurate diagnosis of the disorder. They have located genetic loci on chromosome 21 that may be linked to the familial form of the disease. In fact, β-amyloid protein (a fragment of the APP), which accumulates in plaques in the brains of SDAT patients, maps also on chromosome 21, and several mutations in the gene coding for the APP protein are associated with an aggressive form of the SDAT.

In the spring of 1993, a surprising discovery was reported by Schächter and his collaborators (1994) who were working on cardiovascular diseases, a leading cause of death in the western world. They were interested in possible associations between longevity and genes encoding apolipoproteins. One of these apolipoproteins, apoE, had been shown earlier to have a major impact on total **LDL-cholesterol** levels in the serum. The finding made by these scientists was that a particular form of ApoE is a genetic risk factor for late-onset Alzheimer's, the most common form of the disease (more information is provided in Chapter 5, pages 91–92).

Recently, researchers reported the discovery of a gene, called S182 (now called Presenilin-1 (PS1)) on chromosome 14 that is responsible for 80% of the cases of familial Alzheimer's disease (Sherrington et al, 1995). But this did not account for all cases of SDAT, and in some families in which SDAT was inherited as an autosomal dominant trait, the role of these SDAT loci was excluded. Another collaborative team has revealed the presence of a gene on chromosome 1 that could account for the remaining familial Alzheimer's cases (STM2, now called Presenilin-2 (PS2)) (Levy-Lahad et al, 1995a, b). Family members with SDAT all have the same mutation in the gene, while unaffected family members do not. These two genes, Presenilin 1 and 2, are closely related, which suggests that the proteins they encode have similar functions. The presence of **missense mutations** in SDAT patients in two similar genes indicates that these mutations are pathogenic. The goal is now to find out the putative function of these genes and to screen all families with familial SDAT to determine whether all have defects in only the currently identified genes or whether still other genes are involved.

These findings will help us better understand the causes not only of familial SDAT, but also of the more common sporadic form of SDAT that affects people over 65 years. This example illustrates also the speed with which complex issues like the identification of the genetic basis of SDAT are solved (for a review, see Iwasaki et al, 1996; Sandbrink et al, 1996).

Diseases Due to Environmental Factors

Long-term exposure to many environmental factors can influence the development of cancers. As early as 1775, London surgeon Percival Pott described the development of scrotal cancer in chimney sweeps. Today, many cancers have been associated with environmental causes: skin cancer and leukemia among radiologists, osteosarcomas in workers who paint watch dials with illuminating radium-containing dyes, lung cancer in tobacco smokers, skin cancer after UV radiation.

Medical treatment may contribute also to the induction of malignancy. Diethylstilbestrol (DES), formerly used for the prevention of miscarriage, was associated with an unusual type of vaginal cancer in young women whose mothers were exposed to the drug during pregnancy. Viruses are also important causes of cancer. For example, Epstein-Barr virus causes Burkitt's lymphoma and nasopharyngeal carcinoma in southeastern China, human papilloma virus causes cervical carcinoma, and hepatitis B, liver cell carcinoma.

Environmental factors are also postulated to play a role in the progressive neurodegenerative disorders described above (SDAT and PD), although the environmental effect is not yet established definitely. Exposure to some chemicals can result in neurological syndromes similar to naturally occurring neurodegenerative diseases. For example, exposure to pesticides has been associated with Parkinson's disease, while aluminum has been implicated in Alzheimer's disease.

Chapter 3

Structural and Physiological Changes Occurring with Age

Old age often is not a period of peace and happiness because elderly people are faced with physical and psychological changes resulting in successive losses and dependence. They more or less adapt to these new situations as long as the changes are not incapacitating. Visible structural changes, i.e., to the face, hair, skin, and body shape, are accompanied by many changes in internal organs. For example, the percentage of fat increases over age, while that of water decreases (Fryer, 1962). The loss of bone mass leads to a higher risk of fracture, and the lower rate of drug elimination by the liver exposes elderly patients to frequent adverse drug reactions. Often, elderly people have difficulty adapting to the rapid tempo of modern life because of the slowing down of their cognitive processes. Their body's capacity to adapt to internal and external demands is diminished, and their **functional reserves** are depleted. This causes an overall fragility, which can lead to death in the absence of specific diseases, a condition called apocrypha. Apocrypha is observed in diet-restricted rodents who enjoy a longer life span. These animals simply die, and careful necropsy of these animals does not reveal gross pathological abnormalities (Finch, 1990). This chapter examines the modifications occurring with age at the level of the different physiological systems and describes the consequences of these changes.

Major Functional Systems

In young subjects, the major functional systems, i.e., the neurological, the cardiovascular, and the pulmonary systems have the capacity to perform in response to unusual and excessive internal and external demands, by virtue of so-called functional reserves. With age, these functional reserves diminish to the point of impairing the equilibrium of regulatory systems. Changes in the functional reserves of the main physiological functions have been measured in several long-term longitudinal studies of aging, the best-known being the Baltimore Longitudinal Study on Aging (Shock, 1983, 1985; Tobin, 1984; Shock et al, 1984). This study represents a remarkable achievement. It was initiated in 1958 with the goal of studying aging in individuals who were leading successful lives and were free from disabling diseases. People who enrolled in the study were tested and examined throughout their entire lives. The study comprises a large number of comprehensive analyses of physiological, psychological, and social aspects of aging. It is still being carried out and continues to provide valuable information on all aspects of human aging.

In one of these studies, repeated measurements of major physiological functions were performed on male volunteers, ranging in age from 18 to 103, who were free of significant diseases within the limits of clinical judgement. Four variables, pulmonary, renal, and cardiovascular functions, and glucose metabolism were checked on a yearly basis. The results of these studies are shown in Figure 9. Pulmonary functions were impaired with age (Figure 9a). However, there were no differences between young and old subjects as far as their blood content of oxygen, carbon dioxide, and electrolytes were concerned. As for renal functions (Figure 9b), a highly significant decrease with age in the capacity to filter creati-

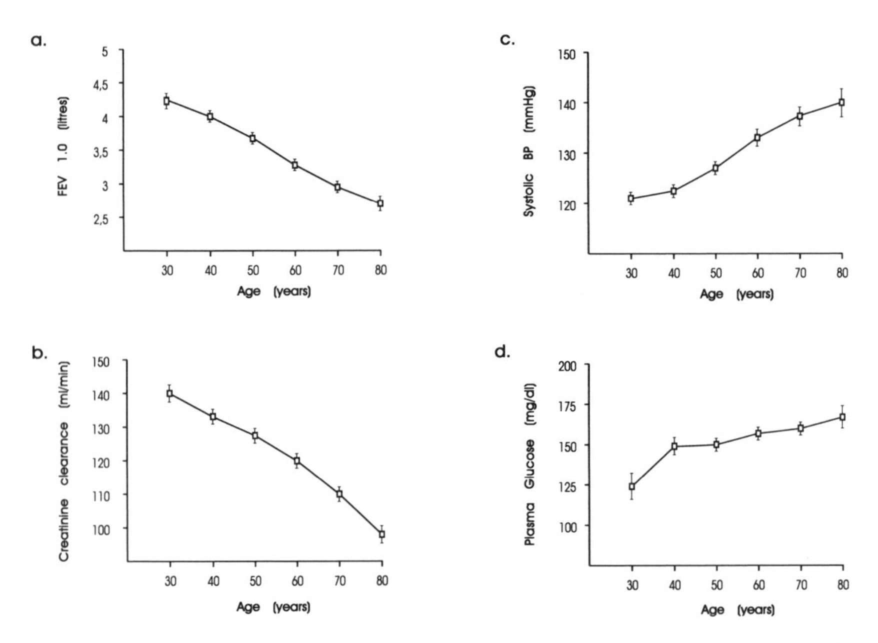

Figure 9. Results on more than 500 subjects from the Baltimore Longitudinal Study on Aging: the Evolution of four physiological functions with age.

(a) Pulmonary functions (maximum expiratory volume of air per second, FEV: forced expiratory volume in one second).

(b) Renal functions (creatinine clearance, an estimate of the glomerular filtration rate).

(c) Cardiovascular functions (systolic blood pressure, the result of complex interactions between the hormonal, neural, renal, and vascular systems).

(d) Oral glucose tolerance test (changes in blood glucose after a high oral dose of glucose). (Redrawn from Tobin, 1984.)

nine (i.e., creatinine clearance) was observed, with a concomitant increase in creatinine blood concentration. There was a progressive and significant increase of the **systolic blood pressure** with age (Figure 9c). The oral glucose tolerance test (Figure 9d) showed that the changes included reduced capacity of blood glucose to return to normal values after glucose administration, but there was little change in fasting serum glucose: there were no age-related effects during the first 40 minutes following glucose administration, but after two hours, the 20-year-old subjects had the lowest concentration of glucose, while the highest concentration (implying a poor regulation of glucose) was observed in 80-year-old people.

These findings reflect a general decrease in physiological functions with age in the absence of diseases or medications.

During the course of the Baltimore Longitudinal Study, about one fifth of the volunteers died. When physiological test values between volunteers who died and those who were still alive were compared, the subjects who died displayed lower physiological performances (Figure 10). This

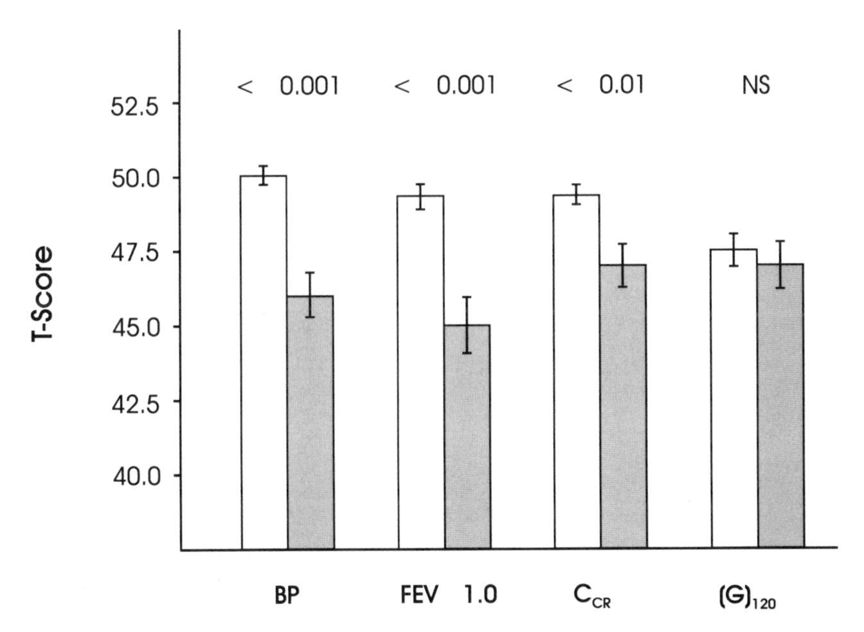

Figure 10. Relation of performance levels to survival on four physiological tests.

The scores for individuals who died (cross-hatched bars) were compared to those for individuals who survived (open bars).

BP: systolic blood pressure. FEV1.0: forced expiratory capacity in one second. C_{cr}: creatinine clearance. $[G]_{120}$: glucose concentration 120 minutes after oral glucose administration. Results are T scores, a normalization that enables the presentation of different data on the same scale: the mean and the standard deviation values of all variables are normalized to 50 and 10 respectively. (Redrawn from Tobin, 1984.)

strongly suggests that the quality of the functional reserves has a predictive value for the length of the survival period at a given age.

The usual deterioration observed with age may in fact be the necessary precondition for the development of a pathology, and it is often difficult to distinguish normal aging changes from pathological disease states. The results of the Baltimore Longitudinal Study are a good illustration of this intricate interdependence between aging and diseases in humans. For example, an increased rigidity of the arterial wall over time is a usual change due to aging, but the resulting structural changes may bring about dynamic changes in blood flow and favor the deposition of lipid plaques within vessel walls. It is now well established that the accumulation of plaques depends both on the individual's diet and exercise regime and on his genetic background, which may determine the ability to metabolize, transport, and excrete lipids such as LDL-cholesterol, for example. The establishment of coronary artery disease depends ultimately on the sequence of these various factors: normal aging changes that put the individual at greater risk, the individual's environment, and his or her genetically determined physiological response mechanisms. Many other age-related diseases are also likely to be a result of the combination of these factors.

Connective Tissue

Skin: With age, the epidermis, the superficial layer of the skin, does not change much, but its attachment to deeper layers of the dermis becomes looser. The dermis loses **collagen** and **elastin**, and its vascularization decreases. The loss of muscle mass also increases the tendency of skin to wrinkle. Thinning of the skin and accumulation of **lipofuscins** (aging pigments) and melanin are characteristic in elderly people.

Bone: Over the years, there is a decrease in total bone mass, called *osteoporosis*. Osteoporosis is an abnormal balance between removal of old bone and replacement with new bone. This is because osteoclastic activity, i.e., bone destruction, becomes predominant over osteoblastic activity, i.e., bone synthesis. Osteoporosis is accompanied by a loss of calcium salts and a high risk of fractures. Many factors are implicated in the development of osteoporosis. Reduction in osteoblast (cells that synthesize bone) activation, decline in physical exercise, decline in growth hormone and **somatomedin C** secretion, increase in parathyroid hormone, and decline in vitamin D levels are all causal factors in the loss of skeletal bone. Clinical studies of osteoporotic postmenopausal women have shown decreased blood levels of the physiologically active vitamin D metabolite 1,25-dihydroxyvitamin D3. This substance, which regulates calcium uptake from the intestine, was thought to be a major factor in the decreased absorption of calcium by the intestine. Indeed, studies in ovariectomized dogs showed that vitamin D reversed bone loss by stimulating osteoblasts (Malluche et al, 1988). However, similar studies carried out with women did not yield clearcut results (see Part II, Chapter 4, page 176).

Until recently, the cellular and biochemical changes that mediated osteoporosis were unknown. Early observations pointed to a potential role of the cytokine interleukin-6 (IL-6) in osteoclast (cells that resorb bone) development. IL-6 is produced by bone marrow stromal cells and osteoblastic cells, both of which influence bone resorption. Later studies showed that IL-6 was inhibited by 17β-estradiol in vitro. Furthermore, 17β-estradiol suppressed osteoclast development in cultures of mouse bone cells. Recent in vivo studies confirmed these observations and showed unambiguously that the loss of **estrogen** in mice induced by ovariectomy stimulated the production of osteoclasts, through an increase in the production of several cytokines and, specifically, IL-6 in the bone marrow (Jilka et al, 1992; Horowitz, 1993). These observations point to a beneficial role of estrogen in bone renewal.

Cartilage: There is loss of cells in cartilage and changes in matrix composition with increasing age. Chondroitin molecules are smaller and therefore contain less water and more calcium. The calcification makes them more rigid.

The Nervous System

Most elderly people suffer some limitation in brain functions, secondary to anatomical and functional changes. The impairment with age of neuronal functions and of intracellular signalling mechanisms of brain cells can affect sensory, motor, cognitive, and emotional functions. This often has a profound influence on the quality of life of older individuals and may prevent them from leading independent lives (Shock, 1962).

Anatomical and Physiological Changes

Neuronal cells (neurons) do not divide during adult life, and their final number is reached around late adolescence. It is a popular belief that people lose hundreds if not thousands of neurons each day. The consequence of such neuronal loss should be a diminution of the size and weight of the brain with age. Indeed, the mean brain size is about 10% lower at the age of 80 (Arking, 1991). The gyri (foldings of the brain cortex) that separate the ridges of the cortex become wider. The situation is, however, more complex: neuron loss occurs at different rates in different areas of the brain; some areas remain intact whereas other areas lose close to 50% of their neurons. For example, the number of cells in the vestibular nucleus (neurons in the brainstem involved in posture) does not change with age. The same is true for cells in the inferior olive that are involved in motor control in the cerebellum. On the other hand, the cortex is the most dramatic example of a brain area that loses a considerable number of neurons (Brody, 1955, 1978). The different cortical regions are not affected uniformly; the associative cortex, where cells channel and process information from various modalities (Arking, 1991) seems to lose more connections than unimodal primary cortices, e.g., those that receive visual or auditory information. There is also a 30% cell loss in the hippocampus of old people (see Chapter 2, pages 32–40). While the number of large-size neurons decreases with age, that of small-size neurons and glial cells (cells that feed and support neurons) seems to increase. One should not forget that neuronal loss is not unique to old age; the greatest rate of neuronal death in the cerebral cortex, the site of higher intellectual functioning, occurs between birth and age 30. Some of this loss can be termed selective cell death and is part of the normal developmental process of the brain. With age, the number of neuronal connections decreases, and the cross-talk between neurons becomes less efficient. There is a marked decrease in neurotransmitter output and in the thickness of the extracellular sheath between neurons and an increase in neurite plaques and fibrillary tangles. These physiological changes may explain, in part, the slowing in the processing of information by the CNS and the slowing of many other physiological functions in old people.

It is important to remember that the brain has the remarkable capacity of generating new connections (dendrites) between neurons (i.e., plasticity through sprouting) throughout life. Neurons are constantly going through the dynamic process of destroying and constructing dendrites, the neuronal appendages that function mostly to receive signals from other neurons. Thus, the number of connections between neurons changes with time, but this capacity does not seem to be much affected with age, at least in laboratory animals (Arking, 1991). The previously accepted observation that neurons lose their dendrites and therefore abolish interneuronal connections has been questioned. Coleman and his colleagues (1982) noticed that one of the effects of old age is dendritic growth, or increased arborization of dendritic spines in certain regions of the brain. He concluded that during early old age, both growth and decline occur concurrently. Thus, the brain eludes any systematic pattern of aging because of its resiliency, and one cannot draw firm conclusions about cognitive function from the physical changes observed in old brains. Neuronal loss is not necessarily accompanied by a loss of function, and this is a puzzling observation for neuroanatomists and neurophysiologists. Of course, there are physical changes in the brain that directly affect behavior, intelligence, sleep patterns, and memory, but people with severely damaged brains are sometimes able to live independent lives, while others with obvious behavior and memory problems turn out to have brains that look normal to a pathologist. (Often, though, there is a loss in white matter, the myelin coating of axons that is necessary for electrical transmission between neurons, and as a result, connections between cortical areas do not function well.) There is still no consensus on which changes are the result of aging and which changes are due to disease (Marantz-Henig, 1987). The challenge is also to find out if there is any relation between cognitive function and memory decline and brain atrophy or perturbed cerebral blood flow; if brain changes precede memory changes, or, conversely, if memory loss precedes changes in brain structure and blood flow; and, finally, if one can predict who will develop memory or cognitive problems. To answer some of these questions, a special study (as part of the Baltimore Longitudinal Study on Aging) was designed and is being carried out: 90 women and 90 men between 60 and 85 years of age will be subjected to a series of tests every year during 9 years. These tests include: magnetic resonance imaging (MRI) tests to investigate brain structural changes, positron emission tomography (PET) tests to measure regional cerebral flow at rest and during activity, and memory and cognitive tests. Investigators hope that this study will give clues about the relation between brain structure and cognitive performance.

Cognitive Capacities

It is generally accepted that aging is accompanied by a decline in cognitive abilities. Elderly subjects asked to undergo tests that measure several cognitive parameters, such as strategies to analyze new situations, to make decisions, and to memorize new material, tend to perform less efficiently than younger subjects. However, when these elderly people are given more time to complete their tests, their performances often turn out to be close to those of young people. After a detailed analysis of cognitive abilities in elderly people, Zec (1995) notes that the causes of the decline can be grouped in three general categories: disuse, disease, and aging per se. The contribution of each of these factors to cognitive decline may explain the significant variability of performances observed among elderly individuals. Zec concludes that disuse, the cessation of use of certain skills and abilities, contributes to a great extent to cognitive decline, and that practice can prevent disuse from occurring. Practice includes training of intellectual flexibility, of reasoning, of memory tasks. Such measures can substantially improve memory and alertness in the elderly.

Difficulties in speech are frequently observed among elderly people. They also may lose fluency in their oral or written expression. This is due to memory loss as well as difficulty in organizing complex sentences. This is not necessarily a handicap as long as these limitations are overcome by using simpler grammar, shorter sentences and paraphrases.

Sensory Systems

Hearing

Impairment of hearing is common, often because of the loss of hair cells that transmit sound in the cochlea. The other causes of age-related deafness are vascular diseases (hypertension, atherosclerosis of inner-ear arteries), and the deleterious effects of the cumulative exposure to noise over a lifetime. Sometimes, the defect does not consist in peripheral hearing loss, but is due to a difficulty in the analysis (ordering and sequencing) of speech sounds by the brain.

Taste and smell

Impairment in taste and smell is frequent with age and occurs for causes that are poorly understood. This has implications for the nutritional status of the elderly.

Vision

Over the years, people progressively lose their capacity to read material close to their eyes. This is due to thickening and rigidity of the lens. Elderly people also detect a smaller spectrum of colors. The major age-related disorders that cause visual disability or blindness in millions of older people are macular degeneration, **glaucoma**, and cataract.

Cellular and Intracellular Modifications

Most neural signals are mediated by specific molecules, the neurotransmitters, that bind to receptors present on target cellular membranes. Following binding of the neurotransmitter to its receptor, different cell events are triggered, depending on the type of neurotransmitter or target cell. Some neurotransmitter-receptors coupled to GTP-binding proteins activate effector systems such as adenylate cyclase or phospholipase C. This intricate sequence of reactions is schematically illustrated in Figure 11.

Activation of the receptors produces so-called second messengers: cyclic adenosine monophosphate (cAMP) or inositol trisphosphates (IP_3). The former phosphorylates and activates specific intracellular enzyme systems, and the latter releases calcium ions from intracellular stores, thereby triggering various biological reactions (secretion, neurotransmission, muscle contraction, cell division). Calcium also plays an important

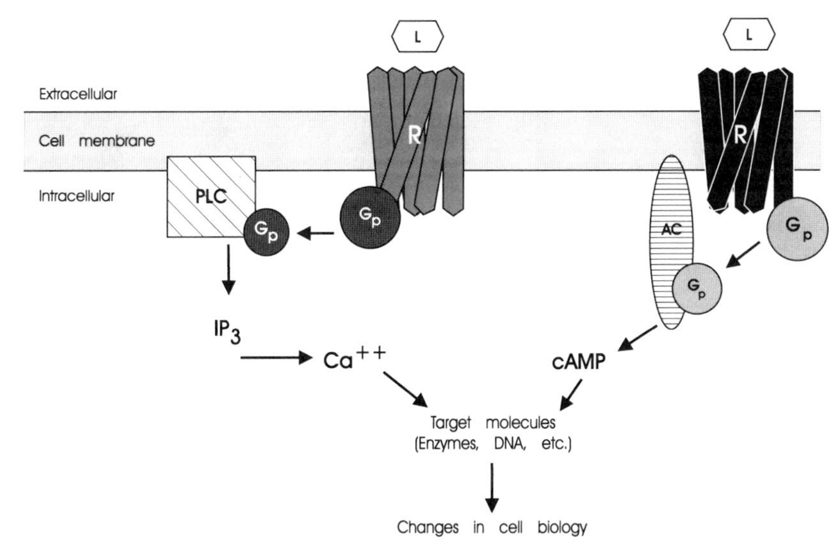

Figure 11. Simplified diagram of two important signal transduction pathways.
See text for explanation. L = ligand; R = receptor; AC = adenylate cyclase; PLC = phospholipase C; IP_3 = inositol triphosphate; G_p = GTP-binding proteins; PKC = phosphokinase C; cAMP = cyclic adenosine monophosphate; Ca^{++} = calcium.

role in the regulation of voltage-operated and receptor-operated cell membrane channels (Roth, 1989).

Aging seems to lower the overall efficiency of the signal-transduction systems of neurons, and all steps of these systems seem to be affected. The capacity to synthesize neurotransmitters is also affected in specific brain nuclei, for example in neurons containing acetylcholine, dopamine, or other neurotransmitters. The synthesis of given neurotransmitters and their respective receptors decreases. A loss of dopamine D_2 receptors in the corpus striatum of aged mice, rats, rabbits, and humans and a loss of β-adrenergic receptors from several other regions of the brain are seen (Roth and Hess, 1982). Calcium fluxes in and out of cells can be affected by aging. In a recent study, Pagliusi and her colleagues (1994) analyzed the ratio of **messenger RNAs** encoding glutamate receptors, which form calcium permeable and calcium impermeable channels, in rat hippocampal neurons, a brain region important in memory function and most vulnerable to aging processes. They found that with aging there is a loss of calcium impermeable channels in these neurons, and therefore a relative increase of calcium permeable channels. They suggest that this could alter calcium homeostasis and cause neuronal death. Interestingly, the observed deterioration in calcium-dependent responsiveness with aging can be partially or fully reversed, provided sufficient calcium is moved to the specific sites of action. Such findings open the way to the development of novel therapeutic strategies for neurological and possibly psychiatric diseases. These and many other modifications of the neuronal signal transduction system are considered normal physiological changes that occur with aging. But these changes at the subcellular level are not trivial and can have serious consequences. They may lead not only to impaired cognitive abilities, but also to neurological or neurodegenerative diseases.

The Endocrine System

The endocrine system is a group of organs that secrete hormones into the bloodstream in order to control other organs and tissues. The endocrine system is, in some respects, a form of diffuse nervous system transmitting its messages more slowly than the central nervous system, and in a more nonspecific manner. Important modifications in the synthesis of hormones have been reported with age. These changes are at the basis of the *endocrine theories* of aging (see Chapter 4), postulating that endocrine systems can speed up the aging process. Good illustrations of these theories are provided by the sudden aging of the marsupial mouse *Antechinus stuartii* and of the Pacific salmon. In marsupial mice, males get involved in fierce fights to have access to females and copulate. They accomplish their courtship and reproduction with extraordinary efforts that are accompanied by a striking rise in plasma androgen concentration and progressive breakdown in disease resistance. They die at about one year of age, soon after the mating period (Bradley et al, 1980). Pacific salmon engage in an exhausting journey to accomplish their single reproductive act and die of physical exhaustion and of hyperadrenocorticism shortly after mating (Robertson and Wexler, 1960). Significantly, when marsupial mice are castrated, their survival is increased dramatically, and they live up to three years. Similarly, castrated salmon will live up to four years longer than salmon that go through their single reproduction phase, essentially doubling their life span (Robertson, 1961).

Endocrine Hypothalamus

The hypothalamus is a neuroendocrine organ located in the brain that controls the synthesis and liberation of several hormones by the pituitary gland. Hypothalamic and pituitary hormones regulate body metabolism and temperature, hunger and thirst, sleep, sexual desire, and reproduction. The hypothalamus also plays an indirect role in the regulation of the immune system. As is the case with many hormones, the concentration of neurohormones from the hypothalamus declines with age.

Glucocorticoids

The effects of excess glucocorticoids on the death of hippocampal cells have been discussed earlier (see Chapter 2, pages 32–40). The elevated secretion of glucocorticoids is implicated in ulcers, loss of muscle and bone mass, and aging.

Dehydroepiandrosterone

Recently, scientists proposed the attractive hypothesis that aging might be due not so much to the presence of death hormones, as to the absence of certain life hormones. A candidate life hormone that received a lot of attention is dehydroepiandrosterone (DHEA), a steroid secreted by the adrenal gland. Levels of DHEA decrease soon after maturity and continue to decrease progressively throughout life in humans of both sexes (Orentreich et al, 1984). The causes and consequences of this decrease are unknown, but it is clearly not due to a loss of the adrenal cortex volume. DHEA shows remarkable effects in inhibiting disease, including cancer and kidney and vascular diseases in rodents. DHEA plays many different physiological functions which justify the interest in this hormone. More information on the effects of endogenous or exogenous DHEA in rodents is given in the second part of this book (see Chapter 5).

Reproductive System

The reproductive system undergoes obvious changes with age, one of which is a decrease in the synthesis of sexual hormones from about the fourth decade of life. In women, the production of estrogens by the ovary ceases at menopause. The mechanisms for this are not yet established but probably involve changes in the physiology of internal biological clocks (see below). When estrogens are no longer secreted, the levels of the **hypophyseal** luteinizing hormone (LH) increases since this hormone is the messenger that induces the secretion of estrogens. Estrogen deficiency has many negative effects, such as osteoporosis, involution of genitalia, and hot flashes, a transient vasodilation on the face and thorax, due to the dysregulation of hypothalamic thermoregulatory centers.

In men, two important functions of the testicles are the synthesis of sperm for reproduction and of testosterone, a steroidal hormone with a number of functions. It ensures sperm production, maintenance of secondary sex characteristics, such as deep voice, muscle mass, hair, and, last but not least, several aspects of behavior. In men, the level of testosterone decreases steadily with age, although this is not found in all men. In animal studies, the level of testosterone has been correlated with low sexual activity in several species; but such a relation is far from being demonstrated in men.

Biological Clocks and Circadian Rhythmicity

Circadian rhythmicity, i.e., diurnal rhythms in physiological functions, is a universal feature of eukaryotic organisms, from algae to man. It is probably the consequence of evolutionary adaptation to light/dark and other cycles from the environment. Scientists have postulated that a common mechanism in the form of a central neural oscillator is responsible for the regulation of all physiological rhythms. In the mid-sixties, Richter (1965) performed a series of lesions of the nervous system and succeeded in localizing the circadian clock to the anterior hypothalamus. Subsequently, Moore and Lenn (1972) identified the suprachiasmatic nucleus (SCN) as the precise anatomical site for the circadian clock in mammals. The SCN is located in the anterior hypothalamus on both sides of the third ventricle, immediately above the optic chiasm. Specific lesions of the SCN in rodents produced arhythmicity of a variety of physiological functions, including corticosterone secretion, drinking, locomotor activity, and sleep-wake behavior. Additional evidence supporting the role of the SCN as the central neural oscillator is provided by experiments in which the SCN was isolated either by deafferentation of the nuclei or removal of the SCN. In both instances, neuronal electrical activity of the SCN continued to exhibit an endogenous rhythmicity. Recent studies on the neural connections of the SCN, neuropeptide distribution, and specific binding of melatonin (see below) confirm that the SCN plays a similar role in humans to that in other mammals.

Melatonin

The pivotal hormone involved in circadian rhythmicity is melatonin, a serotonin derivative synthesized and released during the night by the pineal gland. Melatonin is considered by many as the hormone reflecting the length of nights (i.e., the duration of nocturnal secretion of this hormone is proportional to the length of night). Melatonin has two major biological effects in mammals. It regulates the reproductive alterations that occur in response to variation in day length in seasonal-breeding mammals, and it is involved in the adaptation of biological circadian rhythms as a function of light (Reiter, 1984). Thus, injections of melatonin in animals at a given time have been shown to synchronize circadian rhythms by acting on the release and uptake of neurotransmitters from the SCN.

Although the exact role of melatonin in humans is not yet firmly established, it probably plays a role in the measurement of day length and controls puberty and various menstrual disorders. Its overall function can be seen as one of coordinating and optimizing the responses of the organism to environmental demands. In addition, melatonin is thought to have profound effects on homeostasis, overall metabolism, and **immune surveillance** (Caroleo et al, 1992). Since it is lipid soluble and can easily cross cell membranes, in particular the **blood-brain barrier**, researchers have postulated that melatonin could have other physiological actions that do not require interaction with receptors. In particular, melatonin can inhibit the growth of cancer cells in vitro and in vivo (Janiaud, 1987). It also has been shown to inhibit DNA damage by chemical carcinogens in vivo (Tan et al, 1993b). This suggests that it can protect cells against oxidative stress and damage caused by free radicals. Recently, Reiter and his colleagues (Reiter et al, 1993; Tan et al, 1993a, see Chapter 5, pages 113–119) have reported that melatonin possesses a potent, endogenous hydroxyl radical scavenging ability in vitro (hydroxyl radicals are the major molecular species mediating oxygen toxicity in organisms).

Because of the multiplicity of effects of melatonin and because it can penetrate into cells and bind to both extracellular and intracellular targets, scientists have postulated that there may be more than one site of action for this hormone. In the last years, significant findings have been made in the discovery of more than one type of melatonin receptor (for review see Dubocovich, 1995). First, the circadian and the reproductive effects of melatonin appear to be mediated by a high-affinity membrane binding site, later called ML1 receptor, found in the hypothalamic SCN, the pars tuberalis of the pituitary, the **retina**, and in other sites in the brain. The pituitary has the highest concentration of melatonin receptors in mammals, and it is the only site containing melatonin receptors in all seasonally breeding animals (for review see Weaver et al, 1991). Recently, the gene encoding the melatonin receptor in several animal species has been cloned, and the predicted amino acid sequence exhibits seven hydrophobic segments, which makes this receptor a member of the superfamily of seven transmembrane-domain, G-protein-coupled receptors. A second binding site with different affinities for melatonin and other agonists, termed ML2, has been identified in a few tissues, mainly in hamsters, but its function remains unclear. Apart from its actions on cellular membrane receptors, melatonin interacts with intracellular targets. It has been shown to act as a calcium-calmodulin antagonist, which may have important consequences for cellular homeostasis. A recent exciting finding is that the action of melatonin as a free radical scavenger can be mediated by receptors located on the nuclear membrane. A putative specific receptor for melatonin, structurally related to the retinoid receptor, has been identified on nuclear membrane and is called RZRβ (Carlberg et al, 1994). This receptor could be the mediator of the direct genomic action of melatonin and of its transcriptional effects. RZRβ receptors can be detected in the pineal gland, in the visual system, and in the SCN.

The finding of these different melatonin receptors raises the important issue as to which of these receptors are involved in the circadian effects of the hormone.

The pineal gland, and, more specifically, melatonin have been implicated in the process of aging and age-related diseases. This is substantiated by the fact that, in humans, melatonin production in the organism is gradually reduced throughout life to the point at which melatonin rhythm is barely discernible in very old individuals. These changes have been attributed to changes intrinsic to the pineal gland (a gland that calcifies in many people in their thirties). Two theories have been formulated to explain the role of melatonin in aging. Some investigators believe that the decrease of melatonin output is responsible for the dysychronization of other circadian rhythms and that this contributes to the onset of age-related diseases. Others think that the reduced melatonin production acts as a trigger for the activation of genes that determine aging at the cellular level. The ability of melatonin to scavenge free radicals and to enhance the activity of some antioxidative enzymes to further reduce oxidative damage is perhaps one of the strongest indications that melatonin plays an important role in aging. These actions are of particular importance in the nervous system, which is highly susceptible to damage by free radicals. Diminished protection due to lower levels of melatonin could be an important cause of neuronal damage by free radicals.

Whatever the mechanisms involved, there are now a number of compelling arguments indicating that melatonin reduction may contribute to aging. The effects of exogenous melatonin are further discussed in the second part of this book (see Chapter 5).

Biological Rhythms

An extensive literature documents changes in the properties of the circadian system with age. Some scientists even believe that the deterioration of biological rhythmicity is the most important factor in aging. While they do not rule out the fact that the deterioration in rhythmicity may be a consequence of a broader neurodegenerative process, they think that it is the loss of synchrony between all biological rhythms that is responsible for the aging process. They claim also that alterations in biological clocks can trigger the onset or contribute to the maintenance and further development of diseases. The existence of a link between disruption of the circadian system and life span has been documented in insect species (Hayes et al, 1977) and in rodents. Tapp and Natelson (1986) compared the life spans of hamsters with inherited heart disease, reared in different regimes of light-dark cycles. They observed that hamsters lived longer when they were kept in constant light as opposed to standard cycles of 12 hours light–12 hours dark. Median life span was 12% longer in hamsters kept in constant light. The change in circadian rhythms was proposed as a likely explanation for the observed increase in life expectancy. In mice, it is the loss of circadian rhythmicity in activity, more than loss of activity itself, that seems to be detrimental to a long life span.

In aged humans, rhythms of many physiological parameters are perturbed, including blood pressure and urinary concentration of electrolytes. Effects of age on the secretion of virtually all the anterior pituitary hormones are manifest. However, there seems to be little alteration in the rhythmicity of the HPA axis with age (see Chapter 2, pages 32–40). In healthy aged people, no significant change in the ACTH rhythms or cortisol secretion has been observed (Blichert-Toft, 1975). HPA rhythms seem to be more robust than other rhythmic behaviors.

The change observed most consistently in circadian organization with age (reported in humans and in animals) is a reduction in the amplitude of rhythmic variation and a shortening of the period of the circadian clock to less than 24 hours. It is possible that the changes in circadian amplitude are causally linked to night-sleep disruption and daytime somnolence in elderly people (Prinz and Halter, 1983; see below).

Rhythmic changes in hormone secretion and behavior in aged humans have been extensively described, but the importance of these variations, i.e., the significance of a disruption of endogenous rhythms, remains speculative. In contrast to animals, there are no data in humans documenting adverse health effects resulting from disruption of physiological endogenous rhythms, with the exception of sleep perturbations.

Sleep

A majority of elderly people experience disturbances in normal patterns of sleep. The major change is a fragmentation of sleep cycles in association with disrupted brain electrical activities and breathing patterns (Richardson, 1990; Carskadon et al, 1982; Dement et al, 1985). They suffer increased wakefulness during their sleep phases and increased napping during the daytime phases. There is also evidence for an age-related advance in circadian sleep-wake rhythms in elderly people: bedtime and awakening time shift to earlier hours (Miles and Dement, 1980). The biological rhythms of other physiological parameters associated with sleep, such as those of body temperature, REM sleep, and cortisol are also shifted to earlier hours. These perturbations are clearly disadvantageous for elderly individuals (Kamei et al, 1979). Sleep perturbations should be taken seriously because they affect the health and quality of life of older people, and, by extension, they could affect their longevity. The age-related diminution in melatonin levels had been thought to be the cause of the deteriorations in sleep-wake cycles (Carskadon et al, 1982). However, the role of melatonin on sleep pattern should be evaluated with caution because the reduction in physical activity in elderly people could also explain the disturbances in their sleep-wake rhythms.

Immune System

The role of the immune system is to protect us from invading microorganisms, from antigens, and from cells that undergo cancerous transformations. This is done through a set of defense mechanisms that are capable of distinguishing self from nonself (Möller and Möller, 1978). The immune system is one of the best studied physiological systems, and it is therefore a particularly attractive model to explore the effects of aging at the genetic, cellular, and developmental levels.

The cells of the immune system are made up mostly of **B and T lymphocytes**, which are divided into several subtypes characterized by unique cell surface receptors. B lymphocytes, originating in bone marrow, are responsible for humoral immunity. This involves the synthesis and secretion of specific antibody molecules into the blood and lymph circulation. Antibodies bind to the antigens that triggered their formation and inactivate them. Cellular immunity is mediated by T cells. In the course of their development, prothymocytes migrate from the bone marrow to the **thymus**, where they mature and differentiate. To date, a large number of T-cell subpopulations have been identified. Some of them are involved in the stimulation of B lymphocyte growth and differentiation, and they are called helper T cells and express surface receptors called CD4. Others can directly recognize and destroy tumor cells or virus-infected cells. They are called cytotoxic T cells and express CD8 cell surface receptors. There are also suppressor T cells that control the extent of T-helper and T-cytotoxic activity in order to stop additional antibody response and avoid **autoimmune** reactivities. In many situations, T cells do not react directly with the foreign substance but via interactions of their specific receptors with antigen-presenting cells, (e.g., **macrophages**). Following interaction with the antigen, antigen-presenting cells produce interleukin-1 (IL-1). IL-1 induces the expression of important **lymphokines** by T cells, including interleukin-2 (IL-2), interferon-γ (Inf-γ), and other lymphokines. These are necessary for maximum production of antibodies by B cells. They are also necessary for those T cells that have interacted with the specific antigen to induce cell proliferation and production of IL-2 receptors (IL-2R) and other receptors (see Figure 12).

The effects of aging on immunological functions are complex, and there is some controversy among scientists about which immune cells and immunological reactions are altered with aging. However, it is generally accepted that the consequences of the deterioration of the immune surveillance system are an increased vulnerability to infections and an increased incidence of autoimmune diseases and neoplasms (Keast, 1981). The term immunological senescence (Thoman and Weigle, 1989) has been used in the literature to describe the decrease in immune efficiency with age. The major changes include atrophy of the thymus, a generalized loss of potency of lymphocytes in responding to foreign antigens, and a decreased ability to respond by a delayed-type hypersensitivity reaction to various stimuli.

The involution of the thymus is considered an important factor in immunological senescence (Weksler, 1983), but the actual effects of thymus involution on immune responses are far from clear.

Gonzalez-Quintal and Theofilopoulos in 1992 addressed the question of the consequences of involution of the thymus and the resulting dramatic reduction in thymic output on immune responses. They investigated specifically whether aging affects the positive and negative selection of T-cell repertoires or destabilizes the peripheral T-cell repertoires. Positive selection is the process by which immature T-cell clones, expressing receptors for self-major histocompatibility complex (MHC) and antigenic peptides presented by thymic epithelial cells, mature and are therefore phenotypically and functionally modified. Negative selection is the deletion of the positively selected cells with high self-affinity from the thymic repertoire prior to exportation to the periphery. Negative selection is a self-protective reaction in that it allows **self-tolerance**. Gonzalez-Quintal and Theofilopoulos carried out careful analyses of stimulated T and B cells in young and old mice of different MHC haplotypes after stimulation with a **superantigen** and found an astonishing stability of the T-cell repertoire. A reduction in the output of the thymus apparently had no effect on this repertoire. This is consistent with the finding that most mature T cells, once they are exported to the periphery, reside in the long-lived recirculating lymphocyte pool. The contribution of newly emerging thymic cells is not significant. The pool of mature T cells in adult mice is largely self-renewed, and the continuous input of virgin cells from the thymus late in life apparently has little effect on its overall composition. Other studies have shown that an involuted thymus can still provide an appropriate environment for homing. Furthermore, aging and thymic involution are not associated with leakage to the periphery of T cells reactive to self-superantigens. These results contradict those of previous studies in which mice thymectomized early in life have been found to have inadequate intrathymic clonal deletion of self-reactive T-cell clones, indicating that a defective thymus would be responsible for the observed autoimmune manifestations.

Gonzales-Quintial and Theofilopoulos also point out that it is only the responses to superantigens that are unaffected by age. Interestingly, endogenous and exogenous superantigens engage in a signal transduction pathway that appears to be different from that engaged by conventional antigens in aged mice. The differential responses of cells from aged mice to the two types of antigens raise the intriguing possibility that the signal transduction defect may be selective. In contrast to the response of mice to superantigens, it seems that, in general, the responses to a variety of conventional antigens are reduced in aged mice, as measured by both in vitro and in vivo assays. However, conclusions concerning the decrease in immunologically competent T cells with aging are sometimes conflicting. For example, many groups report that aging affects the ability of T cells in mice, humans, and rats to proliferate in response to foreign antigen (Makinodan and Kay, 1980). Other studies report that the interaction of T lymphocytes with foreign stimuli appears to be intact. Likewise, some researchers have found a decrease in the density of lymphocyte surface markers with age, as illustrated in Figure 12, while others have found no differences between young and old animals. And in humans, several investigators have reported a decrease in both the absolute number and the percentage of peripheral T lymphocytes with age, while others saw no modifications.

As described above, T cells are a heterogeneous population of cells, and it is likely that aging might preferentially affect a specific subset of T cells. Studies on T-cell heterogeneity in young mice have revealed that differences in the expression of surface receptors between the naive T cells (T cells that have left the thymus recently) and the memory T cells (T cells that have undergone several cycles of proliferation after stimulation by an antigen). These surface markers have been used to estimate the proportion of naive versus memory T cells in young and old mice. It has been found that memory T cells made up about 20% of the peripheral T cells in young mice, but 50% of the population of T cells in older animals. The shift from naive to memory affected both helper (CD4) and cytotoxic (CD8) cells (Miller, 1994). This accumulation of memory T cells in older mice could account for most of the decline in the proportion of cells that can secrete or respond to IL-2. Moreover, as discussed below, memory T cells show a weak stimulation of the intracellular signalling system in response to antigen.

Defects in the intracellular signalling pathways following foreign stimuli have also been claimed to be responsible for the observed immunological senescence. In support of this hypothesis, scientists have observed a reduction in lymphokine production and a substantial decrease in the level of cytosolic calcium after stimulation of T cells in old mice (Figure 12; see below). This defect is present mostly in the memory T cells (Miller, 1994). But again, there are contradictory results. These may be due to the use of different animal species. For instance, decreases in messenger RNA for c-myc, IL-2R expression, and IL-2 production have been reported in some, but not all species. Decreases in IL-2 production occur in mice and humans, but not in rats; decreases in Inf-γ occur in humans, but not in mice or rats. In humans, one individual might show a decrease in IL-2 that

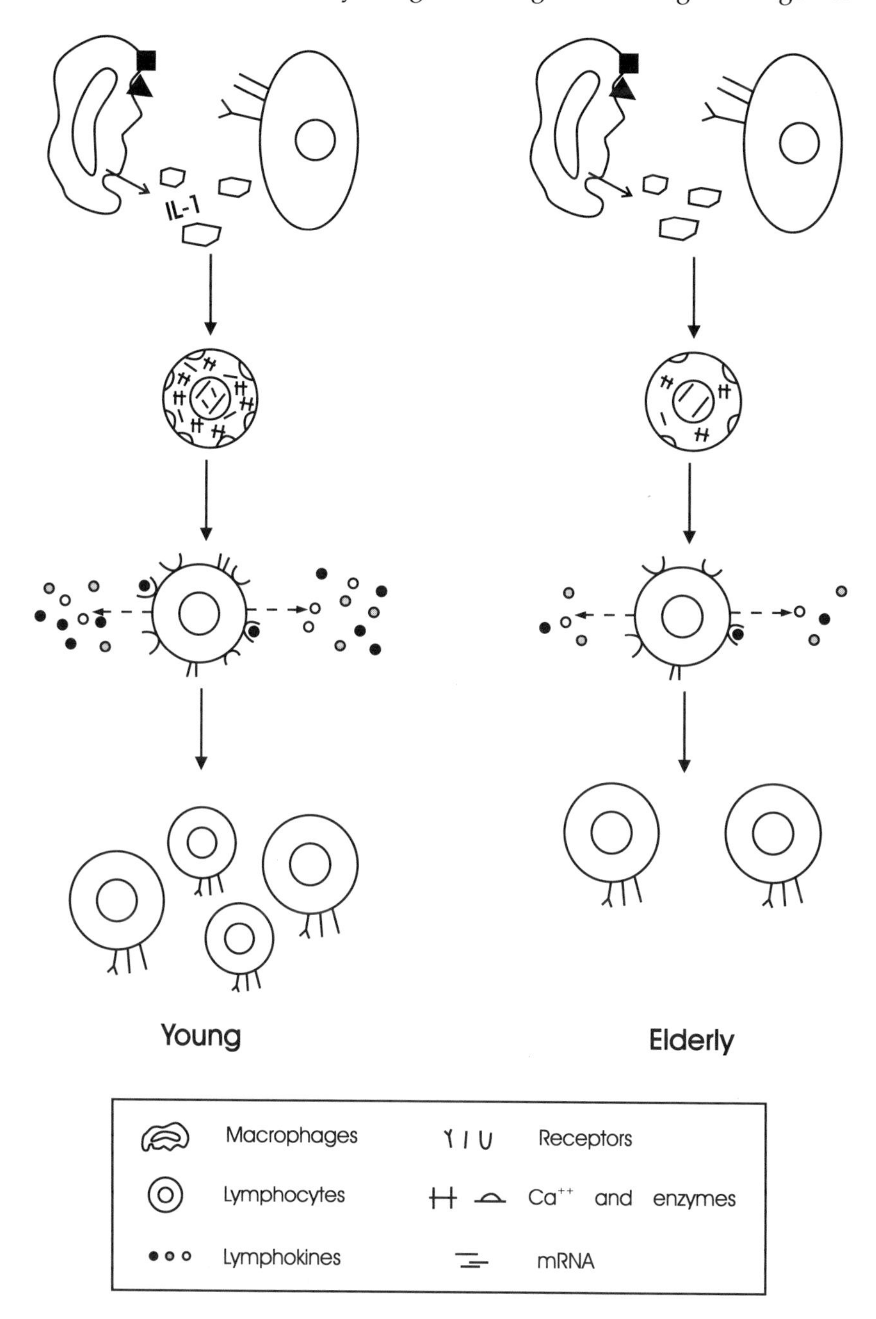

Figure 12. Diagram of the activation process of T cells from young and elderly individuals.
This diagram shows the loss of cell surface receptors and intracellular second messenger molecules and the decreased number of stimulated immune cells in elderly as compared to young individuals. (Redrawn from Murasko and Goonewardene, 1990.)

would precede changes in IL-2R, while another might show the opposite, i.e., a decrease in IL-2R before a decrease in IL-2; in a third individual, it is the expression of another immunological marker that might be defective.

The production of antibodies by B cells is also affected by aging. Several studies have demonstrated that qualitative changes in the antibody response to an antigen like phosphorylcholine, a natural epitope of certain pneumococci, occur in aging mice. The antibodies produced by an aged immune system are structurally different from those observed in a young immune system (Nicoletti et al, 1991; Nicoletti et al, 1993; Yang et al, 1994). There is a change in the usage of V genes (genes encoding the variable region of light and heavy chains of immunoglobulins) encoding specific antibodies in aged mice, which results in the production of antibodies with lower affinities for the antigen. This may explain the lower protective capacity of aged immune systems. Yang and his colleagues (1994) propose that the choice of different V gene repertoires may be influenced by a random, aged-related event, such as DNA methylation (see Chapter 5, pages 110–111). Random methylation could influence the rearrangement, transcription rate, and the pairing of specific light chain and heavy chain variable regions.

These various studies challenge the accepted idea that the immune system of older animals or elderly individuals is incompetent. However, many controversial issues have not yet been solved and additional work remains to be done to specify the role of a shift in different T-cell subsets, the role of a decrease in intracellular signalling, or the role of T-cell subsets that have not yet been identified, in the **functional decline** of the immune system. An important question concerns the causal relation between the rate of immune senescence and the rate of aging processes and whether, as is generally accepted, the greater vulnerability of elderly people to infections, cancer, and degenerative diseases is due to immunosenescence. Of interest, food restriction, the most convincing means of prolonging the life span of all animal species so far studied, has been shown to retard immunological senescence (see Part II, Chapter 1).

Aging of Intracellular Functions

The intracellular message transduction systems are ubiquitous. Activation of these systems following the binding of hormones, neurotransmitters, or growth factors to their respective receptors is a crucial step in the response of cells to specific stimuli. Different cascades of intracellular events can be triggered, depending on the type of stimulating agent or target cell. As described earlier in Chapter 3 (Figure 11), many receptor types are coupled to molecules that bind GTP. These so-called **G-proteins** activate or inhibit adenylate cyclase or phospholipase C. These in turn lead to the synthesis of second messengers such as cyclic adenosine monophosphate (cAMP) or inositol trisphosphate (IP_3). Cyclic AMP phosphorylates and activates several intracellular enzymes, whereas IP_3 releases calcium ions from intracellular stores. The intracellular concentration of free calcium seems to be a common pathway by which cells translate the stimuli received from hormones or neurotransmitters into various reactions. These include protein synthesis, secretion of hormones, neurotransmission, muscle contraction, and cell division. Calcium also plays an important role in the regulation of voltage-operated and receptor-operated channels (Roth, 1989). The changes in intracellular calcium increase the synthesis of proteins from so-called early genes, i.e., genes that are expressed within minutes after cell stimulation. Thereafter, the proteins synthesized from early genes (*c-Fos*, *c-Jun* and others) bind on chromosomes and derepress the synthesis of cell-specific structural and functional proteins.

With aging, important changes occur in the efficacy of the transduction system to transfer signals from the cellular membrane to the nucleus. An overall decline in the concentration of most second messengers is observed in almost all types of cells. It is as if signals triggered by receptor stimulation were decreased, thus diminishing the whole cascade of intracellular events and nuclear gene expression. In senescent fibroblasts cultured in vitro, the transcription of the early gene *c-Fos* is nearly completely repressed. Moreover, epidermal growth factor and phorbol esters, which induce the transcription of *c-Fos* in cells in early passages, fail to induce *c-Fos* messenger RNA in senescent cells. A reduction of early-gene expression may be a causative factor in inhibition of cell growth. It may also explain the decreased expression of other transcriptional factors.

The consequences of aging on the transduction pathways in immune T cells were elegantly described by Miller (1994). The proportion of T cells able to generate a rapid intracellular calcium response to strong polyclonal activators like Concanavalin A was found to decline with age. The low intracellular calcium signal seemed to be due in large part to the inability to extrude calcium ions from cellular stores. However, there was also a diminished capacity to respond to activation by increasing extracellular calcium influx. The decline in calcium generation was shown to affect T-cell functions such as proliferation and cytotoxic and helper functions. Further analysis showed that different subsets of T cells differed in their level of resistance to increases of intracellular calcium. Thus, memory T cells in mice of any age were found to generate smaller calcium responses than naive T cells. It is likely therefore, that the accumulation of memory T cells in old mice might contribute to T-cell hyporesponsiveness. Protein kinases, key enzymes that phosphorylate proteins, were also found to be less efficient in aging T cells.

It is likely that similar age-related modifications of the transduction pathways occur in other cell types and may contribute to cellular aging in other tissues.

Homeostasis and Multisystem Regulation

Multicellular organisms are composed of specialized cells which form various tissues and systems. These functional systems cannot work independently, and to ensure a state of equilibrium overall of the thousands of physiological variables, they must be coordinated. Equilibrium is determined basically by cellular activity, but it is also influenced by the composition of the milieu that surrounds cells and by the external environment. Claude Bernard (1813–1878) was the first to draw attention to the "relative constancy of the internal environment". This concept, later called homeostasis by Walter Cannon, is the ability of the body to adapt constantly to external factors and to internal cellular and physiological demands. Homeostasis is a dynamic and fragile state of life order. The neurological, endocrine, and immune systems play a crucial role in ensuring stable homeostasis and are closely linked with and dependent on one another. Often, they are grouped into a single system known as the neuro-endocrine-immune system (see below).

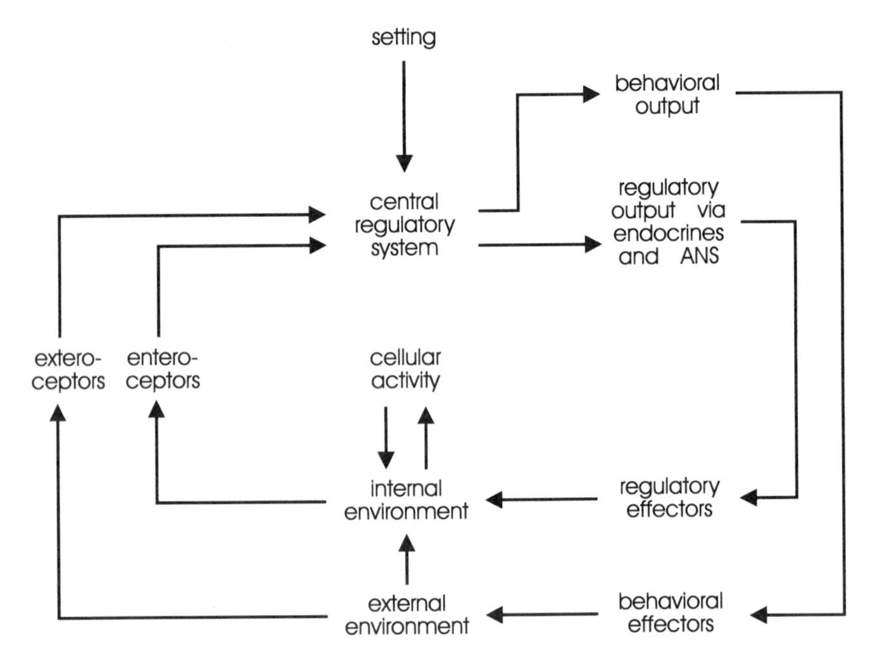

Figure 13. Model of interactive systems operating to maintain stable homeostasis of the internal environment.
The different systems are regulated through feed-back loop arrangements. Each loop contains sensors, a control center, and effectors. The information from sensors is processed by control centers that compare the newly received information with a target value. Behavioral and regulatory signals are then sent to the body via the endocrine and **autonomic nervous system** (ANS). The combination of all feed-back loops results in permanent oscillations in the levels of biological variables. (Redrawn from Kenney, 1985.)

Figure 13 presents a model of the complex interactions between the internal and external environment.

With aging, the complex regulation between these important systems is affected. Nathan Shock (1984) has suggested that "... impaired effectiveness of these coordinating mechanisms may be the primary factor involved in aging." Each element of the complex loop arrangements shown on Figure 13 can change. There may be a reduction in the number of sensors, decreased accuracy of the control centers, or errors made by the effectors. The resulting impairment in the internal environment leads to a loss of normal cell activity and healthy tissue function. A constant and progressive deterioration at any level of these complex interactive systems may decrease the "functional reserve", and this may favor the development of diseases.

If impairments in homeostasis can lead to the occurrence of age-related disease, what are the limits between loss of homeostasis and development of disease (Dilman, 1984)? For example, the distinctions between normal aging and Alzheimer's disease is often difficult to make. The cytological changes observed in Alzheimer's patients are prevalent in patients under 75 years, compared with normal people in this age group, but after 75 years, the number of neurocytological changes become increasingly similar in Alzheimer's patients and in nondemented people, although they are located in different cortical areas. Finch (1990) added further support to homeostatic decline as a major cause of mortality. He has observed a higher mortality rate with age in all species irrespective of the specific diseases that affect these species. Moreover, the age-related mortality rate is similar throughout all human or mouse populations, despite the different distributions of disease. This strongly suggests that factors other than disease influence age-related mortality.

Two examples illustrate these observations. First, Simms and Berg (1957) studied age-related diseases in a carefully maintained rat colony and showed that the onset of specific types of tissue damage increased exponentially after the rats had reached the age that was characteristic for each type of disease. The incidence of diseases anticipated the increase in mortality rate by several months and was attributed therefore to the gradual loss of homeostatic balance. The second example concerns the impairment of homeostasis in elderly people. Very often, it is difficult to assign a single major cause of death since there is an increased incidence with age of multiple disorders such as vascular diseases, perturbed regulation of glucose, or abnormal blood pressure. These perturbations may interact in subtle ways and finally cause death. It seems accurate to say that the impairment of many aspects of homeostasis leads to death in many elderly persons. Kohn (1982) suggested that: "lesions that would not cause death in younger persons are commonly accepted as a cause in the aged ... the final disease that goes on the death certificates is arbitrary." The example proposed by Fries and Crapo (1981) is a good illustration of the sudden breakdown of homeostasis, in the absence of a specific major cause: "An 88-year-old woman is admitted to the hospital from a nursing home with an infection of the bladder. When given intravenous fluids she develops heart failure, and the diuretics given to remove the fluid of the heart failure result in the deterioration of her kidney function. She is transferred to the intensive care unit where oxygen is given . . . She begins to bleed slightly from a stress ulcer in the stomach, and, during passage of a stomach tube to remove the blood, she vomits, and a small amount of vomitus enters the lung, where an aspiration pneumonia develops. Despite oxygen treatment, she becomes comatose, a tracheostomy is performed, and a respirator used to assist her breathing. Several other adverse events occur, the family is eventually told that there is no hope, and she dies in her seventh week in the hospital." Establishing the nature of homeostatic events and the causes of the age-related decrease in functional reserve will undoubtedly produce a major breakthrough in our understanding of the phenomenon of aging.

Neuro-endocrine-Immune Connections

The multiple interactions between the brain, the neuroendocrine system, and the immune system are now widely recognized. So great is this interconnection that these systems are often grouped into a single system and referred to as the neuro-endocrine-immune system. At present, there is evidence that neuro-endocrine-immune interactions play an important role in inflammatory diseases, affective disorders, and other stressful situations (De Wied, 1969; Weihe et al, 1991). A functional impairment or degradation of any one of the three systems has important repercussions on the function of the other two. The failure of the neuro-endocrine-immune system is considered an important cause of homeostatic disturbance of the organism, and this, as we have seen, may play a key role in senescence and death (Finch and Landfield, 1985). A few examples are given below to illustrate neuro-endocrine-immune connections and to suggest how age-related decline of one system could influence the other two.

Lymphoid organs are innervated by a complex neuronal network. The spectrum of neuropeptides present in nerves supplying the different lymphoid tissues and the presence of their respective receptors on cells of the immune system speaks for a specific immunomodulatory function for these nerves (Felten, 1991). Many studies have shown that neuropeptides and classical neurotransmitters released from nerves of lymphoid tissues can cause immune responses in vitro.

The nervous system affects immune responses via two main pathways: (1) an indirect neuro-endocrine communication through the secretion of corticosteroids regulated by the hypothalamo-pituitary-adrenal axis; and (2) a direct neural influence on the immune system mediated by the autonomic nervous system (Bohus and Croiset, 1990). The recent discovery of the complex autonomic and sensory peptidergic innervation of lymphoid tissues clearly demonstrates neuro-immune interactions. More needs to be discovered about the origin, the distribution, and the target of peptidergic nerves in the various lymphoid tissues, as well as about the identity and role of peptides present in these nerves. Findings in this area should help us better understand how aging affects lymphoid organ function.

Some of the neuropeptides identified in nerve fibres innervating lymphoid organs and tissue are substance P (SP), **calcitonin gene-related peptide** (CGRP), vasoactive intestinal peptide (VIP), and neuropeptide Y (NPY). There is evidence suggesting that these neuropeptides act as neurotransmitters with cells of the immune system as targets (Weihe et al, 1991).

The control of the thymus over the neuro-endocrine-immune system emphasizes its role in the maintenance of the organism's homeostasis. It synthesizes hormones called thymosins, which are thought to influence the aging process. Thymosins are directly linked to important hormonal systems in the brain. They stimulate the pituitary gland to produce hormones that regulate the reproductive endocrine system, ACTH, growth hormone, prolactin, and β-endorphins. This thymus-brain loop may play a key role in the aging process. The involution of the thymus with age would decrease the levels of thymosins and thyroxine, a hormone that governs the rate at which cells convert food into energy, and lead to the deterioration of the brain.

In addition to its role in immune functions, the thymus also exerts a regulatory influence on the adrenergic system, which is an important component of the homeostatic network. In fact, involution of the thymus in aging mice, or removal of the thymus in young mice, causes alterations of β-adrenergic responsiveness and decreases levels of β1-adrenoreceptors in submandibular glands and brain cortex (Fattoretti et al, 1982; Piantanelli et al, 1985). The expression of β-1 receptors, it has been found, is not definitively impaired and can be restored by grafting a neonatal thymus. The thymus also controls the regulation of α1-adrenoceptors (Rossolini et al, 1991).

Psychoneuroimmunology

In the last decade, the influence of social environment and stress on the neuro-endocrine-immune system has been recognized as a major factor in the incidence and evolution of diseases. The study of the role of psychological factors in the etiology of diseases is a new discipline in science called psychoneuroimmunology. Psychoneuroimmunologists try to understand the way in which psychological factors influence this integrative system and how the destabilization of one component of the system affects the others. Studies in psychoneuroimmunology address problems concerning the influence of psychological factors on disease development, as well as the role of psychological support in preventing progression of diseases.

It has been suggested that psychological factors are implicated even in the etiology of cancer, but the possible interaction of environmental carcinogens with psychological factors are exceedingly difficult to study, among other reasons because of the time factor: for example, childhood emotional disturbance may affect the tendency to develop cancer in adulthood. However, how does one examine this tendency? The nature of early child-parent relationships and the incidence of cancer in adulthood might be explained by the fact that, early on, difficult family patterns may make people defensive, repressed, and poor at communication. Bahnson (1979, 1981) reports that some cancer patients saw their parents as cold and uninvolved with them as children. Another study reports that breast cancer patients perceived their childhood relationships with their parents as stressed or lacking the normal parents' affectionate behavior (Wrye, 1979). Such studies are, of course, very difficult to confirm because of the subjective nature of patient reports.

One of the mechanisms by which psychological factors might contribute to the incidence of disease is through the immune system. Greene reported that subjects experiencing many changes in their lives in the previous year had a reduced lymphocyte cytotoxicity to virus inoculation and a lower mitogenic response (Greene et al, 1978). The recurrence of clinical manifestation of *Herpes simplex* virus is another example of the relation between events in life and mood and the proportions of T lymphocytes (both helper and suppressor cells). *Herpes simplex* virus remains in the body after the primary infection and is prone to recur and cause clinical symptoms. People who are subject to more turmoil in their lives show lower immunocompetence and an increased likelihood of *Herpes* recurrence (Kemeny et al, 1989).

Even though the role of psychological factors in the etiology of many diseases is generally recognized, researchers and physicians are more skeptical about the possible beneficial therapeutic effects of psychological interventions on diseases like cancer (Walker and Eremin, 1995). Which interventions might be useful, which harmful? The lack of firm evidence about the impact of psychological interventions may be due to different levels in different individuals of host immune resistance and endocrine responses. The relationships of these various factors are, of course, exceedingly complex.

Chapter 4

Theories of Aging

The descriptions provided so far have given a flavor of the complexity of the process of senescence. In an attempt to understand the various components of aging, and to analyze them critically, scientists have proposed testable hypotheses and constructed theories about the possible mechanisms of aging. These theories can serve as the basis for experimental testing. As Einstein stated: "It is the theory that decides what we can recognize." The term theory comprises an organized set of observations and concepts that might account for a given process. This is the case for some theories of aging. Other so-called theories of aging could rather be described as detailed observations of the consequences of biological and biochemical processes.

The phenomena of aging are so complex that a large number of theories have been advanced, each one claiming to decipher the fundamental mechanisms of biological aging. Unfortunately, a few among this wealth of theories have been propounded at the expense of scientifically proven facts. Here is a summary of the best of these theories of aging, each of which provides its own survey of the causes and mechanisms of aging.

Theories of aging have been classified according to different criteria. Hayflick grouped them according to the physiological levels analyzed: he wrote of organ-based, physiologically based, and genome-based theories (Hayflick, 1985). Hart and Turturro (1983) classified them on the basis of the size of physiological entities, from the smallest entity, the cell-based theories, to the next larger functional entity, organ-system-based theories, ending with population-based and integrative theories. Esposito (1987) wrote of causal, systematic, and evolutionary theories. Two other criteria of classification are based on whether aging arises from stochastic or random processes, or from programmed or predetermined processes, and whether the mechanisms of aging operate within most (or all) cells, i.e., intracellular theories, or within the regulatory systems that operate between groups of different cells, i.e., intercellular theories (Arking, 1991).

Stochastic Theories

Stochastic theories propose that aging is caused by the accumulation of insults or damage caused by the environment. These affect the organism to such a degree that its functions deteriorate to the point of no longer being compatible with life. This is best illustrated by the *somatic mutation* theory which states that senescence is caused by progressive genetic damage in the form of mutations accumulated because of direct ionizing radiation or radiomimetic agents. This theory is supported by the observation that exposure to ionizing radiation shortens life span but has been criticized on the grounds that the mechanisms of life span shortening by radiation are not necessarily related to normal mechanisms of aging. To test the theory further, scientists have estimated the occurrence of spontaneous mutations by assessing amino acid substitutions, errors in DNA synthesis, and resistance to cytotoxic purines. The results of these studies are also controversial. Regardless of the extent of DNA modification, the capacity for repair remains quite efficient and does not alter markedly throughout life.

At present, there is little evidence that accumulated point mutations or other types of errors in genomic information constitute a basic mechanism in senescence (Burnet, 1974). A convincing argument against this theory is that the differences in life span in different animal species cannot be explained by random genomic damage. Few experiments have been conducted in the past decade to further support or disprove the somatic mutation theory.

It has been revised, and the new version states that mutations in genes that specifically regulate the fidelity of DNA replication, transcription, or translation induce errors in protein synthesis and lead to alterations in many important regulatory as well as functional or structural proteins and finally cause cellular breakdown.

A second example of a stochastic theory is the *error catastrophe* theory. This theory differs from the somatic mutation theory in that errors in the transfer of information are postulated to occur at sites other than in DNA itself. It was proposed originally that the ability of a cell to synthesize functional proteins depends not only on the intact genetic information for the various polypeptide sequences but also on the competence and fidelity of the protein-synthesizing machinery. If errors are introduced into a protein molecule involved in synthesis of the genetic material or in the protein-synthesizing apparatus, it could cause additional errors in other protein molecules. The number of proteins containing errors would thus expand and lead eventually to collapse of cell functions and the death of the organism. This theory is reasonable and provides testable predictions. One of them is that proteins obtained from cells of old donors will exhibit a higher frequency of errors than proteins isolated from cells of young donors. Recent advances in techniques of molecular separation of the whole pool of cellular proteins allow investigators to detect proteins that contain a change in either charge or molecular weight or both, indicative of errors in synthesis. Thanks to these new technical procedures, the error catastrophe theory is one of the few theories of aging that has been critically tested in a wide variety of species and cell types. In all cases studied, there has been no evidence of the predicted electrophoretic heterogeneity characteristic of synthesis errors in the proteins obtained from older donors, indicating that no differences in error rate between young and old subjects have been detected. On the contrary, all cells show remarkably good fidelity in synthesis. It is therefore unlikely that **transcriptional** or **translational** errors contribute to the mechanisms of senescence (Rothstein, 1987). In spite of the excellent fidelity of the protein-synthesizing machinery over time, some modifications in proteins do occur with aging, and scientists have proposed a modified version of the error catastrophe theory: it is thought that old cells have a decreased capacity to remove damaged proteins and that the ensuing accumulation of defective proteins could result in impaired functional capacity.

The *DNA damage-repair* theory is a stochastic theory related to the error catastrophe theory. The theory states that cellular mechanisms that repair the damage caused to DNA become deficient (Arlett, 1986). This results in poor or altered transcriptional efficiency and ultimately leads to cellular breakdown. This theory is compelling since it is known that a decrease in the ability to repair damage to the genetic material is indeed associated with aging. However, there is no experimental support for this theory.

Finally, a set of closely related stochastic theories have been proposed: *the cross-linking, the altered proteins,* and *the wear-and-tear* theories. All of them stipulate that insults or damage and injuries accumulate during life and decrease the efficiency of the organism. According to the *cross-linking* theory, chemical reactions that result in intra- and intermolecular bonds, or cross-linking, in important macromolecules increase with age. Evidence of cross-linking has been shown so far for **collagen** and **elastin**, two molecules that do not readily turn over. Cross-linking is not restricted to proteins; DNA, for example, undergoes cross-linking as well. However, this particular theory is too simplistic, and although quite popular when proposed years ago, cross-linking is not now viewed as a major cause of aging. The *altered proteins* theory shares the ideas of protein cross-linking, stated above, with an emphasis on the detrimental role of cross-linking on enzymes: altered enzymatic activity would be responsible for a decrease in the cell's efficiency. The *wear-and-tear* theories are probably the oldest precursors of the concept of failure to repair. They propose that organisms are constantly exposed to infections, wounds, and injuries that cause minor or major damage to cells and tissues. The progressive accumulation of this damage contributes to the age-related decline in functional efficiency. Today, there are at least two arguments to counter these theories. First, animals raised in protected environments in which they suffer a minimal amount of damage caused by the environment not only age but also fail to show any increase in their maximum life span. Second, recent discoveries in cell and molecular biology have provided other challenging explanations for why cells and organisms age (see Chapter 5). As a result, modern reformulations of the failure-to-repair theories, which involve all the intracellular stochastic theories, are more convincing in explaining particular aspects of aging than the original wear-and-tear theories.

Programmed Theories

Programmed or genetic theories state that aging is genetically controlled. One of the strongest arguments in favor of such views is that maximum life span is undoubtedly species-specific. Differences in life span are far greater between species than within species. The hereditary diseases of precocious aging, i.e., progerias, are also an argument in favor of a genetic basis of aging. As discussed earlier (see Chapter 2), patients with progerias suffer early in life from some of the commonly recognized aging changes. This implies the existence of specific genes that regulate the aging process. Such genes have recently been identified and their products are being characterized (see Chapter 5), providing the basis for the *death genes* theory.

As part of the genetic theories of aging, the *selective* theory proposes that senescent cells express specific membrane proteins that are recognized by phagocytic cells, and thus, old cells are readily eliminated. Recent findings add increasing support to this theory. A related theory, the *telomere theory of aging* states that the gradual degradation of the repeated arrays of identical DNA motifs located at the ends of chromosomes leads to chromosomal instability and ultimately causes cell death. Recent evidence about age-related shortening of telomeres and the presence of telomerase (the enzyme that repairs telomeres) in cancer cells but not in normal somatic cells, strongly supports this theory.

Dysdifferentiation and *waste accumulation* theories are also classified among the genetic theories. They postulate that a general failure of normal cell physiology results in either faulty gene activation or repression mechanisms, which decrease cellular efficiency, or in the incapacity of the cell to get rid of its waste products of metabolism. Both events lead to cellular suffocation and, eventually, to cell death.

The *neuroendocrine* and the *immunological* theories are among the most important programmed theories. This group of theories implicates the neuronal, the endocrinological, and the immunological systems in the aging processes. Experimental results supporting these theories have been discussed in Chapter 3. The neuroendocrine theories propose that a general functional failure in neurons and in the synthesis of their associated neurohormones are the central cause of aging. As discussed in Chapter 2, the hypothalamic-pituitary-adrenal axis (HPA) is thought to be one of the primary regulators of aging. Functional changes in this system are believed to have strong repercussions throughout the organism. Even though the importance of the neuroendocrine system is universally acknowledged, some critics point out that organisms lacking a complex neuroendocrine system also age and in ways analogous to that of higher vertebrates. Decrements in hormonal functions have led scientists also to formulate an *endocrine* theory of aging, postulating that endocrine systems are responsible for speeding up the aging process or for causing cell death. The functional capacity of the immunological system also declines with age and the immunological theories state that as T-cell function is reduced with age, there is a lowered resistance to infectious disease (Walford, 1981). The fine tuning of the immune system is also believed to be altered with age, as evidenced by the striking age-associated in crease in autoimmune diseases. As discussed in Chapter 2, the genes of the major histocompatibility complex play an important role in regulating life span (Walford, 1979). Interestingly, this gene locus also regulates **superoxide** dismutase and oxidase levels, a finding that connects the immunological theory of aging to the free radical theory of aging, described below.

In some respects, the *free radical* theory belongs to the random damage theories, but other aspects of this theory are more suggestive of the theories that favor developmental and genetic aspects of aging. The main idea of the free radical theory is that changes with aging are due to damage caused by atoms or molecules with an unpaired electron called free radicals. These are highly unstable molecules, and in normal conditions they are rapidly destroyed by protective enzymes. A decline in the efficiency of these enzymes can lead to an accumulation of free radicals and can damage important biological structures or interfere with vital functions (see Chapter 5, pages 113–119). This theory has increasing appeal, even though the role of free radicals in aging has been difficult to demonstrate. Of course, the idea that free radicals represent a single basic mechanism leading to all the changes observed with aging is unlikely to be correct, but more and more observations support the idea that the ability of the organism to control the detrimental effects of free radicals is an important factor in longevity.

A completely different approach to understanding the aging process is the *evolutionary* theory of aging. Wallace, a contemporary of Darwin and codiscoverer of natural selection, made the following statement: ". . . when one or more individuals have provided a sufficient number of successors, they themselves, as consumers of nourishment in a constantly increasing degree, are an injury to those successors. Natural selection therefore weeds them out, and in many cases favors such races as die almost immediately after they have left successors" (quoted in Weismann, 1891). All evolutionary theorists agree that the ultimate cause of aging is the declining force of natural selection with age.

Overview of Aging Theories

None of these theories has been fully confirmed, but many are plausible and supported by experimental data. A few theories have now been proven wrong. For example, the ideas put forward by the metabolic theories, proposing that longevity is inversely proportional to metabolic rate (or the rate of some physiological variables such as heart rate) were disproved on the basis of a more critical analysis of life span data in many species.

The large number of theories that have been proposed indicate that gerontologists are faced with an incredibly multifaceted problem and cannot yet reconcile the observed facts into a single theory. Each theory covers one aspect of the aging process, often in the field that corresponds

Table 7. Overview of theories of aging

Stochastic and programmed theories of aging.
A summary of most currently accepted aging theories is given below with a brief description of the concepts on which each theory is based and an evaluation of their validity.

Theory	Assumptions	Current status
STOCHASTIC THEORIES		
Somatic mutation	Somatic mutations destroy genetic information and decrease cell's efficiency.	Recent findings support this theory.
Error catastrophe	Erroneous transcriptional and/or translational processes decrease cell's efficiency.	Original theory disproved, but modifications of the theory are valid.
DNA damage DNA repair	Damage to DNA is repaired constantly by various mechanisms. Repair efficiency is positively correlated with life span and decreases with age.	Recent findings support this theory.
Altered proteins	Cell's efficiency is affected by conformational changes (cross-linking) of proteins and enzymes.	Proved.
Cross-linking	Chemical cross-linking of important macromolecules (e.g., collagen) impairs cell's and tissue function.	Proved.
Wear-and-tear	Accumulation of insults and injuries during daily living decreases the organism's efficiency.	Probable.

Table 7. *Continued*

Theory	Assumptions	Current status
PROGRAMMED OR DEVELOPMENT/GENETIC THEORIES		
Genetic theories	Senescence is due to programmed changes in gene expression, or to expression of specific proteins.	Proved.
Death genes	Existence of genes that determine cell death.	Supported by recent genetic findings.
Selective death	Cell death is induced in cells bearing specific membrane receptors.	Supported by recent findings.
Shortening of telomeres	The length of telomeres shortens with age, in vivo, and in vitro, leading to chromosomal instability and cell death.	Recent findings support of this theory.
Dysdifferention	Errors in gene activation-repression mechanisms, resulting in the synthesis of excess, insufficient, or unnecessary proteins.	Possible.
Waste accumulation	Accumulation of waste products of metabolism affect cell's efficiency.	Proved in some cases.
Neuroendocrine	Failure of neuronal and endocrine systems to maintain homeostasis. Loss of homeostasis leads to senescence and death.	Proved for the female reproductive system and other specific situations.
Immunological	Given alleles of the immune system would extend and others shorten longevity. These genes are thought to regulate a wide variety of basic processes.	Supported by recent findings.
Metabolic theories	Longevity is inversely proportional to metabolic rate.	Disproved.
Free radical	Longevity is proportional to free radical damage and to **antioxidant** capacity of organisms.	Increasing. experimental evidence.
Clock of aging	Aging and death result from a predetermined biological plan.	Recent findings support this theory.
Evolutionary	Natural selection eliminates individuals as soon as they provide successors.	Proved in some cases.

to the individual's scientific background and expertise (Cristofalo, 1990). The complexity of the aging processes implies the existence of many mechanisms that can be incorporated into more than one valid scheme, and many of these mechanisms interact. For example, an individual with long-life genes may also be subject to free radical damage as well as other wear-and-tear injuries that reduce the genetic life expectancy. At any rate, it is easier to construct theoretical explanations of aging than to test them critically and, often, technical difficulties in devising experiments are immense.

Table 7 gives an overview of the many theories of aging.

Chapter 5

Mechanisms of Cellular Aging

Cells are the fundamental units of living beings. Tissues, organs, and functional systems result from the assemblage of specialized cells interacting with each other. Studying cells outside their natural environment, i.e., isolated from their tissue and cell-type interactions, helps determine whether aging mechanisms also operate at the cellular level. Research has focused on how aging affects different types of cells and what intracellular modifications can be observed over time. The findings have profound implications for the theories of aging (Peacocke and Campisi, 1991).

Aging at the cellular level turns out to be complex. Cells age because of the functional deterioration of their internal constituents, called intrinsic factors, and/or because of extrinsic factors, i.e., changes in their environment. The existence of genes that shorten or prolong life span has now been firmly established, and the evidence for genes that provoke cell death have added weight to the programmed theories of aging. Aging can no longer be considered solely as a passive phenomenon, as proposed in the wear-and-tear theories, whereby cells, organs, and systems slowly deteriorate over time as a result of repetitive exterior attacks. The existence of death genes indicates that aging is, at least in part, an active intrinsic process. In addition to death genes, antigrowth genes and long-life genes have been identified. Other types of intrinsic factors are somatic mutations in particular genes. These give rise to modified proteins which can lead to disease and premature death. Cellular aging can occur also at any level in the organism's many intracellular biochemical reactions. These epigenetic modifications can perturb gene expression. For example, aging can affect normal chemical modification of DNA such as methylation, or transcription processes. Poor transcriptional efficiency can lead to a suboptimal concentration of essential proteins, such as hormones, neurotransmitters, or enzymes and this may perturb the metabolism of the cell. Erroneous intracellular mechanisms, such as errors in the overall metabolism of proteins, lipids, and sugars occur during aging.

The influence of intrinsic factors on cellular aging processes seems to outweigh that of extrinsic factors. Yet, cells are very much affected by environmental factors; free radicals and radiation are probably the two most harmful. Free radicals profoundly perturb cellular functions and are possibly a leading cause of aging processes. The generation of free radicals is due essentially to oxygen, the very molecule essential for life; for this reason, free radicals are discussed arbitrarily in the chapter concerned with extrinsic factors. However, as was suggested in Chapter 4, free radicals are also generated by intracellular biochemical reactions and could have been discussed equally well in the chapter on intrinsic factors.

Intrinsic Mechanisms

The Influence of Genes and Chromosomes on Health and Longevity

The Discovery and Role of Death Genes and Long-Life Genes in Invertebrate Species

Analysis of the life spans of animals shows that the range of maximum life span between closely related species (different species of mammals for example) far exceeds the variation in maximum life span between individuals of the same species. The maximum life span of a mouse is ten times longer than that of a fruit fly, and humans live about thirty times longer than mice. How can those differences be explained?

The gradual deterioration in the synthesis of control mechanisms or damage of intracellular components due to outside factors are believed to occur passively and at random. It is not surprising, therefore, that inbred mice of different genotypes differ widely in the onset of age-related diseases and dysfunctions. What is astonishing, however, is that their *mortality rate doubling time*, or MRDT (defined as the time necessary to observe an increase of twofold the mortality rate) are quite similar. The increase in the mortality rate has been considered the major determinant of the maximum life span. This is also true in human populations. The fact that in all ethnic groups, a similar MRDT is observed in spite of different distributions of age-related diseases suggests that life span is regulated by specific genetic determinants. The recent discovery of genes, identified in several species, that accelerate or cause death, has substantiated these hypotheses. These so-called death genes open the door to new fields of research in the discovery of aging mechanisms. They encode products that display a large variety of physiological functions. Although homologous genes have not yet been demonstrated in humans, it can be inferred from animal studies that they may well exist. The discovery of genes that have the potential to prolong rather than shorten life span is a pleasant counterbalance to the discovery of death genes, and these so-called long-life genes will also be discussed. The roles of both sets of genes and their possible interactions are key elements in understanding the complex mechanisms of senescence.

Death and Long-Life Genes in the Nematode Caenorhabditis Elegans

A prerequisite for the identification of death genes and long-life genes is to study simple species, and the nematode *Caenorhabditis elegans (C. elegans)*, a small soil worm, is an ideal model. The life span of this worm is about 20 days, and its rapid senescence is an advantage in studying the effects of genes and of genetic manipulations. These worms reproduce largely through hermaphroditic self-fertilization which leads to populations of highly inbred individuals. A variety of mutants that affect development, behavior, morphology, fertility, or cell lineage have been isolated, and these are relevant for research on aging. In some mutants, the relative timing of events in larval development is affected, and in others, modification in cell death mechanisms has been observed (Ellis and Horvitz, 1991). The selection of *C. elegans* populations with different life spans is an additional demonstration that the rate of senescence is under genetic control. Different approaches have been used to identify the genes influencing senescence and life span. One of them involves producing mutants with the use of the mutagen ethylmethanesulfonate (EMS). This treatment has produced worms that have up to a 50% greater mean life span and a twofold greater maximum life span than their progenitors (Friedman and Johnson, 1988a, 1988b). This lengthened life span is not associated with reduced food intake, an important point to emphasize, in view of the increase of life span observed in many species, including *C. elegans*, under dietary restriction (see Part II, Chapter 1). Genetic analysis of mutants with increased life spans has led to the identification of a gene, designated *age-1*. The modification or the suppression of this gene product results in life span prolongation. *Age-1* specifies sperm activation; it is recessive and tightly linked to a gene locus influencing fertility, *fer-15*. Scientists failed to separate the two loci by genetic crosses and hypothesized that *age-1* and *fer-15* could be one and the same gene, suggesting that one of the functions of the wild-type gene product of *age-1* could be to increase fertility and another could be to shorten life span. Thus, *age-1* may be one of the regulators of the aging process, at least in this species (Johnson, 1990). Mutations in other genes called clock genes (*clk-1*, *2*, and *3*) also have been found to affect nematode life span. The analysis of worms that carried mutations in two clock genes or in one clock gene and other genes known to prolong life span, has led to the surprising finding of worms with specific combinations of mutations that live up to five times longer than normal. Each phase of the life cycle of these mutant worms is extended. Moreover, they eat and defecate less frequently, and they move more slowly. The functions of clk proteins are not yet known, but they could act as a regulator of aging in nematodes (Lakowski and Hekimi, 1996).

Horvitz and his colleagues isolated a series of *C. elegans* mutants in which migration and outgrowth of specific neurons were affected (Yuan and Horvitz, 1992; Ellis et al, 1991). In this small worm, as in all animal species, specific mechanisms exist, called programmed cell death (PCD), that are destined to kill cells that have presumably not migrated to their final destination or established proper connections with other cells. PCD is regulated by 11 genes. In *C. elegans* it is a complex process involving three successive steps: killing, engulfment or phagocytosis, and degradation (Ellis et al, 1991). Two mutants, unable to perform the first step, i.e., cell killing, were isolated: they were called *ced3* and *ced4*. Scientists postulated that *ced3* and *ced4* mutants had deficient calcium binding proteins and proteins active in the process of phosphorylation. Recent findings have shown that there is a mammalian homologue of the product of *ced3*, the protease interleukin-1β converting enzyme (ICE) (Yuan et al, 1993). While searching for cells that could exert a regulatory control over cells expressing *ced3* and *ced4* mutated proteins, Horvitz and his colleagues found another very interesting mutant, called *ced9*. *Ced9* mutants bear mutations in a gene encoding a protein that protects cells from programmed cell death (Hengartner et al, 1992). Mutants that have a decreased *ced9* activity die early in embryogenesis because of massive cell death due to the activity of *ced3* and *ced4*. *Ced9* thus acts as a control gene. When it is active, it inhibits the function of *ced3* and *ced4*, and when it is not, *ced3* and *ced4*, uncontrolled, cause the massive cell death. Interestingly, the nucleic acid sequence of the *ced9* gene is similar to that of *bcl-2* oncogene, a gene isolated in human B cell lymphoma (see below) that specifically antagonizes programmed cell death (see below). Not only are the two proteins encoded by *ced9* and *bcl-2* genes structurally similar, but they were found to have a strong similarity in their biological functions. This has been demonstrated by the ability of human *bcl-2* to rescue the *ced9* deficient phenotype after **transfection** in nematodes.

The identification and the characterization of all the proteins involved in PCD is important and will yield valuable information on their cellular function. This will contribute to the identification of related genes in mammalian species. Indeed, whether homologous genes display the same functions in mammalian species and in humans in particular, is an important issue.

Death and Long-Life Genes in the Fruit-Fly Drosophila melanogaster

The fruit fly *Drosophila melanogaster* is another useful model to study the genetic components of longevity. As in the case of *C. elegans, Drosophila* mutants that display quite different life spans were isolated after treatment with the mutagen EMS (Mayer and Baker, 1985; for review, see Rose and Graves, 1990). Among the various mutants generated, some carried a mutation in the gene for superoxide dismutase. Homozygote mutants for this gene develop normally, but they have a short adult life of 10 days versus 60 days for wild-type flies (Phillips et al, 1989). A closer analysis showed that these mutants have an increased sensitivity to substances generating free radicals, such as paraquat and transition metals. Further, they lack motile sperm at adult-stage, which may point to an important role of superoxide dismutases in protecting against DNA damage during gametogenesis.

Two other short-lived mutants, called drop-dead (drd) and dunce have mean life spans of a few days and about 15 days, respectively. The defects were identified in both types of mutants. Drd mutants undergo a rapid degeneration of brain cells (Hotta and Benzer, 1972). Scientists have not found the cause of this neurodegeneration yet, but they think that it could reflect an acceleration of a normal process of neuron atrophy and death. Dunce mutants are useful for understanding the physiological processes associated with cAMP metabolism and how these processes may underly the shortened longevity of dunce male and female flies. Mutations alter or abolish cAMP-specific phosphodiesterase activity and this causes measurable changes in the cAMP content of whole flies. In null mutants (in which there is no active cAMP-specific phosphodiesterase), there is a five- to sixfold increase in cAMP concentration. The loss of phophodiesterase has an effect, therefore, on longevity, but this is not the only effect. Cyclic AMP-phosphodiesterase of *D. melanogaster* is also involved in several physiological processes. Dunce mutants exhibit five specific physiological and behavioral defects. Three of them underlie the female sterility phenotype and cause learning deficiencies in both sexes. The other two defects involve an interesting behavioral characteristic. Scientists noticed that the longevity of dunce females in the presence of males is reduced by 50%, compared to control dunce females kept without males. They also noticed that mutant dunce females mate two to three times more frequently than wild-type females (Bellen and Kiger, 1987). Frequent mating and the associated mechanical trauma are factors that have been implicated in the shortened life span.

Accelerated senescence in *Drosophila* is not only the result of mutations in specific genes, but it can also be the result of modifications in the expression of epigenetic factors. While studying protein synthesis mechanisms, investigators noticed by serendipity a sharp decrease in old flies of the expression of a protein, later identified as a translation factor, the **elongation factor** *EF-1*α (Webster and Webster, 1983). The translation machinery is complex and many factors and cofactors interacting in subtle ways are involved in well-defined steps like initiation, elongation, and termination of protein synthesis. Interestingly, the decrease in *EF-1*α preceded the decrease in total protein synthesis. By use of genetic manipulations, Shepherd and his collaborators (1989) introduced additional copies of the *EF-1*α gene in flies. The result was spectacular: a considerable extension of the life span of these flies. What was remarkable was that the manipulation of a single gene that does not necessarily represent the primary cause of senescence induced major changes in the complex translational machinery. The most plausible explanation is that the additional copies of the *EF-1α* gene had complex **pleiotropic effects**.

Candidate Death and Long-life Genes in Humans

bcl-2 Gene of Human Lymphoma

About 10 years ago, changes in the expression of a gene called *bcl-2*, were identified in cells of a cancer in humans known as follicular B cell lymphoma (Tsujimoto and Croce, 1986). In these cells, a chromosomal rearrangement brought the *bcl-2* gene close to the J segment of immunoglobulin heavy-chain genes, and this translocation was accompanied by over-expression of *bcl-2* protein. The gene *bcl-2* was found later to encode a mitochondrial membrane protein (Hockenbery et al, 1990). It was localized recently on nuclear and **endoplasmic reticulum** membranes also (Krajewski et al, 1993). When scientists transfected *bcl-2* gene into mouse cells that were dependent on specific immunological hormones for survival, they uncovered a special function of *bcl-2* (Korsmeyer, 1992). It became apparent that this powerful gene conveyed an immortal phenotype to cells in which it was expressed. However, in contrast to all other oncogenes, *bcl-2* did not induce cellular proliferation; instead, cells went into a quiescent state that persisted even when the hormones were withdrawn. As cells accumulated, the chance that some of them would become malignant increased. Korsmeyer and his colleagues (Nuñez et al, 1991) made the same observations: insertion of active copies of *bcl-2* into murine antibody-producing B cells resulted in the accumulation of B cells, and mice developed cancer.

The implication of these findings is that *bcl-2* can block programmed cell death in B cells and thus extend their lifetime. This characteristic is interesting since one of the crucial features of the immune system is its ability to remember infectious agents and its capacity to mount a secondary immune response even years after a primary exposure to an antigen. The signal that enables certain B cells to acquire this long-term biological memory is unknown at present, but *bcl-2* could be a proto-oncogene responsible for maintaining immune responsiveness. Recently, more pieces have been added to the *bcl-2* puzzle. The *bcl-2* gene product has been found to counteract the toxic effects of hydroxyl radicals thereby protecting senescent cells from oxidative stress. Hengartner and Horvitz (1994) and Yin and his collaborators (1994) have provided further evidence in nematodes of the way in which *bcl-2* protein and other newly discovered members of the *bcl-2* gene family interact to preside over the cellular decision between life and death. These scientists propose that *bcl-2* inhibits damaging lipid peroxidation chain reactions that occur in membranes and thus protects cells from free radical damage. As the gene *bcl-2* is part of a genetic program counteracting the cell-death program and prevents cells from dying on schedule, biologists are now thinking of ways to use *bcl-2* as a therapeutic tool. In particular, they are devising strategies to insert active versions of *bcl-2* into patients affected by nervous system degenerative diseases in which cells die massively, like Alzheimer's, Parkinson's, and Huntington's diseases (see Part II, Chapter 8).

ApoE and ACE Genes

The startling finding of Schächter and his collaborators (1994) on the association of a particular form of apolipoprotein E (ApoE) with Alzheimer's disease has been alluded to in Chapter 2. In 1992, a group of scientists analyzed the genetic components of longevity in humans. Since cardiovascular diseases are among the leading causes of death, they decided to investigate, in over 300 centenarians, the associations between longevity and genetic variations of two genes implicated in the occurrence of cardiovascular disease (vanBlockxmeer and Mamotte, 1992). These genes, coding for ApoE and angiotensin-converting enzyme (ACE), play a major role in lipid metabolism (Schächter et al, 1994). ApoE proteins participate in the formation, secretion, transport, and binding of macromolecular complexes, called lipoprotein particles. They are components of atherogenic triglyceride-rich lipoproteins and of high-density lipoproteins (**HDL** or good lipoproteins) in plasma (see Chapter 5, pages 96–97). ACE is a key member of the **renin**-angiotensin system important in the control of blood pressure, salt, and water homeostasis and control of cell growth. Three common variants (or isoforms) of ApoE proteins have been described, E2, E3, E4, and are encoded by the corresponding alleles called ε2, ε3, ε4 (Saunders et al, 1993; Strittmatter et al, 1993). While the isoform E4 is associated with higher plasma cholesterol levels and is related to an increased risk of **ischemic** heart **disease**, the relatively rare E2 isoform is associated with lower cholesterol levels. Schächter and his collaborators (1994) found a higher frequency of the ε2 allele in centenarians and suggested that ε2 might confer a protective effect by its cholesterol-lowering influence and secondary beneficial impact on ischemic heart disease. However, this is not the only explanation. ApoE (and particularly the ApoE4 variant) has been found to be directly implicated in late-onset Alzheimer's disease (SDAT). ApoE4 protein binds abnormally to some intracellular proteins, resulting in their aggregation and the formation of intracellular neurofibrillary tangles, and it also binds extracellularly to β-amyloid protein (Schächter et al, 1994). These researchers suggest that SDAT develops because the ApoE4 variant is detrimental to the neurons and deprives cells of their normal protection against neurodegeneration. It may be that ApoE4 binds to other extracellular or intracellular components that might be involved in coronary vascular disease. More research is needed to uncover all the steps of this pathophysiological process. What is interesting is that Strittmatter and Corder and their respective groups have suggested that people who carry the gene for ApoE4 protein would be more likely to suffer SDAT (Corder et al, 1993; Strittmatter et al, 1993). Their proposal is supported by another recent study in which the authors evaluate the role of ApoE in predicting the outcome of memory-impaired individuals; they have concluded that possession of the allele Apo ε4 is the strongest predictor of clinical outcome (Peterson et al, 1995).

The greater representation of the D allele of ACE in very old people is somewhat paradoxical because that same D allele is predominant in subjects with myocardial infarction. Schächter and his collaborators (1994) speculate that, apart from its adverse effect on cardiovascular disease, the D allele might have some as yet unknown beneficial effects, for example, on repair of damaged tissues or resistance to neoplasia or infection.

Antigrowth Genes

The processes of cell growth and proliferation appear to be controlled by a large number of genes that act in a positive or in a negative way. A wealth of information exists about growth factors (considered as positive factors), but little is known about negative regulators. The gradual loss of the capacity of cultured animal or human cells to proliferate was a hint that negative regulators existed and played an important role in aging. Scientists therefore searched for molecules that prevented the continuous proliferation of cells in culture. McClung and his colleagues (1989; Nuell et al, 1991) reported the isolation of a gene from rat liver encoding a protein that they called prohibitin. Injection of prohibitin messenger RNA into actively growing cells inhibited cell proliferation while maintaining their viability. Unlike other antiproliferative proteins, prohibitin exerts its effects on normal cells at physiological levels and without concomitant changes in cell shape. Prohibitin is widely expressed in different tissues and highly conserved in evolution. Recently, the human equivalent of the rat prohibitin gene has been characterized and mapped to chromosome 17q21, in a region containing a gene responsible for hereditary breast cancer (Sato et al, 1992). The interesting properties of prohibitin led scientists to postulate that changes in the expression of its gene could play an important role in cellular senescence. Indeed, DNA analysis of part of the prohibitin gene isolated from breast cancer tissues revealed the presence of somatic mutations in the gene of several patients. Scientists hypothesized that this gene could act as a tumor suppressor gene and could be associated with tumor development and/or progression in at least some breast cancers.

The role of prohibitin might be extended to other cancer cells. In fact, many types of cancer cells, grown in culture, are immortal, that is, they escape normal cellular aging and continue to divide indefinitely. Possibly, changes in the expression of the prohibitin gene could be one of the key factors in the uninterrupted growth of cancer cells. Scientists are now looking at the regulation of prohibitin gene expression in normal as well as in cancer or aging cells. If, indeed, overexpression of prohibitin is involved in age-related loss of cell growth, then finding a way to control the expression of this gene could perhaps minimize age-related pathology. However, such a control should be exercised with caution—inhibition of prohibitin might destabilize the balance between anarchical cell growth and total inhibition of proliferation and might, paradoxically, promote an increased incidence of cancers.

The prohibitin gene is one of many genes expected to have a role in decreased cell proliferation. The isolation and characterization of genes encoding proteins with analogous functions will help scientists better understand the causes of abnormal cell proliferation as well as the basic mechanisms of cell division or division prevention.

The Role of Specific Chromosomes in Cellular Aging

When cells isolated from different tissues are cultured, they undergo a limited number of divisions, enlarge in size, and undergo cellular senescence resulting in cell death. In contrast, most tumor cells (cells transformed with chemical carcinogens, viruses, or oncogenes) can be grown indefinitely in culture; they have escaped senescence and are termed immortal. Scientists have taken advantage of the phenotypic properties of normal and transformed cells to search for specific genetic determinants of aging. One of their approaches is based on the study of the proliferative potential of **hybrid cells**. Most hybrids of human cells with a finite life span and immortal cells with an indefinite life span have been found to age, indicating that senescence is dominant over immortality. On a few occasions, certain hybrids of two different immortal human cell lines have been found to have a finite proliferative capacity (Pereira-Smith and Smith, 1983, 1988). This suggested that chromosomal rearrangement or loss of chromosomes might occur during hybridization between two types of cells, which would result in the ability or inability of cells to grow indefinitely. Sugawara and his colleagues (1990) studied the chromosomal content of hybrids between normal human diploid fibroblasts and immortal Syrian hamster cells. As observed previously, the majority of these hybrids exhibited a limited life span, similar to that of human fibroblasts, which confirmed that senescence is dominant in these hybrids. A few hybrids, however, did not senesce. **Karyotypic** analyses of these clones revealed that both copies of human chromosome 1 had been lost (all other human chromosomes were maintained in at least some of the immortal hybrids). The application of selective pressure to retain human chromosome 1 in the hybrids resulted in an increased percentage of hybrids that senesced. Moreover, the introduction of a single copy of human chromosome 1 into the hamster cells by microcell fusion caused typical signs of cellular senescence, whereas transfer of other human chromosomes did not cause senescence. Immortal cells arose, therefore, as a consequence of recessive changes in the growth control mechanisms of normal cells. These experiments indicate that cellular senescence results from a genetic program by which specific genes on chromosome 1 limit cell proliferation. The implication is that genes determining aging are *not* randomly distributed on all chromosomes. Scientists are now trying to identify and localize these putative aging genes.

The Role of Telomeres and Telomerase in Cancer and Aging

Of the various cytogenetic alterations that are involved in cellular senescence of normal cells and the continuous proliferation of cancer cells, changes in the length of telomeres is considered one of the most important.

Human telomeres are tandemly arranged repeat arrays of the sequence motif TTAGGG, located at the end of chromosomes. The length of these repeats is several thousand bases long. Not only are telomeres necessary for DNA replication, but they protect chromosomes from DNA degradation and rearrangement. Telomeres are maintained by telomerase, a ribonucleoprotein enzyme that overcomes the inability of DNA polymerase to replicate telomeric regions (Greider and Blackburn, 1985). Interestingly, in most somatic human cells, telomerase activity is repressed, leading to telomere loss during cellular replication in vivo and in vitro (Harley and Villeponteau, 1995). It is estimated that chromosomes lose stretches of approximately fifty nucleotides from the end of telomeres with each cellular division. The gradual degradation of telomeres is thought to lead to chromosomal instability. Scientists who subscribe to the *telomere* theory of aging believe that telomere shortening is the main cause of cell aging and death. Supporting this theory are observations that telomeres are shorter in fibroblasts from an old donor than in fibroblasts from a young donor (Kruk et al, 1995) and that, in contrast to somatic cells, telomere length is maintained in essentially all human cancer cells, due to the presence of the enzyme telomerase. Recently, scientists have found evidence that telomerase plays a role in the uncontrolled growth of tumor cells: inhibiting telomerase activity by injecting sequences complementary to the telomerase messenger RNA into tumor cells leads to a shortening of telomeres. More importantly, tumor cells, that are normally immortal undergo a number of cellular divisions and die (Feng et al, 1995). Conversely, when cells are immortalized in vitro, telomeres maintain a stable length.

However, the relation between the length of telomeres and the presence of telomerase is not constant. An analysis of telomeric repeats in nonagenarians has revealed a surprising stability (Luke et al, 1994). In another study, considerable variation of telomere length has been observed among individuals of the same age, indicating that these differences might be genetically determined (Slagboom et al, 1994). Moreover, the length of telomeres is shorter in the liver of patients with chronic hepatitis or liver cirrhosis compared with that in the normal liver and tends to decrease with the progression of the disease (Kitada et al, 1995). The same observation has been made in cancer cells, where chromosomes have shorter telomeres than do normal cells. To explain this contradictory finding, the authors proposed that early cancer cells do not turn on telomerase expression until later in the progression of the disease when telomere loss generates a selective pressure for cell immortality.

Differences in telomerase activity between species have been reported. In mouse, for example, telomerase activity is detected in somatic cells, whereas it is absent from most human tissues (Prowse and Greider, 1995). Furthermore, a tissue-specific regulation of telomerase expression has been observed in mouse during development and aging in vivo.

The causal relation between telomeres and aging is not yet established. Evidence is still needed to show that the decline in telomere repair with aging has functional significance to an age-related decline in genomic stability. Telomere shortening could be a consequence of aging, like grey hair or loss of muscle. Since telomerase expression is unique to cancer cells, it represents an attractive therapeutic target. Researchers are now actively developing inhibitors of telomerase for the treatment of a broad spectrum of human malignancies.

Human Genetic Diseases Due to Somatic Mutations

In recent years, researchers discovered that many diseases are caused by single defective genes. In these so-called genetic diseases, single point mutations, **amplifications** or deletions of DNA sequences in specific genes lead to abnormal structural or enzymatic proteins. This induces modifications in basic cellular functions and eventually results in pathological conditions. The number of diseases found to be due to inborn errors is increasing as scientists succeed in detecting the molecular defects of each disease. A few examples of human genetic diseases are described below.

Diseases Caused by Mutations in Genomic DNA

LDL/HDL Disease

Familial hypercholesterolemia (FH) is an inherited disorder that causes blood cholesterol levels to be abnormally high. To search for the biochemical cause of FH, Brown and Goldstein (1986) carried out research on a spontaneously hypertensive animal model, the Watanabe rabbit. They showed that in these rabbits, the high blood cholesterol is due to a deficiency of low-density lipoprotein (LDL) receptors. LDL receptors remove cholesterol from the blood so that it can be degraded by the liver. The same defect is present in human patients.

FH exists in two clinical forms, the less severe heterozygous form and the more severe homozygous form. FH heterozygotes, who carry a single copy of a mutant LDL-receptor gene, are quite common, accounting for one in every 500 people among most ethnic groups throughout the world. These individuals have a twofold increase in the concentration of LDL particles in plasma. They begin to suffer from heart attacks at 30 to 40 years of age. Homozygote individuals are rare, and account for about one in a million. They inherit two mutant genes at the LDL receptor locus. Consequently, their disease is much more severe than that of heterozygotes. They have six to ten times the normal concentration of plasma LDL from the time of birth, and die at an early age (often in their twenties) of heart attacks. The severe atherosclerosis that develops in these patients in the absence of other risk factors is a proof that high levels of plasma cholesterol can produce atherosclerotic plaques leading to heart attacks and strokes in humans.

Cholesterol molecules are transported as spherical particles, called LDL particles, containing cholesteryl ester molecules in a core. This is shielded from the aqueous plasma by a hydrophilic coat composed of phospholipids and unesterified cholesterol molecules and one molecule of a protein called apoprotein B-100. Cholesterol passes the cellular membrane via binding of LDL to lipoprotein receptors of which the prototype is the LDL receptor. These receptors bind LDL and carry it into the cell by receptor-mediated **endocytosis**. An interesting property of this receptor is its ability to cycle many times in and out of the cell allowing large amounts of cholesterol to be delivered to tissues, while keeping the concentration of LDL in blood low enough to avoid the formation of atherosclerotic plaques. In the cell, cholesterol is used for the synthesis of cell membranes, bile acids, and steroid hormones. When the LDL receptor does not function properly as a result of genetic defects or in response to regulatory signals, cholesterol is not transported normally into cells and it builds up in plasma. The mutations detected in the LDL-receptor gene have been divided into four classes defined by the localization of the genetic defect that disrupts specific functions of the receptor.

Class 1 is the most common class of mutant alleles. Mutated genes produce either no LDL receptor proteins or only trace amounts. Molecular analysis of one of these alleles has shown that the gene contains a large deletion. In class 2 mutations, the receptors are synthesized but transported slowly from the endoplasmic reticulum to the Golgi. The exact molecular defect of this class of mutations has not been determined. In class 3 mutations, there is a suspicion that amino acid substitutions, deletions, or duplications in the cystein-rich LDL-binding domain might be the cause of the defect. LDL receptors are processed and reach the cell surface but fail to bind LDL normally. Finally, in class 4 mutations, the rarest, receptors reach cell surface and bind LDL but fail to internalize or cluster in coated pits. All class 4 mutations involve alterations in the cytoplasmic tail of the receptor.

The studies of Brown and Goldstein (1986) emphasized the clinical importance of the LDL receptor in the uptake of LDL; about two-thirds of the removal of LDL from blood is mediated by these receptors and occurs mostly in the liver. The reason that high levels of plasma LDL are so common in western industrialized societies is not clear at present (nor are the mechanisms producing these high levels), but extensive evidence implicates two major factors, diet and heredity.

The mutations in the LDL receptor gene serve as examples of a genetic defect, but it should be emphasized that there are important alternative pathways by which LDL is removed from blood, independent of the LDL receptors.

Cystic Fibrosis

Cystic fibrosis (CF) is a complex inherited disorder that affects children and young adults (Boat et al, 1989). The frequency of the disease varies among ethnic groups and is highest in individuals of Northern Europe where about 1 in 2500 new-borns is affected. Significant numbers of affected individuals are also found in the Ashkenazi Jewish population and in American blacks. CF is inherited in an autosomal recessive fashion. Heterozygotes with one normal CF allele and one mutant allele are asymptomatic carriers. In the early fifties, the finding of excessive salt loss in the sweat of affected children led to the measurement of sodium and chloride in sweat, which served as a diagnosis of the disease. The clinical symptoms of CF are characterized by damage to the respiratory tract, with predominant obstruction of the airways by thick, sticky mucus. Often, airways become infected. There is also a pancreatic deficiency caused by obstruction of the pancreatic ducts and loss of exocrine function. As a result of considerable improvement in current treatments (particularly thanks to the administration of antibiotics), individuals born today with CF are expected to survive until they are about 40 years old.

In the last 10 years, impressive breakthroughs have led to the discovery of the primary cause of the salt transport abnormality. First, scientists have demonstrated that the normal efflux of chloride ions across respiratory epithelial cell membranes in response to elevated adenosine 3′,5′ monophosphate (cyclic AMP) is lacking in cells derived from patients with CF (Sato and Sato, 1984). These studies point to a deficiency of protein kinase A (PKA) to activate a chloride conductance. The second finding is the localization of the gene for CF on chromosome 7 (achieved using the techniques of genetic linkage analysis of affected individuals) using a panel of polymorphic DNA **markers** and DNA cloning techniques. Subsequently a combination of **chromosome jumping** and **chromosome walking** was used. A large group of scientists combined their efforts and identified a candidate transcript, which is expressed in sweat glands, lungs, and pancreas (Rommens et al, 1989). The proof that the identified gene is involved directly in the pathology of CF came with the finding that the genes characterized in affected individuals contain mutations that are not present in the genes of normal people. In individuals with CF, a 3 base-pair deletion, resulting in the loss of a single amino acid, has been identified.

Surprisingly, this particular mutation (referred to as DF508) is responsible for 70% of all CF mutations. The other 30% represents about 170 different mutations. DF508 mutation is associated with severe CF with almost universal pancreatic deficiency and a high risk of *meconium ileus*. Unlike DF508 mutations, a few other mutations are associated with very mild expression of the disease, including some that result in normal sweat chloride values. The protein encoded by this gene is found to have a striking homology with a superfamily of proteins involved in active transport across cell membranes, called traffic adenosine trisphophatases (ATPases) (Ames et al, 1990). The CF protein, called cystic fibrosis transmembrane conductance regulator or CFTR protein, functions as an ion channel, most likely as a chloride channel. It is possible that the CFTR protein may have other important functions such as cAMP-mediated endocytosis and exocytosis. The expression of the CF gene is restricted largely to epithelial cells, where a low-level transcription is detected. High levels of CF RNA are found, however, in pancreas, salivary glands, sweat glands, intestine, and reproductive tract.

Today, therapies focus on the pulmonary complications of CF. The new approach using gene therapy raises some hope in the medical community. Scientists recently attempted to correct the ion transport defect in transgenic mice carrying the CF gene by transferring the normal CFTR coding region into their CF cells (Drumm et al, 1990). Even though this new therapy is still at a very early stage, medical doctors have now taken a similar approach with human patients and have started clinical trials in severely affected patients. Additional information on the techniques used in gene therapy is given in Chapter 8, Part II.

Malignant Hyperthermia is a life-threatening disorder that manifests itself only in conjunction with general anaesthesia. Abnormalities in the calcium release channel on the skeletal muscle sarcoplasmic reticulum are the cause of both the porcine and human syndromes (MacLennan and Phillips, 1992).

Gaucher Disease is characterized by the accumulation of glucocerebrosides, leading to enlargement of the liver and spleen and to bone lesions. It is due to an inherited deficiency of the enzyme glucocerebrosidase. Many mutations have been detected, but four of these account for over 97% of the mutations in Ashkenazi Jews, the group in which Gaucher disease is the most common (Beutler, 1992).

Epidermolysis Bullosa is a hereditary disease of the skin, characterized by blisters following minor trauma. Recent findings reveal that epidermal basal cells are fragile, due either to mutations of genes encoding keratin intermediate filament proteins or to mutations of the gene encoding type VII collagen, which is the major component of anchoring fibrils (Epstein, 1992).

Retinitis Pigmentosa is a term used to describe a genetically and clinically heterogeneous group of human inherited retinopathies. The prevalence of these diseases is about 1 in 3000. The first molecular defects to be identified were anomalies of rhodopsin which were caused by various mutations in the corresponding gene. Then, lesions in a second protein, peripherin, were found of which a whole spectrum of mutations have now been described (Kajiwara et al, 1993; Nichols et al, 1993; Wells et al, 1993). The mutations of the gene, called RDS/peripherin (RDS stands for retinal degeneration slow) cause a whole array of retinopathies with diverse symptoms. All of these lesions are the result of mutations in the same gene. There is evidence though that additional genes are implicated in retinopathies (Humphries et al, 1992). Recently, mutations in other proteins also implicated in phosphotransduction, the phosphodiesterases (PDE), have been suspected as the cause of retinitis pigmentosa. The PDE in rods of the retina is composed of two larger catalytic subunits, α and β, and two smaller inhibitory subunits, γ. In 1991, Pittler and Baehr showed that the recessive disease in mice, called retinal degeneration, characterized by a loss of rods' photoreceptor cells, is due to a **nonsense mutation** in the β gene of PDE. Recently, an analogous mutation has been reported in the same gene in Irish setter dogs that causes rod/cone dysplasia (Suber et al, 1993). Concomitantly, Dryja and his collaborators (McLaughlin et al, 1993) found three patients with recessive forms of retinitis pigmentosa, who have a nonsense mutation in the β subunit of PDE and one patient with a missense mutation in this gene. The gene encoding the β subunit of PDE is therefore an additional gene in which mutations have been found to be the cause of autosomic recessive retinitis. The mechanisms by which a loss of PDE causes retinitis is still unclear. It is likely, however, that the defects in rhodopsin and PDE β subunit are only the first genetic mutations discovered in the cascade of phototransduction. Other candidate genes probably exist and may be identified by screening large numbers of patients.

The Genetic Origins of Neoplasia

The last fifteen years have witnessed important breakthroughs concerning the molecular and genetic basis of the origin of neoplasia. Thanks to the huge number of observations gathered by scientists, it is now possibe to have a more extensive picture of the field. Mutations in a variety of genes have been shown to contribute to the origins and progression of cancer in humans. Several of these genes can be grouped in three categories: the oncogenes, the tumor suppressor genes, and the mutator genes. Genes called proto-oncogenes become oncogenes after mutations. Many oncogenes have been identified, but only a small subset of these contribute to the origin of human cancers. At present, eight to ten tumor suppressor genes have been identified. An example of one of them, p53, is given below. Genes belonging to the third category, the mutator genes are responsible for ensuring the fidelity of DNA replication. When these genes are modified and fail to function, an increase in mutation rate is observed.

The p53 Gene and Cancer

The small protein called p53 could turn out to be a very important molecule of our body (p stands for protein, and 53 refers to the molecular weight of the protein: 53 kilodalton; p53 protein is 393 amino acids long). It has been discovered that p53 supervises the evolution of cancerous cells, slows down their anarchical growth, and may even cause tumor regression. It could also play a role in cell aging by eliminating old, nonfunctioning cells. The product of p53 gene was first described as a cellular protein that forms a complex with large-T antigen in virus SV40-transformed cells. Later, wild-type p53 gene product was shown to activate transcription in vitro (Farmer et al, 1992), and to be involved in the control of cellular proliferation. Today, p53 protein is best classified, like the retinoblastoma gene product (see below), as a member of the tumor suppressor family. P53 protein behaves as an anti-oncogene: injection of normal p53 in transformed cells inhibits their uncontrolled proliferation. The discovery of mutated forms of p53 led scientists to postulate that if normal p53 is involved in controlling the growth of tissue by activating genes involved in growth inhibition, mutated forms of p53 could interfere with this process and promote tumor formation. To date, p53 gene mutations have been found in inherited and in spontaneous cancers. They are the most common mutations in human tumors, and have been detected in colon, lung, breast, brain, bladder, and uterus cancers, as well as in some forms of leukemia and cancers of the lymph nodes (Hodgkin's disease). About half of the people diagnosed with cancer each year worldwide have p53 gene mutations in their tumors. These mutations differ depending on the cancer-causing factor. For example, p53 mutations in lung cancers of workers in uranium mines are different from those who smoked tobacco. The same is true in colon cancers; specific mutations are observed in people living in China or Africa where the level of food contamination by **aflatoxins** is high, while other mutations are seen in people living in areas where there is a high contamination with hepatitis B and C viruses.

Inherited mutations of this gene have been recently identified in families with Li-Fraumeni syndrome, a rare familial syndrome characterized by an unusually high incidence of sarcomas, premenopausal breast cancers, brain tumors, leukemias, and adrenocortical carcinomas (Srivastava et al, 1990). Two groups have reported recently an inherited defective p53 gene in patients with sarcoma and in children and young adults with secondary malignant neoplasms (Malkin et al, 1992; Toguchida et al, 1992). Even though the incidence of a mutant p53 gene in these patients is low, it suggests that an inherited defective gene may play a role at times when youngsters who have been treated for cancer develop a second unrelated malignancy. Germline mutations of the p53 gene are rare among patients with sporadic sarcoma but may be common in patients with sarcoma who had either multiple primary cancers or a family history of cancer.

Mutations in p53 include large deletions or insertions as well as **point mutations**. The interesting observation is that the majority of somatic point mutations cluster in five separate regions of the gene that were all highly conserved during evolution. The seven germline p53 mutations are even less diverse: all are amino acid substitutions occurring between **codons** 242 and 258. One can only speculate that mutations located in other areas in the p53 gene would be lethal during embryogenesis and that only a limited number of subtle changes are nonlethal. The important message is that mutations in p53 are associated with instability in the rest of the genome because it supposedly causes multiple genetic alterations that can lead to cancer. The recent discovery that p53 is an integral component in one pathway of programmed cell death (apoptosis) means that an inactivated p53 could inhibit programmed cell death. This in turn could lead to an increase in proliferating cells and to the probability that these cells become transformed (Clarke et al, 1993; Lowe et al, 1993). There are still many questions to be solved about the functions of p53 gene product, but there is little doubt that this gene plays a central role in the origin of a number of human cancers.

Other hereditary mutations of tumor-suppressor genes have been correlated with a genetic predisposition to cancer, and several of these genes have been linked to distinct familial cancer syndromes. The best-studied example concerns patients with retinoblastoma who have a hereditary form of the disease. These patients have a high risk of multifocal retinoblastoma and other tumors, and they may pass the disease to their progeny as an autosomal dominant trait.

Diseases Caused by Amplification of Triplet Nucleotides

Dynamic mutations consist of a class of genetic human diseases characterized by the pathological expansion of normal repeats of specific trinucleotides (**triplet repeats**). Four diseases belonging to this category have been known for some time: fragile X syndrome, spinal and bulbar muscular atrophy (SBMA), myotonic dystrophy (DM), and Kennedy's disease. Recently, amplifications of triplet repeats have been identified in three other diseases, Huntington's disease (The Huntington's Disease Collaborative Research Group, 1993), spinocerebellar ataxia type 1 (Orr et al, 1993), and hereditary dentatorubral-pallidoluysian atrophy (DRPLA) (Koide et al, 1994). Five of these disorders are described in more details.

Spinal and Bulbar Muscular Atrophy (SBMA) is an X-linked recessive genetic disorder characterized by progressive muscular weakness of upper and lower extremities that starts in adulthood and is secondary to neural degeneration (Harding et al, 1991). Affected men have reduced fertility and excessive development of the mammary glands (or gynecomastia); female carriers have few or no symptoms. The molecular defect has been localized tentatively on the androgen receptor (AR) at the CAG repeat sequence (La Spada et al, 1991); it is a polymorphic repeat of the codon CAG, encoding a stretch of sequential glutamines, which is amplified to approximately twice the normal size. The same mutation is observed in all SBMA patients (Figure 14a and b).

Fragile X Syndrome is also an X-linked recessive disorder. The clinical and physical features are a moderate to severe mental retardation, large head, long face, large ears, and large testicles. It is one of the most common forms of mental retardation, with an estimated incidence of 1 in 1250 men and 1 in 2500 women (heterozygotes). Fragile X syndrome has been reported in a broad spectrum of ethnic groups. The gene causing Fragile X syndrome has not been easy to localize. First, a constriction at a precise site in the X chromosome was identified by cytogenic studies of cells from affected individuals. Then, it was discovered that, by culturing leukocytes of affected individuals in media deficient in folate or containing the folate inhibitor methotrexate, one could induce the break in the Fragile X region. The Fragile X site (Xq27) was therefore defined as a folate-sensitive site. Finally, identification of the gene causing Fragile X syndrome was facilitated by the development of somatic cell hybrids. The Fragile X gene, called FMR-1, contains a highly polymorphic CGG repeat (Fu et al, 1991). The range of CGG repeats is 5–54 in the normal population with a mean incidence of 29 repeats. The study of Fragile X families facilitated the correlation of a phenomenon called anticipation with the molecular events of CGG amplification. Anticipation is defined as the appearance of the disease in increasing severity or earlier onset in successive generations with a heritable disorder. For example, normal transmitting men carry CGG repeats slightly above the normal range, from 52 to 200 repeats but still below those found in affected men (above 200). These men transmit the gene containing these repeats to their progeny with small changes in the repeat number. The progeny (men or women) of a transmitting man with a woman who carries similar premutation alleles (an allele with a phenotypically silent mutation) will carry large expansions of the repeat region (250–4000 repeats) and will show signs of mental retardation (Figure 14a and b).

Myotonic Dystrophy (DM) is an autosomal-dominant disease characterized by myotonia, cardiac arrhythmias, cataracts, male balding, male infertility, and other endocrinopathies. An unstable sequence of GCT repeats in the gene associated with DM, designated myotonic protein kinase or MT-PK gene, was found by several groups (Aslanidis et al, 1992; Buxton et al, 1992; Harley et al, 1992) (Figure 14a and b). The same phenomenon of anticipation occurs in DM families: the severity of the clinical symptoms in DM families increases over successive generations and is correlated with the observation that the triplet repeat, GCT, undergoes progressive expansion. Surprisingly, a recent study reported the reduction in size of the DM triplet repeat mutation during transmission in three patients. In one case, the number was reduced to within normal range and correlated with a delay in the onset of clinical signs of DM (O'Hoy et al, 1993).

Triplet repeats associated with diseases can occur at different position in genes: as shown on Figure 14a, the CGG repeat of FMR-1 is located upstream of the initiation AUG codon, and the GCT repeat of MT-PK lies in the untranslated sequence following the coding sequence. Even though the sites containing these triplets are highly polymorphic in length, it is

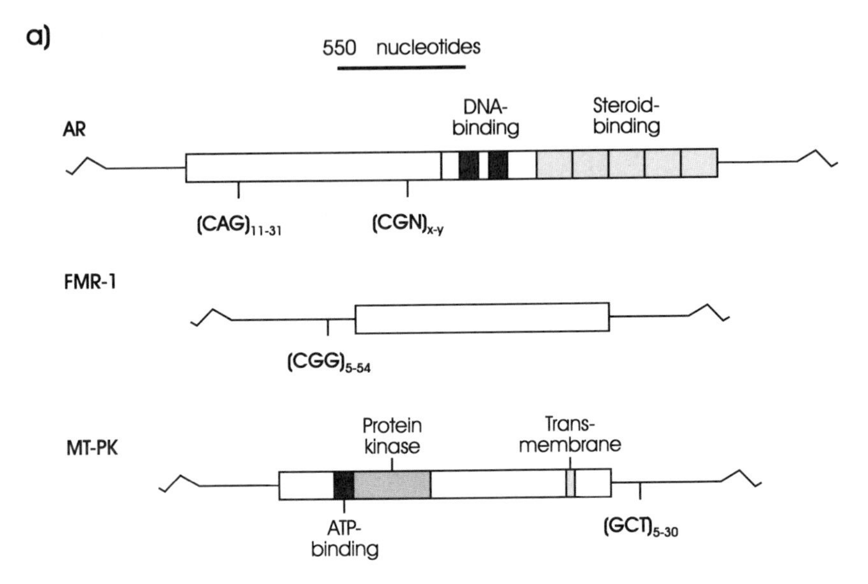

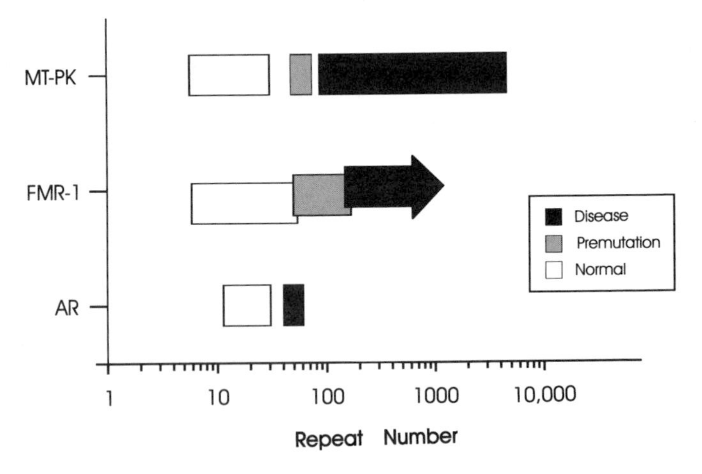

Figure 14. Triplet repeat mutations.

(a) Diagram of three mRNA molecules containing polymorphic triplet repeats.
Boxes represent the coding sequence of the gene. The localization of the trinucleotide repeat is indicated with the range of the number of repeats seen in normal individuals. Some functional domains are indicated. AR: androgen receptor; MT-PK, myotonin protein kinase; FMR-1, Fragile X gene.

(b) Variation of polymorphic triplet repeat number in three inheritable diseases.
The range of repeat numbers is shown for normal individuals, those carrying premutations and those affected with spinal and bulbar muscular dystrophy (AR), Fragile X syndrome (FMR-1), and myotonic dystrophy (MT-PK), respectively. Notice that the number of triplet repeats that determines a pathological state is specific for each disease.

(Redrawn from Caskey et al, 1992.)

the overall number of triplet repeats that distinguishes the normal person from those manifesting the disease. Figure 14b represents schematically the point at which the number of repeats results in disease (Caskey et al, 1992).

Huntington's Disease (HD) is a progressive neurodegenerative disorder, manifested by distinctive choreic movements, that affects one individual in 10,000 in most populations of European origin. It is inherited in an autosomal dominant fashion. Patients affected by HD have motor disturbance, cognitive loss, and psychiatric manifestations, starting in the fourth to fifth decade of life and worsening over the next 10 to 20 years until death. Currently, there is no treatment to either delay or prevent the onset and progression of this devastating disease. HD patients display distinctive neuropathological lesions, with selective loss of neurons particularly in the caudate and putamen. There is currently no clue as to the biochemical basis for this neuronal loss. The genetic defect causing HD was assigned to chromosome 4 more than 10 years ago (Gusella et al, 1983). Since then, scientists from six independent research groups in Europe and in the United States have striven to refine the HD gene localization, using a variety of genetic approaches. The search for the HD gene has been a long and arduous gene hunt. They finally identified a large gene, called IT15 (for interesting transcript 15), that encodes a previously unknown protein of about 348 kilodaltons (The Huntington's Disease Collaborative Research Group, 1993). The IT15 reading frame contains a polymorphic (CAG)n trinucleotide repeat with at least 17 alleles in the normal population. Whereas the number of repeats varies from 11–34 CAG copies in the normal population, the length of these repeats on HD patients' chromosomes increases to a range of 42 to over 86 copies. The number of repeats of the CAG trinucleotide seems to be correlated to the age at which the first symptoms appear. This protein is expressed in many tissues; yet, cell death in HD is confined to neurons of specific regions of the brain. The challenge now is to find the normal function of the Huntington protein and to discover the mechanism by which the increase in the trinucleotide repeat leads to the characteristic neuropathology of HD. Now that the gene for HD has been identified, it is possible to test whether an individual has inherited the defective gene. The ethical issues raised by this diagnostic test are important, because people with hereditary predisposition are still left with the prospect of no efficacious cure after onset of the clinical symptoms.

Spinocerebellar Ataxia Type 1 is an autosomal dominant progressive neurodegenerative disease, characterized by ataxia, ophthalmoparesis, and variable degrees of motor weakness. A selective loss of neurons in the cerebellum and brain stem and degeneration of the spinocerebellar tracts are observed. In common with all other triplet repeat diseases, there is a strong correlation between the number of CAG repeats and the age of onset of clinical symptoms (Orr et al, 1993).

The discovery of the mutational mechanism of triplet repeat amplification found in the above disorders has stimulated the development of systematic methods to detect new trinucleotide repeats in other diseases. Undoubtedly, as more refined molecular analyses are carried out, the number of diseases belonging to this class is likely to increase. For example, various neurodegenerative diseases such as hereditary spastic paraparesis, bipolar affective disorders, hereditary essential tremor, and zonular cataract could be the result of the phenomenon of anticipation described above.

Diseases Caused by Mutations in Mitochondrial DNA

Mutations in mitochondrial DNA (mtDNA) have attracted a lot of attention in recent years, because many of them cause defects in a biological process called oxidative phosphorylation (OXPHOS). The biochemical reactions of OXPHOS generate mitochondrial ATP and are therefore a primary source of energy for tissues and organs, including the brain, muscle, heart, kidney, liver, and pancreatic islets (Wallace, 1992). A number of mtDNA mutations have been identified, including base substitution mutations in both **transfer RNA** (tRNA) and protein coding genes and mtDNA insertion and deletion mutations. Human mtDNA is a closed circular molecule, containing about 16,500 base pairs, located within the mitochondrial matrix. It encodes mitochondrial **ribosomal RNA**s, tRNA, and genes encoding structural and enzymatic proteins, many of which are involved in generating energy for the cell. The genetics of OXPHOS is very complex and involves proteins encoded by mtDNA as well as nuclear DNA (nDNA). For example, the expression of mtOXPHOS genes requires functional mitochondrial replication, transcription, and translation systems. All the polypeptides necessary for these processes are encoded by nDNA. The biogenesis of OXPHOS therefore requires hundreds of nuclear and mitochondrial genes. Defects in any of these genes can result in a deleterious OXPHOS process. Several diseases result from mutations in mitochondrial genes (Table 8).

Diseases caused by mtDNA insertion-deletion mutations are interesting because they show a relation between the quantitative reduction in OXPHOS and the severity of the disease phenotype. Diseases in this category are generally spontaneous, which suggests that the deletions are of somatic rather than genetic origin. Patients suffering from these diseases have paralysis of the eye muscles, ptosis (droopy eyelids), and mitochondrial myopathy. Patients who are only mildly affected show

Table 8. Examples of diseases resulting from mutations in mitochondrial genes

---

MERRF: myoclonic epilepsy and ragged-red fiber disease. MERRF is the result of an mtDNA mutation in the tRNA gene carrying the amino acid lysine. This mutation leads to a defect in mitochondrial protein synthesis and to deficiencies in some respiratory processes. Patients suffering from this disease have mitochondrial myopathy and uncontrolled movements, neurosensory deafness, myoclonic epilepsy, progressive dementia, hypoventilation, cardiac insufficiency, and renal dysfunction.

MELAS: mitochondrial encephalomyopathy, lactic acidosis, stroke-like symptoms. MELAS results from an mtDNA mutation in the tRNA for leucine. Patients suffer from mitochondrial encephalomyopathies.

LHON: Leber's hereditary optic neuropathy. This disease is caused by mtDNA missense mutations in electron transport genes. Individuals with the most common mutation are usually normal throughout childhood but have an increased probability of going blind with age.

---

these symptoms and are designated CEOP (chronic exernal opthalmoplegia), whereas severely affected patients are designated KSS (Kearns-Sayre); in addition, these patients suffer from retinitis pigmentosa, hearing loss, heart conduction defects, ataxia, and dementia.

Deletions in mtDNA are highly heterogeneous in both size and position (Figure 15). Patients with sponteaneous cases of CEOP and KSS display only one type of deleted mtDNA, which suggests that the disease is the result of a single deletion event occurring early in development. Furthermore, in these patients, the proportion of mutant and normal mtDNAs varies markedly among tissues. It is possible that the severity of the disease reflects the proportion of the deleted mtDNAs among tissues.

Apart from the above diseases, defects in OXPHOS are thought to result in a large number of other tissue-specific diseases and could be the underlying cause of diverse clinical problems like ischemic heart disease, late-onset diabetes, Parkinson's disease, and Alzheimer's disease. They are even considered as one of the main causes of aging (Wallace, 1992). Scientists postulate that, at birth, each individual is endowed with an initial OXPHOS capacity. As this capacity declines with age and falls below the energetic threshold of an organ, disease symptoms appear. Several observations support this hypothesis:

(1) Patients with OXPHOS defects show early symptoms characteristic of degenerative disorders. It is as if these individuals start at lower OXPHOS level and cross a presumed threshold earlier within their lifetime.

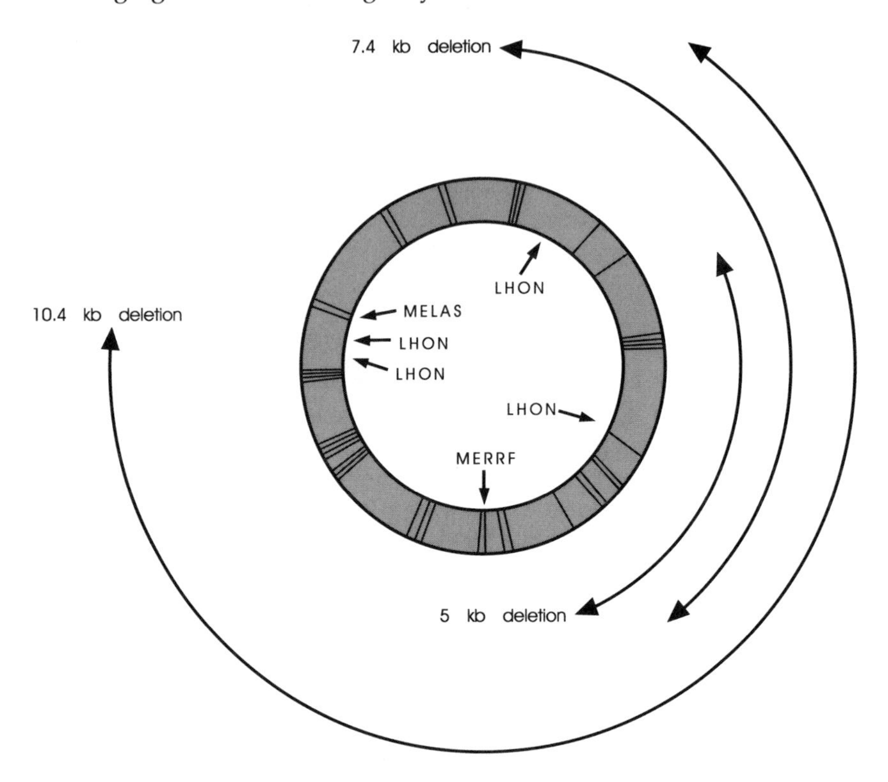

Figure 15. Mitochondrial DNA mutations.
Diagram of human mtDNA with the location of major disease mutations. Location of some base substitution mutations are shown by arrows inside the circle representing mtDNA and carry the abbreviated name (see text) of the disease caused by these mutations. Deletion mutations are drawn outside the circle. Mitochondrial genes are delineated by bars in the circle. (Redrawn from Wallace, 1992.)

(2) In ischemic heart disease, OXPHOS could be chronically inhibited because atherosclerotic plaques occlude coronary arteries, which leads to a deprivation of substrates and oxygen to heart mitochondria. At least some forms of cardiomyopathy have been associated with increased secondary damage to mtDNA and a compensatory induction of OXPHOS gene expression.
(3) Some cases of diabetes mellitus are thought to result from OXPHOS defects. A chronic deficiency in OXPHOS appears to inhibit insulin production by pancreatic islets.
(4) Aging seems to be associated with the appearance of damaged mtDNA in tissues, and this observation correlates with a decline in OXPHOS capacity.

The role of mitochondrial dysfunctions may be particularly important in relation to the nervous system. Scientists have observed for a long time that the brain is subject to region-specific metabolic deficiency. Recently, two groups reported an age-related increase of mtDNA deletions in multiple regions of the human brain (Corral-Debrinski et al, 1992; Soong et al, 1992). There was a significant increase of mtDNA damage around age 80. All regions of the brain, except the cerebellum, accumulated damage more than the heart; the three regions with the highest levels were the caudate, putamen, and substantia nigra, all regions characterized by high dopamine metabolism. Interestingly, the breakdown of dopamine by mitochondrial monoaminoxidases produces H_2O_2, a potent oxidant leading to free radical formation (see pages 113–119). Furthermore, the preferential and progressive increase of iron deposition in caudate, putamen, and motor cortices with age could also be due to decreased OXPHOS processes.

A recent study in rats provides evidence that sensitivity to the neurotoxin 3-nitropropionic acid (3-NP), which blocks mitochondrial energy production in the cell, increases with age. After treatment of young and old rats with 3-NP, Bossi and his collaborators (1993) noticed that 3-NP had no adverse effects on young rats, but caused striatal lesions or death in most animals older than 16 weeks. The increased sensitivity was attributed to changes within the neurons of older animals and, specifically, to the decline in mitochondrial efficiency. The authors postulated that in specific neurodegenerative diseases, a genetic predisposition for an energy failure in cells may be precipitated by an age-dependent mitochondrial decline. Several studies support this hypothesis: the brains of patients with Parkinson's disease show mitochondrial DNA deletions. Furthermore, most patients with Alzheimer's disease have been reported to have OXPHOS defects in mitochondria in platelets; platelets taken from Huntington's Disease (HD) patients reveal a marked decrease in the activity of some enzymatic processes called mitochondrial complex I (NAD dehydrogenase) relative to age-matched controls.

Thus, a decline in mitochondrial function with age, as exemplified by 3-NP sensitivity, accelerated in some way by the HD gene, is likely to produce progressive degeneration in brain areas containing the most metabolically active neurons. These and other age-related degenerative diseases could all fit the OXPHOS paradigm of neurodegeneration, namely the inability of energy-depleted cells to maintain an ionic balance and correctly generate, transport, or break down proteins. The neuropathology of HD, Alzheimer's, and Parkinson's patients could therefore all be related to an impairment of cellular function due to reduced mitochondrial efficiency. Scientists are now making diagnostic tests of cellular mitochondrial function. If future research confirms the important role of mtDNA defects, mitochondrial replacement therapy through in vivo gene transfer or treatment with cellular energy enhancers might be possible. In the meantime, DNA analysis will be useful in screening asymptomatic subjects to identify mutant OXPHOS genes that predispose individuals to numerous diseases. Mitochondrial DNA mutations and their associated diseases are representative of a large number of possible somatic mutations produced by oxidative damage.

DNA Methylation

The molecular events that determine transcription are of fundamental interest to gerontologists, since regulation of gene expression is known to have an impact on aging and on aging disorders. Over-expression of amyloid proteins or inappropriate expression of oncogenes are good examples. Genetic factors that influence gene expression, but that do not directly involve changes in the genetic code, could play a role in aging. One of these molecular modifications is DNA methylation (Catania and Fairweather, 1991). Up to 5% of total cytosine residues in mammalian DNA are methylated at the 5′ position to form 5-methylcytosine (5mC). This is the only permanently modified base in DNA of higher eukaryotes. Methylation occurs on both strands, is symmetrical, and 5mC residues are always flanked by guanine residues at the 3′ side (CpG). The putative roles of these methylations are numerous: they include strand selection in mismatch repair, recognition sites for restriction endonucleases, alterations in DNA conformation and **chromatin** structure, and modifications of specific DNA-protein interactions. More importantly, DNA methylation is involved in the regulation of gene activity. Changes in DNA methylation, in particular demethylation of CpG dinucleotides in vertebrates, is related to increased levels of transcription (Mays-Hoopes, 1989).

Age-related DNA demethylation has been described by Vanyushin and his collaborators (1973), who found marked tissue differences in rats, with more demethylation in brain than in liver. Wilson and Jones (1983) observed an age-related decline of 5mC in lung and foreskin fibroblast cultures. Similarly, others have reported that demethylation reduces the growth potential of fibroblast-like cultures (Fairweather et al, 1987).

Not only demethylation, but hypermethylation have been found. Hypermethylation in DNA noncoding sequences or spacer sequences from brain, liver, and spleen have been observed in CBA mice at their midlife stage. Increased DNA methylation has been reported also in ribosomal RNA gene clusters. Consistent with the inverse relationship between methylation and transcription, the ribosomal gene cluster appears to be selectively inactivated (Swisshelm et al, 1990).

Overall, the data on DNA methylation indicate that the age-related changes can be bidirectional (demethylation and hypermethylation) and that there seems to be considerable sequence selectivity (this consequently causes changes in gene expression). The possibility that age-related demethylation predisposes cells to tranformation is an attractive idea. It is supported by experiments showing that methyl-deficient diets are carcinogenic (Wilson et al, 1984). However, the link between DNA methylation and cancer is far from demonstrated.

DNA and Protein Glycation

Nucleic acids and proteins can be modified by the addition of sugars to their free amino groups and this results in structural and functional alterations of the molecule. The nonenzymatic glycation of biologically important molecules has become an increasingly important area of research in *diabetes mellitus* and normal aging processes. For example, monosaccharides such as D-glucose or D-galactose trigger a chain of chemical events that produce substances called advanced glycation end products or AGE; these products are yellow-brown and fluorescent and are capable of covalent intramolecular cross-linking of proteins and of cross-linking between different proteins. The alteration of proteins by glucose affects many of their properties: tertiary/quaternary structure, antigenicity, solubility and susceptibility to degradation, enzymatic activity, aggregation, and turn-over rate. Many important proteins, such as collagen, contain a large amount of glucose and are found to be cross-linked in older and diabetic individuals compared with normal individuals (Schnider and Kohn, 1981; Kohn et al, 1984). Cross-linking of collagen decreases its elasticity and such molecular change could be the basis for thickening of the basement membrane in the kidney mesangial matrix, for example, and could lead to kidney failure in diabetes; it may also be a factor in age-related reductions of kidney function. It is thought to play a role in the constriction of arteries, in the perturbation of vascular flow, and the restriction of flexibility in tendons.

Myelin is another candidate protein affected by AGE. Modified myelin could lead to demyelination disorders (Vlassara et al, 1984). In fact, experimental diabetes in rats causes abnormalities in hypothalamic luteinizing hormone releasing hormone (LHRH) neurons and slows axonal transport in the sciatic nerve. This suggests that the formation of AGE in the brain could have serious consequences, notably on the neural control of pituitary secretions of growth hormone or autonomic influences on glucose metabolism (Mobbs, 1990).

Interestingly, as will be further discussed in Part II, one of the few efficacious interventions to postpone aging is to lower total caloric intake. Maybe the mechanism by which diet restriction operates is related to the nonenzymatic addition of glucose to long-lived proteins. In support of this hypothesis, Masoro showed that after diet restriction, blood glucose was reduced by about 15%, and he observed a corresponding decrease of glycosylated hemoglobin (Masoro et al, 1989). Cataracts may also involve

AGE in lens crystallins. In this case as well, caloric restriction strikingly retarded the loss of soluble γ-crystallins in normal hybrid mice (Leveille et al, 1984).

Nucleotides and DNA are also affected by nonenzymatic glycation. Addition of sugars to DNA results in mutations through direct damage and in activation of error-prone recombination/repair systems. It also causes increased chromosomal breakage. In vitro, glucose has been shown to induce DNA rearrangements in an *Escherichia coli* plasmid (Bucala et al, 1985).

Garbage Molecules

One of the early observations in biological gerontology was of the age-related accumulation of intracellular pigments, or lipofuscins. Today, lipofuscin accumulation is viewed as a consequence, rather than a cause of aging. Constituents of lipofuscin granules are partly degraded proteins that result from intracellular oxidative activity and represent some sort of garbage for cells. Intracellular protein degradation appears to be carried out by several different pathways, one of which is found in **lysosomes** (Chiang and Dice, 1988). This vital garbage service tends to slow down with age, and cells lose their ability to dispose of damaged proteins. Build-up of lipofuscin molecules eventually paralyzes normal cellular metabolism. This is, however, not a general phenomenon applicable to all types of proteins. Much remains to be learned about the degradation of specific proteins and changes in degradation pathways with age.

Extrinsic Mechanisms

Aging, as seen from the preceding chapter, is to a large extent due to intrinsic factors. But, attacks from the outside, i.e., environmental or extrinsic factors should not be minimized. The difference in longevity between people who live in cities and those who live in the country is a good illustration of their impact. The air is dirty in cities and in areas of industrial activities and is the cause of a higher rate of respiratory diseases. Noise, tense professional working conditions, and hours of daily commuting all contribute to higher levels of stress among city inhabitants and represent potential life-shortening factors. Damage to the organism can also be caused by other types of extrinsic factors. The generation of free radicals is believed to play a crucial role in the process of aging. Radiation is another factor of increasing concern. The generation and detrimental actions of free radicals and the effects of radiations are discussed below.

Free Radicals

The existence of free radicals was first suspected about 100 years ago, when scientists discovered that free radical reactions were responsible for the rancification of fats and oils. Recently, there has been much interest in the roles of free radicals in normal body chemistry, in the pathophysiology of diseases, and, in particular, in aging (Dormandy, 1989; Harman, 1984; Halliwell and Gutteridge, 1985). Free radicals are generated in a variety of ways. They are constantly being produced within the body from normal enzymatic biochemical oxydation-reduction reactions. Sources of free radicals include activated leukocytes and macrophages, the arachidonic acid cascade, nitrous oxide, tissue **ischemia**, and reperfusion after ischemia. Environmental perturbations such as heat, radiation, air pollution, toxic chemicals (tobacco, pesticides, asbestos, cured meats, dietary fats), and even sunlight or heavy physical exercise can generate free radicals.

A free radical is defined as an atom, an ion, or a molecule that contains one or more unpaired electrons (Halliwell, 1981). A compound becomes a free radical either by losing or by gaining an additional electron. This electron imbalance makes the compound highly unstable and ready to combine with other atoms or molecules to share electrons. By so doing, free radicals damage or destroy the electron-sharing partner. Free radicals are highly damaging to cellular components: they cross-link proteins, cause DNA mutations, and oxidize lipids; virtually all cellular constituents are sensitive to oxidant damage, even carbohydrates.

The chemical reactions involved in the generation of free radicals are as follows: aerobic cells have the ability to oxidize substrates in a controlled manner and, thereby, to produce high energy phosphate compounds like ATP. As elemental oxygen is highly electronegative, it readily accepts single electron transfers from cytochromes and other reduced cellular components (Foreman and Fischer, 1981). During the univalent reduction of oxygen and its conversion to water, a number of free radical intermediates are formed: a small portion of the O_2 consumed by cells is reduced to superoxide radicals ($\cdot O_2^-$) (Cadenas, 1989). Further reduction of ($\cdot O_2^-$) produces hydrogen peroxide (H_2O_2), hydroxyl radicals ($\cdot OH$), and water. These chemical reactions are illustrated in a simplified form in Figure 16.

It is ironic that oxygen, the element that above all others keeps organisms alive, also kills them slowly through the formation of free radicals. Transformation of superoxides into oxygen peroxides is activated by a class of enzymes present in most cells, the superoxide dismutases (SOD). SODs are metalloenzymes that depend on zinc and copper for their enzymatic activities.

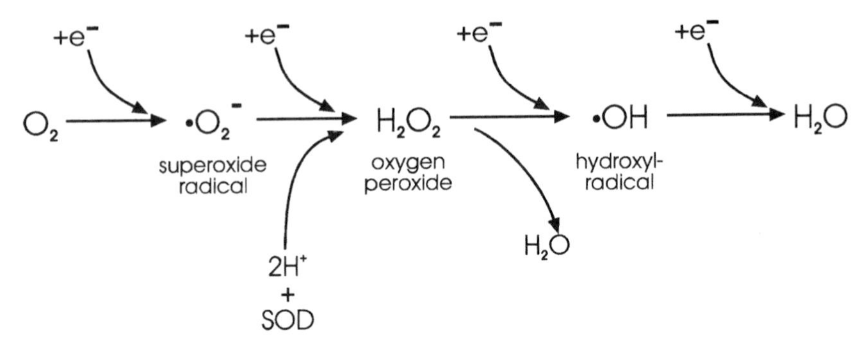

Figure 16. The formation of free radicals.

Although hydrogen peroxide is not a free radical in the true sense, it is still a highly reactive molecule and participates in reactions that result in free radical generation. Once they are formed, the free-radical-damaged products can interact further to form still other free radicals and nonradical oxidants such as singlet oxygen and peroxides.

During evolution, living organisms have been faced with the necessity to destroy free radicals, and they have devised ways to fight oxidative attack. These defense mechanisms are already present in quite primitive species, which further emphasizes their crucial role in the survival of organisms. A variety of enzymes such as the superoxide dismutases (SOD), glutathione peroxidases, and catalases specifically remove superoxides ($\cdot O_2^-$), H_2O_2, and organic peroxides. In addition to enzymes, other low-molecular-weight free radical scavengers and antioxidants, like uric acid, glutathione, and ascorbic acid (vitamin C), are active in the liquid phase of the cell, while α-tocopherol (vitamin E) and β-carotene are thought to protect membranes because of their lipophilic character. All these molecules help prevent toxic levels of free radicals from occurring. The relevance of antioxidant levels for survival is demonstrated by the positive relationship observed between the concentrations of SOD and α-tocopherol in different tissues and the maximum life span potential of many mammalian species, including man (Figure 17) (Cutler, 1985).

In addition to these protective agents, cells contain proteolytic systems (proteases and peptidases). These rapidly degrade proteins damaged by oxygen radicals as well as phospholipases that cut out damaged parts of oxidized lipids in membranes (so that other enzymes can repair injured segments). There are also acetyltransferases, which are thought to replace fatty acids cleaved from lipids, exo- and endonucleases that clip out damaged segments of DNA, glycosylases, and polymerases, that fill in the gaps left by exonucleases and endonucleases, and ligases that seal the repairs.

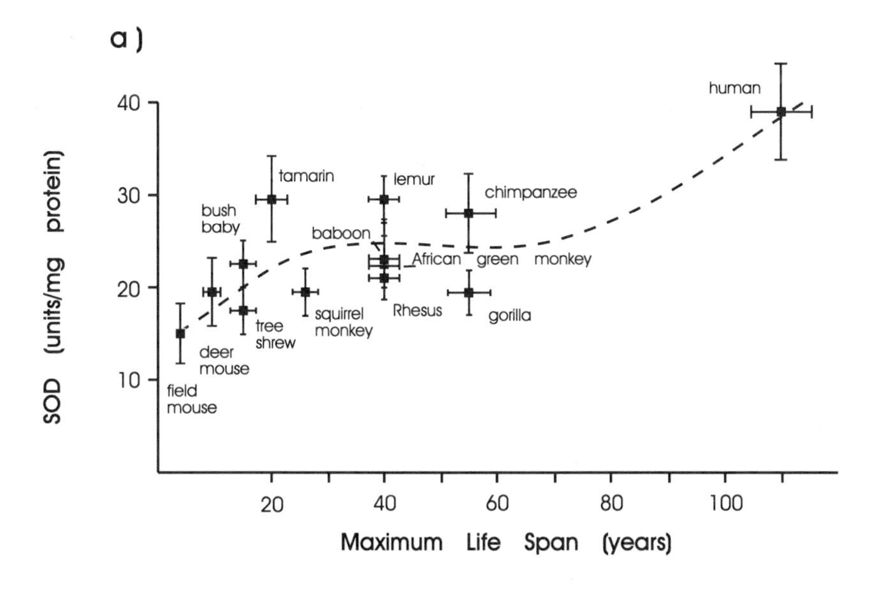

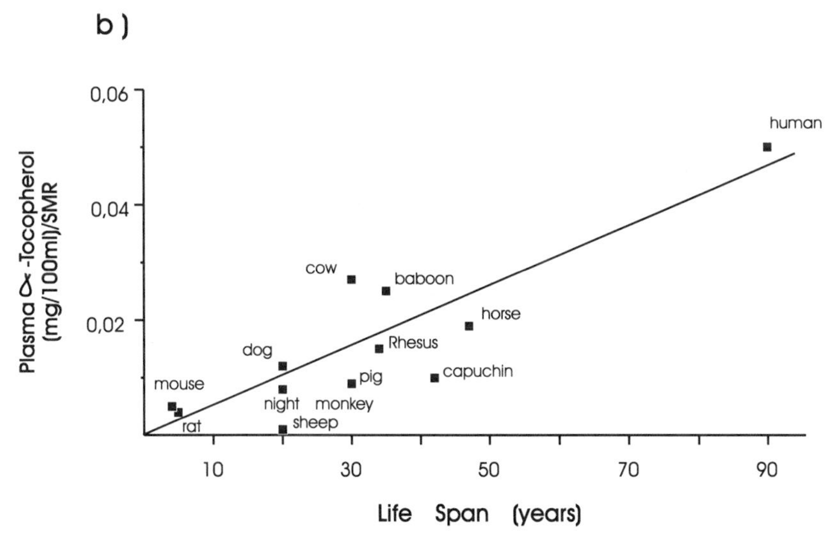

Figure 17. Average antioxidants levels and longevity.

(a) SOD concentration in liver of mammals as a function of maximum life span potential (MLSP).

(b) Plasma levels of vitamin E (α-Tocopherol) per specific metabolic rate (SMR) as a function of maximum life span potential (LSP) in mammalian species.

(Redrawn from Cutler, 1985.)

As long as the processes generating free radicals are counterbalanced by appropriate protective mechanisms, there is no damage or dysfunction of cellular components. If the protective system breaks down or is exceeded, accumulation of cellular damage limits homeostatic mechanisms and leads to cellular breakdown and finally death of the organism. This is the basis of the *free radical* theory of aging, first proposed by Harman in 1956. The original theory stated that cellular damage induced by free radicals was the main cause of aging. Reformulation of the theory states that free radical reactions could lead to specific and subtle changes that directly affect homeostasis. DNA regulatory mechanisms and, in particular, factors that control gene expression may be damaged by free radicals. These alterations would lead to a gradual loss of cellular control over the genome, a process called dysdifferentiation. Observations by independent investigators support the *dysdifferentiation* theory of aging. The appearance of abnormal proteins, nuclear release of unprocessed RNA in aging oviduct tissues (Schroder et al, 1987), and increased incidence of metaplasia are just a few examples of dysdifferentiation. A variation of the dysdifferentiation theory proposes that free radicals stimulate changes in gene expression that cause aging without causing direct damage to cellular components. In other words, rather than loosening the genetic control as suggested by the dysdifferentiation hypothesis, free radicals stimulate changes in gene expression and lead to a form of continuation of developmental processes. These age-associated changes could be due to the actions of oxidants on genetic programs (Pacifici and Davies, 1991; Sohal and Allen, 1990). In support of this, variations in redox potential can infuence chromatin-controlling proteins and the binding properties of nucleic acid binding proteins. Moreover, increases in the rate of synthesis of some proteins have been observed during differentiation and have also been reported during aging. Aberrant gene expression could provide an explanation of some aging processes.

In summary, free radicals may cause aging through direct structural damage as well as through deleterious effects on nuclear control mechanisms. Whatever the mechanisms involved, the free radical theory of aging is attractive because free radicals are ubiquitous and have notably deleterious effects. Almost every type of cell is vulnerable to tissue destruction caused by free radicals or their by-products, and these radicals can attack everything from the cell's membrane to its DNA (Figure 18).

Free radical damage could account for an astonishingly broad range of pathological processes including carcinogenesis, arthritis, adult respiratory-distress syndrome, emphysema, retinopathy of premature infants, cataracts, atherosclerosis due to oxidized LDL cholesterol, muscular dystrophy, and ischemia-reperfusion tissue injury. More importantly, it may be a leading cause of disturbance of general homeostasis and thus be a main factor in all aging processes.

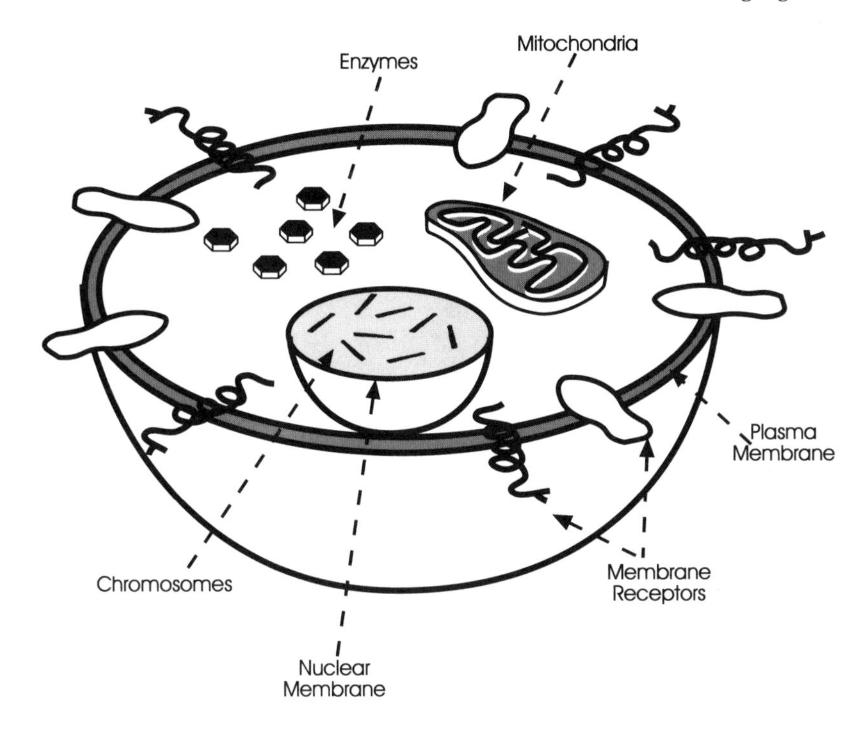

Figure 18. Diagram of model cell and its constituents.
Oxidative attack can impair the functioning of many cellular constitutents:
(a) Oxidation of lipids and proteins damages membranes of cells and organelles, thereby affecting their permeability.
(b) Mitochondrial DNA and membranes are particularly prone to oxidation. When mitochondria, which are themselves a major source of free radicals, are severely damaged, cells may be starved of energy.
(c) Many enzymes can be inactivated by oxidation at their active sites. Free radical attack can also damage structural proteins.
(d) Free radical attack on DNA molecules can interfere with the accuracy or amount of protein made by a cell.

Effects of Free Radicals in the Central Nervous System

Recent discoveries have added weight to the free radical theory of aging and provided evidence that free radicals cause brain damage and neurodegenerative processes. Free radicals are involved in the genesis of brain edema (Ikeda et al, 1989).

Iron injected into the brain causes superoxide formation and subsequent lipid peroxidation of cortical neurons at the injection site. Iron is crucial for the formation of oxygen-free radicals. In fact, hydrogen peroxide is transformed to hydroxyl radicals in the presence of Fe^{2+} as depicted in the reaction illustrated in Figure 19.

$$H_2O_2 + Fe^{2+} \longrightarrow \text{Intermediate complex} \longrightarrow OH^- + \cdot OH + Fe^{3+}$$

Figure 19. The transformation of hydrogen peroxide to hydroxyl radicals.

The liberation of iron from hemoglobin following hematoma in the brain could therefore be a critical factor triggering neuronal death. The same causal relation could exist between free radicals and iron-induced epilepsy, which occurs in people who have experienced head trauma (Jesberger and Richardson, 1991).

A speculative idea proposes that free radicals are implicated in the pathophysiology of schizophrenia. The hypothesized mechanisms involve membrane lipid metabolism. Phospholipase A_2 (PLA_2) plays an important role in the metabolism of phospholipids by initiating the release of arachidonic acid from membrane phospholipids. The metabolism of arachidonic acid by cyclooxygenase and lipoxygenase involves the formation of free radicals, which can lead to lipid peroxidation. Interestingly, studies have reported elevated levels of PLA_2 in some schizophrenic patients, and, curiously, chronic treatment with neuroleptics has been found to lower the PLA_2 levels (Gattaz et al, 1987). The mechanisms by which neuroleptics inhibit the action of these enzymes on the arachidonic acid cascade are unknown.

The idea that continuous microattacks on brain cell membranes could lead to neurological diseases has attracted the attention of scientists for some time. Recently, a startling finding related to amyotrophic lateral sclerosis (ALS) was reported by Rosen and his collaborators (1993). This finding was made possible by virtue of a vast scientific and medical collaborative effort. ALS, also called Lou Gehrig's disease, is a neurodegenerative syndrome with onset during adulthood. It is a degenerative disorder of motor neurons in the cortex, brainstem, and spinal cord. The death of these neurons eventually leads to total paralysis and muscular atrophy. The incidence of the disease is about one in 100,000, and many patients with ALS die within five years from the first diagnostic evidence. About two years ago, a gene responsible for familial amyotrophic lateral sclerosis (FALS) was found on chromosome 21. This was based on the study of 13 families. FALS is an autosomal dominant trait and represents about 10% of the patients affected by ALS. There is a tight genetic linkage between FALS and the gene that encodes a cytosolic, Cu/Zn-binding superoxide dismutase called SOD1, an enzyme that protects cells against the toxic effects of oxygen. Rosen and his colleagues then hypothesized that the SOD1 gene itself could be responsible for the disease. They analyzed SOD1 gene in the affected families and identified 11 single amino acid changes in a very conserved region of the SOD1 gene, which suggests that the mutation affects the function of superoxide dismutase. These changes were not detected in more than 100 chromosomes from normal individuals and are clearly associated with FALS. The authors proposed that the mutant SOD1 protein is not only functionally defective but also inhibits the function of the normal SOD1 protein expressed from the normal allele. The normal and mutant proteins combine to form an inactive heterodimer. The implication is that toxicity by oxygen-free radicals is pathogenic in the motor neuron death in FALS and perhaps in sporadic ALS as well.

Free radical toxicity could be responsible for other forms of neurodegeneration: free radicals have been implicated in the pathogenesis of Parkinson's disease and cerebral ischemic lesions. Defects in a number of different cellular processes, such as iron storage or pentose phosphate pathways, could lead to an increased free radical production in the brain. Once tissue destruction begins, the alteration in ion homeostasis and the disruption of mitochondrial electron transport are likely to result in more electrons being available for generating additional oxygen-derived free radicals and initiating peroxidative lesions.

The development of methods to measure oxidant generation in these types of diseases is important in the further understanding of the diseases. Also, these findings will lead to new ideas for therapeutic approaches. For example, compounds that penetrate the central nervous system and decrease the levels of free radicals would be useful for FALS and ALS patients. Such a compound, the spin-trapping N-tert-butyl-α-phenylnitrone, or PBN, is reported to diminish the increase in oxidized protein and the loss of glutamine synthetase activity that accompanies ischemia/reperfusion injury in the brain. PBN also decreases the level of oxidized protein in aged gerbils. The surprising finding is that older gerbils treated with PBN make fewer errors in the radial arm maze test for temporal and spatial memory than untreated aged controls (Carney et al, 1991). The authors think that oxidation of cellular proteins may be a critical determinant of brain function. This example shows clearly that the age-related increase in vulnerability of tissue to oxidation can be modified by free radical trapping compounds.

Effects of Radiation

Everyone is bathed throughout life with low-level ionizing radiation from natural sources (cosmic and terrestrial radiation) and high-level radiations from man-made sources, such as medical X-ray machines or nuclear plants. If the harmful consequences of high-level radiation on living organisms are obvious, the effects of low-level radiations are not as well known. The question is: how harmful are low-level radiations, and at what point should these radiations be considered hazardous?

Radiation causes electrons in atoms to be dislodged. These electrons interact with molecules of water, for example, to give rise to free radicals.

Free radical damage to DNA is probably the most critical effect of low-level radiation, causing double-strand breaks, deletion of bases, and chemical cross-linking of DNA strands. The damage to DNA could be a triggering factor in many types of cancers: breast cancer, lung cancer, and cancers of the digestive tract, namely stomach, liver, colon, etc. However, even though the prevalent view is that low-level radiation is toxic, there is no clear answer concerning the limit at which low-level radiation becomes harmful.

Recently, scientists came up with surprising findings. They observed that low-level exposure to radiation does not result in the type of damage that would be predicted by extrapolation from higher levels of exposure. They noticed a biphasic response to radiation and low doses of radiation had the beneficial effects of increasing longevity in mice (Congdon, 1987), rats (Carlson et al, 1957), insects (Willard, 1971), and even plankton (Apostol and Clain, 1975). The process by which low doses of an otherwise harmful agent can result in stimulatory or beneficial results is called hormesis. Beneficial effects have also been observed with low-frequency electromagnetic radiation as well as cosmic, gamma, and ultraviolet irradiation (Planel et al, 1987). The explanations for hormesis are numerous: many agents and activities such as mixed-function oxidases, enzyme activity, and DNA repair are thought to be induced by low doses of radiation. These low-level radiations may first trigger a kind of molecular agitation, followed by a feed-back control, which may involve an overcompensation and then fluctuations to a lower value. This overcompensation at low doses could contribute to the observed increased longevity often accompanied by increased growth and fecundity. In multicellular organisms, the beneficial effects are thought to be due to stimulation of the immune system. In fact, changes in the suppressor/cytotoxic T-cell subset of human lymphocytes in culture (Gualde and Goodwin, 1984), changes in natural killer cell activity of human lymphocytes in vitro (Onsrud and Thorsby,1981), and enhanced mitogen-stimulated proliferation of human lymphocytes in vitro have been reported to occur under low-level radiation.

Since low levels of toxic agents like radiation can have beneficial effects, such as increased growth and longevity, throughout the animal kingdom, then perhaps the safe utilization of radiation might be a good way to improve cell function in aged organisms.

Chapter 6

Apoptosis, Diseases, and Senescence

Throughout the different phases of life, there is a delicate balance between proliferation and differentiation of cells and cell death. During ontogeny, the equilibrium is shifted towards cell proliferation, whereas during adult life it is assumed that cells are lost continuously from many normal tissues to balance the cell division that takes place. During senescence, there is a reduction in cell proliferative potential and a gradual loss of cells leading to atrophy and physiological involution of most tissues and organs. The equilibrium is shifted then towards cell death. The mechanism of cell deletion plays a complementary but opposite role to mitosis. It is designed to control cell populations or to eliminate cells that do not function properly. This phenomenon, called apoptosis or programmed or physiological cell death, occurs during normal development and throughout life. The morphological changes occurring during apoptosis are shrinkage and fragmentation of the cytoplasm, formation of apoptotic bodies, and changes in chromatin structure. These apoptotic bodies are shed from the cell epithelium and are taken up by other cells where they are rapidly degraded by lysosomal enzymes. Importantly, removal of apoptotic cells does not cause an inflammatory response. Apoptosis is an active process that requires protein synthesis. Apoptosis is distinguished from another type of cell death, called necrosis, which is characterized by swelling of the whole cell and of intracytoplasmic organelles such as mitochondria. This may be the result of an increased permeability to calcium which may be implicated in the death process. Necrosis occurs mainly during senescence and is, to a large extent, the consequence of cellular injury or exposure to toxins, hypoxia, or other deleterious environmental conditions. Necrosis occurs independently of protein synthesis and seems to be the result of a general cellular breakdown. In contrast to apoptosis, there is nothing to suggest that necrosis is involved in the control of cell populations.

It is now accepted that the mechanisms involved in apoptosis are active beyond adulthood and may be active during senescence. However, the critical question of how apoptosis contributes to the aging of the organism has not yet been answered definitively and the role of cell death and of cell death genes in cellular aging and aging of the whole organism is not entirely clear. For example, a decline in the efficiency of the apoptotic program, designed to eliminate damaged or nonfunctional cells, would result in an increased number of deleterious cells, which might contribute to diseases. Alternatively, the mechanisms of aging could affect the apoptotic machinery itself, and this could result in death of essential cells thereby contributing to diseases. Cancers, autoimmune diseases, or degenerative diseases can all be the result of a defect in the deletion of malignant cells, or of immature lymphocytes that recognize self-antigens, or the death of neurons due to toxic excitatory amino acids, or lack of trophic factors (for review, see Driscoll, 1994; Carson and Ribeiro, 1993).

Aside from the link between apoptosis and diseases, there are many connections between the molecular mechanisms of apoptosis and those of senescence. The lack of growth factors or survival factors causes cell death, but the same phenomenon occurs during senescence. For example, the involution of the uterus and mammary glands at the time of menopause is due to a the lack of estrogen. Another factor that causes cell death is the decreased efficiency of the signalling system that triggers proliferation. In the immune system, a system in which apoptosis plays an important role, some lymphocytes express membrane receptors called Fas that induce apoptosis when stimulated. With aging, there is an increase in the proportion of lymphocytes CD45 RO expressing a high density of Fas receptors, which makes these cells more susceptible to apoptosis. This could be an important factor in immunosenescence (Hanabuchi et al, 1994; Itoh et al, 1991). Similarly, other phenomena that have been described during cell death, such as deficient intracellular signalling (in particular an inhibition or lower expression of protein kinases or PKC) or the alteration of the cell cycle, have been observed in cellular aging.

The finding that all eukaryotic cells have an in-built program of self-destruction that can be activated emphasizes the importance of genes in the process of cell death. Apart from death genes, several regulatory genes encoding proteins that extend cellular life have been identified (see Chapter 5). Almost every month, scientists report on the identification and characterization of new factors that play a role in the process of cell death. New members related to the anti-apoptotic bcl-2 proto-oncogene product and to the pro-apoptotic interleukin-1β converting enzyme (ICE) family of proteases have been found. Interestingly, a protein related to bcl-2, called bax, can form homodimers or heterodimers with bcl-2 protein. When heterodimers are formed, bcl-2 influence on cell survival is counteracted. Thus, the balance between the level of homo- versus heterodimers formed seems to determine the fate of cells. Recently, yet another protein called Bik (bcl-2 interacting killer) was found to interact with proteins belonging to the bcl-2 family and to trigger apoptosis. The picture that emerges is that complex regulatory mechanisms exist in cells, which involve factors that promote cell death or cell survival, and the activation of either cell death or cell survival machinery may depend on intracellular and extracellular signals. A detailed analysis of the molecular mechanisms of apoptosis and senescence might be a way of bringing to light the connections between these phenomena. If apoptosis is found to play an active role in aging processes, then, cell or tissue aging should be seen as the result of an active cellular suicide. The pharmacological manipulation of physiological cell death might be a way to affect the aging process.

At present, pharmaceutical and biotechnology companies are putting emphasis on research on apoptosis in order to find new agents that could induce or prevent apoptosis. In diseases such as cancer or degenerative diseases in which cells are either proliferating in an uncontrolled way or die massively, induction or prevention of cell death programs may provide an alternative therapeutic approach to current treatment of these diseases. The regulation of altered cell death programs in defined subsets of cells could result in the increased life span of the individual. The use of factors that affect cell death programs constitutes a new approach to treating diseases characterized by uncontrolled cell proliferation or cell degeneration.

Conclusion of Part One

The first part of this book contains a detailed description of the causes and consequences of aging. The overall picture is somewhat gloomy, because of the undeniable deterioration that accompanies senescence. We all know that life is a kind of fatal disease that leads all vertebrates to their demise with no hope for immortality. We described the complexity and the variety of manifestations of aging. We emphasized that it does not occur in the same way in all species or in all individuals from the same species; aging is not a singular phenomenon but is characterized by a variety of scenarios. Its mechanisms are complex, and in spite of recent and spectacular discoveries that allow us to understand a lot more about senescence, few fundamental mechanisms of aging have as yet been uncovered.

Why should aging occur at all in nature? Reasonable answers to this philosophical question have been proposed by different researchers. Aging is seen by proponents of the evolutionary theory of aging as a necessary process to guarantee the survival and evolution of the species at the expense of the individual's evolution. Forty years ago, Medawar (1952) and Williams (1957) proposed that the most important determinant of selection in any species is the optimization of reproduction. They suggested that the expression of deleterious genes associated with senescence (and eventually leading to death) would be delayed until after the reproductive period. The link between reproduction and senescence is best illustrated by the example of the Pacific salmon, which dies shortly after the only episode of sexual reproduction in its lifetime. The message is that when individuals no longer play an important role within the group or the species they must be eliminated. Williams proposed a variation of this idea and suggested that genes associated with the optimization of fecundity, expressed early in life, might have deleterious effects later in life (this is the *negative* or *antagonistic pleiotropy* hypothesis). These theories are attractive, and examples demonstrate that they are valid in several vertebrate species. However, the huge differences in the speed with which individuals age after their reproductive period also argue against these theories. Aging is inescapable, but the manifestations of aging depend greatly on the efficacy of protective mechanisms that prolong life. We emphasized the role of functional reserves or homeostasis of all physiological systems. One of the main reason why very few persons reach the maximum life span of 100–120 years, is because of the slow accumulation of cellular changes perturbing the functional equilibrium between physiological systems. The deterioration of normal homeostasis is one of the most important mechanisms of aging, and it explains the high vulnerability of the elderly to diseases. Kohn (1971) proposed that: "There may be a small number of basic aging processes in tissues which cause or predispose to (all of) diseases. Aging processes would then constitute a 100% total disease that everyone has". Finch (1990) also supports the idea that the decline in homeostasis is a major cause of mortality. He writes: ". . . decreased homeostatic powers are a major source of age-related mortality and may decrease independently of specific disease. This would account for the similar mortality accelerations in genotypes and populations with different disease distributions. The presence of a major disease may also decrease the homeostatic reserve and physiological resiliency, however."

At present, most medical interventions aim at the control of signs and symptoms resulting from specific diseases, while cellular and tissue perturbations that do not have obvious clinical consequences are not searched for, identified, or treated. Clearly, establishing the nature of homeostatic events and the causes of age-related decrease in functional reserve will be a major breakthrough in our understanding of the phenomenon of aging. Research projects on ways to detect homeostatic perturbations are much needed. They are as important as efforts to decipher the pathophysiology of the processes that lead to major age-related diseases, like heart failure, cancer, or neurodegenerative diseases.

In fact, demographers estimate that if all cancers, neurological, or cardiovascular diseases were eliminated as causes of death in the population of the United States, about ten years would be added to the average life span. This is surprisingly little in view of the theoretical maximum human life span.

It is worth pointing out that the somewhat pessimistic view that aging is an endless progression of unfavorable events that lead to death should however be tempered by some encouraging observations. If it is true that there is functional decline with age, the concept that all tissues and systems are affected is not correct. Indeed, there are a number of molecular and cellular functions that remain unimpaired throughout life. Aging, the loss in homeostasis, and death are often the result of a few limited deficits in otherwise healthy organisms. One tissue may be losing functional capacity while others stay quite young functionally and never get a chance to age. It is the state of the oldest nonredundant and necessary element or system that determines the biological age of the organism, i.e., its theoretical remaining life span at a given age. Death is due, in these cases, to the failure of the cell or tissue type on which the entire organism depends. The conclusions drawn from developmental biological studies that most somatic cells of young adults, and perhaps even of older adults, are genetically totipotent, support this view.

While research on homeostatic events is multifactorial and necessitates approaching the organism in all its physiological complexity, genetic research is more focused on specific genes that trigger cascades of beneficial or detrimental effects. We described the role of genes that lead to dramatically different life spans, thus emphasizing the role of genetic determinants in senescence. There are genes that speed up or slow down senescence, called long-life genes and death genes. The subtle balance between the level of expression of factors that determine cell survival and of factors that lead to apoptosis is important in aging. Findings in this area will undoubtedly give clues to some of the mechanisms that cause senescence and provide explanations about the underlying processes that determine the maximum life span. The presence of these genes clearly means that senescence in all animal species, including man, is under tight genetic control.

It is not known whether this clock-like genetic program is the same in all species. Will the mechanisms of senescence be traced to a very limited number of highly conserved genes, and will the differences in life spans between species be explained by diversity in these genes, as Finch (1990) proposes, or is there an array of genes that might operate in aging? There is no clear answer at present.

Even though there is a multitude of causes of aging, genes and the loss of homeostasis are probably the most relevant ones. This conclusion is based on impressive research efforts by many authors. If scientists acknowledge the crucial role of genes in determining senescence, they also recognize that genes are not the ultimate answer to explain senescence. Even individuals of highly inbred strains do not have perfectly identical patterns of senescence, nor do they have exactly the same longevity. Many aspects of senescence are therefore not determined by genes only but are the consequences of positive or negative environmental influences.

The second part of this book addresses the issue of how to prolong life, by investigating factors that can affect positively the inescapable phenomenon of aging.

PART II

THE PROLONGATION OF LIFE

Introduction

Living longer has been a goal of humankind since ancient times, and scholars in ancient civilizations promoted healthy living through medications and made recommendations about life-style. Based on recent information from the fields of biology, medicine, gerontology, psychology, and sociology, scientists now believe that the rate of senescence can be slowed and that the time of death can be postponed by many years. But what is the evidence for this? Answers to this question and a review of the measures aimed at slowing the rate of senescence are the theme of the second part of this book. We will discuss many subjects, from experimental attempts to postpone aging in animals, to behavioral, medical, and surgical approaches that might lead to increased life span in humans. These measures cover a broad area, from changes in psychological attitudes to the replacement of defective biochemical or physical components of the body. We will discuss mostly primary prevention measures aimed at the general population, or at individuals at risk and predisposed individuals. Because of their impact on longevity, we cover also recent developments in secondary prevention, aimed primarily at persons whose disorders have been diagnosed.

Humans know that they are doomed to age and die but many accept this situation only with difficulty. The dogma of many religions softens this inevitable fate by postulating second-best immortality, such as the Hindu reincarnation or Christian paradise. In a more prosaic way, Woody Allen gave his opinion about immortality: "Some people try to achieve immortality through their offspring or their works. I prefer to achieve immortality by not dying". Following this line of thinking, a few North American citizens have gone so far as to pay for being deep frozen in liquid nitrogen, where they will wait patiently to come back to life after some future medical breakthrough. Bob Dylan views aging with a more optimistic attitude, when he sings: "I was so much older then, I am younger than that now".

As death is inevitable, most of us do not wish to extend our lives at all costs, but to maintain good general health and an alert mind as long as possible, without suffering from debilitating diseases. Jazwinski (quoted by Keaton, 1992) summarizes this wish in simple words: "The ideal would be to live a long and healthy life and then undergo a rapid demise—to die with your boots on, as they say in cowboy movies". In our Western culture, the desire to postpone disease and death has become an overriding goal, both because of the prominent cultural idea that every individual should have a long and enjoyable life and because recent scientific discoveries might indeed provide new ways to treat diseases and to postpone aging.

In the first part of this book, a number of determinants of aging were examined, from psychological or social issues to molecular changes in cells. The resulting picture is complex and varied. We have seen that heredity is a powerful factor in determining longevity but that other factors, independent of our genes, also determine who will remain younger biologically even as they get chronologically older.

What can one do to retard the damages caused by aging? We will discuss the role of diet, food restriction, antioxidant and vitamin supplements, hormones, and drugs in prolonging life. We will also evaluate the importance of the mind, of psychological factors in the way people age. Mood, social contacts, and sexuality are also important in longevity.

Is there a way to evaluate the biological age of individuals? We have seen in the first part of this book, that testing an individual's functional capacity can give some indication of his or her life reserves. However, it would be useful to be able to test a person's cellular reserves, in other words, to test the efficacy of cells in different tissues: for example, to test how the individual is capable of fighting oxidative damage, or repairing damaged DNA, or regulating the gene expression machinery. Whether this kind of information will be of good predictive value to estimate an individual's residual lifetime is another question.

We all know that life leads all vertebrates to their demise with no hope of immortality; however, there are a number of preventive measures that might delay aging and the onset of diseases. Among all the possible interventions to postpone aging and death, the most relevant ones are changes in life styles, the ingestion of food supplements, novel drugs, and medical screening (preventive medicine). We hope that the implementation of these measures, particularly those concerning diet, exercise, smoking, alcohol, and vitamin and mineral supplements, will help people stay mentally and physically active. Good life styles, by alleviating medical, social and psychological problems of the individual, may also be beneficial to society. Biological approaches for life extension will have their full impact on the fate of the elderly only if they are accompanied by societal attitude changes towards older people.

Finally, the spectacular advances in the identification of genetic mutations that cause diseases have opened the way to new therapeutic approaches. Even though the measures described may seem limited at present, there is hope that a cure will soon be found for some of the severely handicapping and life-shortening human diseases.

Chapter 1

Life Extension

Life Extension in Nonhuman Species

In some species, certain individuals have a particularly long life span compared to other individuals. In the case of the queen bee, her long life is due to environmental and hormonal influences. In laboratory experiments with flies, it is possible to extend the life span by postponing reproduction artificially, or by mutagenesis followed by selection of mutants that have a longer life span (see Part I, Chapter 1 and Chapter 5). These studies and others carried out in mammalian species illustrate the diversity of the different biological and developmental mechanisms that lead to a longer life span.

Pharmacological Strategies

Many drugs have been shown to increase the life span of different animal species raised in laboratory conditions. Here we describe one of these drugs, L-deprenyl. (Findings with several other compounds are described under the chapter of anti-aging medications (Chapter 6).)

L-deprenyl, or selegiline, is a medication for the treatment of Parkinson's disease. It slows down the progression of this severe brain degenerative disorder. In pharmacological terms, it is a selective and irreversible inhibitor of the B-form of the enzyme monoamine oxidase, or a MAOI-B. Research with rats has shown that L-deprenyl increases the life span by about 30% (Kitani et al, 1994). Interestingly, some of the anatomical changes that occur in the brain of aged animals can be prevented by L-deprenyl (Amenta et al, 1994). The inhibition of MAO-B is said to prevent the transformation in the brain of exogenous and endogenous compounds that might lead to neurotoxic metabolites. But there are other possibilities as well. L-deprenyl might activate several of the antioxidant enzymes, such as superoxide dismutase (although this has not been consistently found by all authors); it might also lead to an enhanced level of growth factors in brain tissues and changes in astrocytes, the cells that support neurons. In short, L-deprenyl might have life-extending effects in animals through diverse pharmacological actions, but these may not necessarily be the same in humans.

Many compounds have been found to enhance life span in laboratory animals, but they do not necessarily have the same effects in humans. This observation is also found in other conditions: animal models often suggest that a given experimental drug might be efficacious in humans for the treatment of dementia, while clinical trials demonstrate soon after that this is not the case.

Food Restriction in Nonhuman Species

Of all the measures that are seen to extend life span in a number of animal species, the most dramatic and consistent is through food restriction. The evidence is strong that food restriction (FR) postpones age-related physiological changes, at least in all animal species studied so far. To date, FR is the only intervention shown consistently to extend life span. A number of studies on the effect of FR in long-lived and short-lived rodents, have shown unambigously that animals fed as much food as they ate spontaneously were not as healthy and did not live as long as those whose diet was restricted in calories; the diets were formulated carefully to guard against malnutrition. The onset of age-related diseases can also be retarded by FR, and this is due mainly to restriction in calories, rather because of specific nutrients or contaminants present in the diet. The exact mechanisms by which FR operates are still unclear, but they are not based solely on the slowing of growth and development, the reduction of body fat, or the decrease in metabolism. Scientists believe that FR has a strong influence on the endocrine, immune, and neuronal regulatory systems, that it changes protein turnover, gene expression, and the level of free radical damage.

The first animal study of FR was by McKay and Crowell in 1934 who demonstrated that an adequate but calorie-restricted diet, when compared with a normal laboratory diet, had a remarkable effect on the life span of rats. The mean and maximum life span of male rats almost doubled. These observations have been confirmed and extended by others. The results obtained by Weindruch and his collaborators (1986), as shown in Figure 20, and by Harrison and his collaborators (1984), in Table 9, are especially striking.

Interestingly, as shown in Table 9, FR acts independently of hereditary metabolic changes: food-restricted, genetically obese mice exhibit a maximum life span comparable to normal restricted controls, despite the fact that they still have 3.5 times as much body fat as the controls.

The effects of FR are impressive, and it is likely that FR affects directly or indirectly basic processes involved in the biological regulation of senescence. Therefore, finding out the mechanisms by which FR operates is a way to gain more insight about senescence. A number of studies already provide information on changes induced by FR.

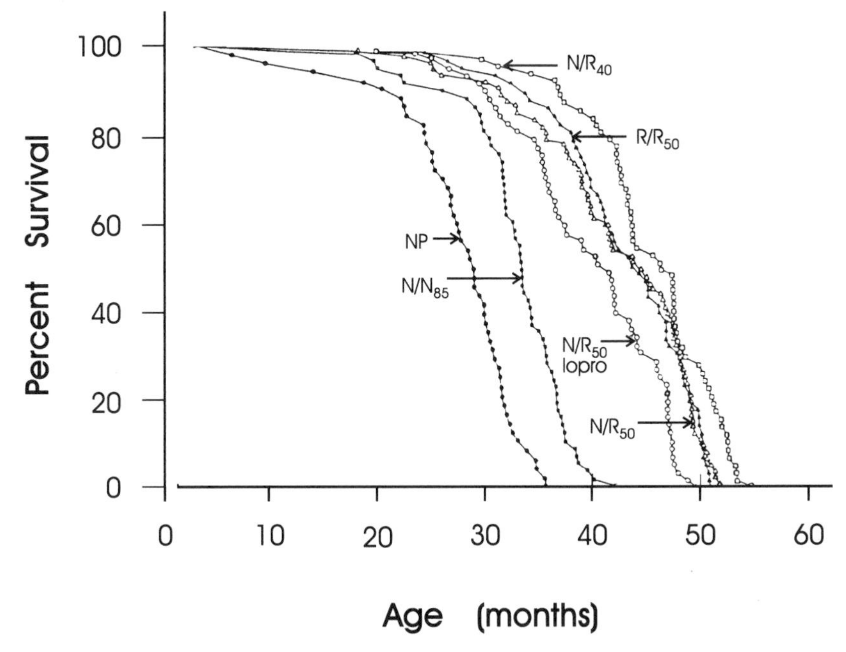

Figure 20. The influence of diet on survival.
Each symbol represents an individual mouse.
Diet groups: NP: nonpurified diet fed *ad libitum* at approximately 113 kcalorie/week;
N/N85: ca.85 kcalorie/week of purified diet fed postweaning;
N/R50: lopro, animals restricted postweaning to a 50 kcalorie/week diet with a decreased protein content;
N/R50: animals restricted postweaning to a 50 kcalorie/week diet;
R/R50: restricted pre- and postweaning;
N/R40: animals restricted postweaning to a diet of approximately 40 kcalorie/week. (Redrawn from Weindruch et al, 1986.)

Effects of FR on Intestinal Cells

Heller and his collaborators (1990) compared cellular changes in the small intestines between two groups of rats, a control group fed an unrestricted diet and another group fed a FR diet. Examining the lining of the small intestine, they found that the number of villus cells (cells with tiny finger-like projections through which nutrients are absorbed and which secrete digestive enzymes) increased with age in animals fed a normal diet. There was also an increase in crypt cells at the base of the villus cells. Medical doctors believe that in humans as well, rich diets cause an abnormal increase in the number of crypt cells, and this could be an important step in the early development of cancer. These changes in crypt cells, characteristic of aging in the small and large intestines, were inhibited by FR. Whereas the increase in villus and crypt cells started at 21 months in rats

Table 9. Effect of genetic obesity and food restriction on aging and longevity of mice

Treatment	Food (g/day)	Body Wt (g)	Fat Wt (%)	Longevity	
				Med.	Max.
Fed obese	4.2	59	67	552	890
Fed normal	3.0	30	22	799	970
Rest. obese	2.0	28	48	814	1300
Rest. normal	2.0	20	13	810	1280

(g = gram; body wt = body weigt; fat wt = fat weight; fed obese and fed normal: rats in these two groups had unlimited access to food; rest. obese and rest. normal: restricted groups were given limited amounts, once daily, 6 days per week; med. and max.: medium and maximum life span respectively).
(Data reproduced from Harrison et al, 1984).

from the control group (fed *ad libitum*), this was only observed from 27 months on in the FR rats. FR also retarded similar proliferative changes in the colon. It was also shown that FR delays the fall in activity of several small intestinal mucosal enzymes (Holt et al, 1991).

Effects of FR on Cellular Membranes and Memory

FR induces changes in cell membrane properties. Tacconi and his collaborators (1991) observed that, unlike aged rats fed a standard diet, those fed a hypocaloric diet presented brain membrane microviscosity and phospholipid composition similar to those of young rats. This is interesting in relation to the deterioration of memory in aged rats. Senescent animals present deficits when they have to perform spatial memory tasks. In particular, their ability to remember internal maps using environmental clues is affected. FR senescent rats performed significantly better. Although there is some controversy as to the type of memory affected, many studies agree with the statement that a life-long hypocaloric diet acts to prevent age-related memory deficits (Pitsikas and Algeri, 1992).

Effects of FR on Immune Functions

As discussed earlier (Chapter 3, Part I), aging is associated with a decline in immune functions, particularly T-cell mediated functions. This decline possibly leads to an increased susceptibility to bacterial, fungal, and viral infections, as well as to the development of cancer and autoimmune diseases. T cells or T-cell subsets from old animals respond poorly to mitogens when compared to cells obtained from young animals, but FR delays the progression of immune deficiency with age. To understand the mechanisms by which FR prevents the loss of immune function, Fernandes and collaborators (1990a) analyzed spleen cells. They reported that FR markedly alters the fatty acid composition of these cells and increases their membrane fluidity. FR also increases splenic intracellular calcium in response to stimulation by Concanavalin A, IL-2 production, and the proliferative response of these cells to mitogens. These cellular modifications may be responsible for increasing the immune activity of T cells and increasing B cell ability to produce antibodies. Prevention of early loss of T-cell functions may be the principal mechanism whereby FR delays the aging process of the immune system and, by extension, of the entire organism.

Effects of FR on Free Radical Generation

FR is believed to protect the "self-regulatory mechanisms that maintain cellular homeostasis" (Yu et al, 1990). This is based on the observation that liver mitochondrial and microsomal membranes from *ad libitum*-fed rats contained a higher amount of lipid hydroperoxide than liver from food-restricted rats. Cells from food-restricted animals showed lower levels of all parameters linked to oxidation: peroxidizability, H_2O_2 production, and lipid hydroperoxides. Cytosols of livers from food-restricted rats had a better protective activity against peroxidation, high concentrations of reduced glutathione, elevated levels of catalases, and reduced glutathione peroxidase activity.

FR has an important effect on cytochrome P-450 enzymes. These play a pivotal role in the metabolism and the detoxification of endogenous steroid hormones and of many drugs and carcinogenic compounds. FR prevents the progressive breakdown of microsomal cytochrome P-450 observed during aging in rat liver tissues. It also prevents the accumulation of several toxic substances in cells (Fernandes et al, 1990b).

Although FR is the most compelling measure that will enhance life span, at least in animal species, it is not without danger. In their early studies, McKay and his collaborators (1935) noticed that animals put on FR immediately after the weaning period, stopped growing and resumed their growth in body size and weight only after they were given additional calories. FR can have other negative effects, including reduced tolerance to cold, chronic increased output of stress hormones, as well as impairment of fertility and reproduction.

It should be underlined that in studies on the effects of diet on longevity, food-restricted rodents were compared to rodents fed *ad libitum*. This comparison is frequently criticized, because *ad libitum* feeding of laboratory animals might not be physiological (as discussed by Masoro (1988)); these animals are believed to die sooner because of an excess of calories, as is the case in Western man. Some scientists even claim that FR reproduces the natural feeding pattern of most animals. These criticisms cannot be ignored since no one knows exactly what the natural nutrition of rodents is. Moreover, laboratory animals are inbred, and it is difficult to make any kind of assertion about normal conditions for inbred animals.

Life Extension in Humans

At the present time, medical interventions to postpone aging or death consist essentially in improving the standards of living and medical services in poor countries and, in richer countries, in counteracting the negative influence of our chromosomes at the phenotypic level, i.e., to treat symptoms of diseases or prevent disease. These interventions have a well-recognized effect on longevity, as seen, for example, in the control of complications from diabetes and high blood pressure or through the prevention of myocardial infarction or of recurrent psychiatric disorders.

Other interventions to postpone death are aimed at the population at large—they include everyone, whether sick or not. The aim is to act on factors that accelerate aging, such as unhealthy life styles, lack of essential nutrients, decreased levels of some hormones and peptides.

Direct modification of the structure of longevity or death genes to increase human life span is still impossible. However, the last years have witnessed fantastic discoveries in the area of the genetic basis of disorders and of genetic engineering techniques and gene therapy. Repairing defective genes can now be envisaged, and there is much hope in the medical community that several diseases causing handicaps may soon find a cure. Moreover, molecular genetics is becoming a powerful tool to identify individuals predisposed to given disorders. This early detection of disorders, at a time when the future patient is asymptomatic, will open new avenues for primary prevention.

The powerful methods of gene transfer techniques and gene therapy may provide means of developing protective procedures against several age-related disorders. These somewhat futuristic molecular approaches applied to the aging process and to age-related disorders may open the way to the prolongation of life that will be as efficacious, or even more efficacious, than the difficult effort to eradicate deleterious human habits through life-style changes.

Facts Versus Hypotheses in Life-Extension Programs

Aging prevention or life extension is a popular theme in the media, but in addition to fascinating, accurate, and serious books (see list of relevant books on aging), sensational and misleading self-help books are also being published. Authors of these books have a biased and incomplete understanding of the mechanisms of aging; they propose conclusions that are quite remote from scientifically proven facts, or construct biological life extension programs based on untested hypotheses, with unknown and potentially harmful consequences. In this book, we take a critical, scientific approach towards life extension methods, rather than recommend untested or irrational life extension programs. However, the distinction between established facts based on good science and irrational hypotheses is not clearcut: recommendations concerning the intake of vitamin supplements differ among United States agencies. For example, the nongovernmental Alliance for Aging Research recently made a recommendation based on a review of more than 200 epidemiological or clinical studies. They advised healthy adults to take daily doses of 250 to 1000 milligrams of vitamin C, 60–250 milligrams of vitamin E, 10 to 30 milligrams of β-carotene to prevent age-related diseases (Voelker, 1995). These doses are much more than the official recommended daily allowances (RDA).

The administration of several endogenous compounds that decline with age might delay the onset of certain decreases in physiological functions. Hormonal replacement for postmenopausal women is a recognized strategy, as Adami and Persson stated (1995): "For now, the aggregate of the epidemiologic evidence is reassuring; we can feel a great deal of confidence that the net effect of hormone replacement with regard to length and quality of life is beneficial. This confidence should guide our counseling of and treatment recommendations to postmenopausal women".

Other potentially useful compounds are melatonin, dehydroepiandrosterone (DHEA), human growth hormone (hGH), and maybe testosterone (for men). Vitamins and drugs acting as antioxidants, as antistress agents, or as cognitive enhancers might also help, as well as compounds that have an impact on the production of energy by cell mitochondria, such as coenzyme Q.

In this book, we give specific recommendations on whether or not to take these compounds on a routine basis to every person who has reached a given age. Our recommendations are based on published information, specifically on the results of several large scale trials (megatrials). Several other megatrials are ongoing, and the conclusions of such studies may impose modifications of proposed replacement therapies. Indeed, more clinical trials are needed to evaluate life-extension programs, and this is urgent in the view of preventive medicine specialists.

What these trials will have to decide is whether one or a combination of life-extending compound(s) is more efficacious and, of course, which compounds are the most effective. One answer seems already clear—no single compound or group of compounds would be efficacious for all people. For example, vitamin E seems to protect against cancers of the lower portion of the digestive tract, while β-carotene seems to be more useful in preventing cancers of the upper digestive tract and respiratory airways.

People who wish to start a life-extension program should consult with their physician, of course. They should see such a program as wide ranging, with global measures concerning diet, physical exercise, and mental well-being, as well as specific measures, such as vaccines against the flu, especially for elderly subjects. The goal of such programs should be not only to extend the duration of life but the number of years of healthy life. The individual and society should have this goal, as illustrated by the United States program entitled *Healthy People 2000*. The good news is that goals for the better health of the population are being met—people are exercising more, smoking less, and following the various other objectives of preventive medicine (McGinnis and Lee, 1995).

Medical Screening and the Prevention of Diseases

Medical screening of the population, through clinical and biological testing, is enabling physicians to detect clinical diseases, or the predisposition to develop diseases, more effectively. Its usefulness increases as new discoveries are made on ways to diagnose diseases at an early stage. The aim is to find predispositions or unhealthy life habits that may play a significant role in limiting a person's potential longevity. Examples of preventive screenings include mammography for breast cancer; testing for blood in stools and colonoscopy for digestive cancers; rectal examination and blood tests for prostatic carcinoma; thoracic X-ray for lung carcinoma; blood lipids, exercise tests, and other functional tests for heart disease, etc.

Medical screening leads to the early treatment of conditions such as skin, breast, or colon cancer and to preventive cardiac bypass of coronary arteries. These are the most publicized conditions, but there are other, less recognized ones, for example, sleep apnea. Its prevalence is said to be between 0.5% and 10% of the adult population. It is defined as repeated short nocturnal periods when breathing is interrupted, either because of brain mechanisms (central apnea) or, more frequently, because of peculiarities in the anatomy of respiratory pathways (obstructive apnea). Sleep apnea leads to daytime sleepiness and is more frequent in those who are overweight or who snore. It is among the factors that negatively influence life span, predisposing to cardiovascular problems, such as heart attack and stroke, or to dementia (McNamara et al, 1993).

Medical screening should be standard practice. For example, rectal examination and prostate specific antigen testing for the detection of prostatic cancer should be performed in men over 50. However, there is controversy about the benefit-to-risk ratio of some of the screening; e.g., repeated mammography in women, which could increase the risk of radiation-linked cancers.

Screening for the prediction of potential diseases raises financial, societal, and ethical issues that cannot be underestimated. Predictive tests are not 100% infallible: false positive or false negative results can occur, and the prediction of unfavorable events can cause psychological trauma. The physician who informs a person that she or he has a high risk of early development of a given disease is being helpful to that person only if primary prevention measures or efficacious therapies exist for the disease. For example, elderly persons who suffer from an age-related cognitive decline might well evolve toward a true Alzheimer's disorder, but they might also be progressing toward stabilization of their mental functions. It now may be possible to predict which of these evolutions will happen by using a test for Apo ε4 status (Peterson et al, 1995) and also one of the new brain imaging techniques (Small et al, 1995).

Measuring the Individual's Biological Age

Longevity and the rate of senescence of a given person depends on genetic, epigenetic, and environmental factors. In older populations, the consequences of these factors become quite obvious, and there is a wide range of biological, physical, and mental fitness: a person with a chronological age of 70 might be biologically or psychologically much younger, or older, than another person of the same age. Research to develop methods to measure biological age and quantify the rate of senescence, and thus to make a prognosis about potential longevity, is an area that has gained popularity in recent years. Prognoses are based on many factors. In the previous section, we mentioned screening tests for early detection of disease and identification of predispositions to diseases. Medical screening has a major role in predicting residual lifetime. However, these evaluations are not foolproof. Everyone knows of a person who was diagnosed with a severe disease by his physician and condemned to live only a few months—but he lived several additional years! Conversely, a middle-aged man is told after his yearly medical check-up that he will live to 100, but he dies a week later.

Predictions of potential longevity are based on a multifactorial approach including, for example, familial history of cardiovascular disorders or cancers, blood lipid levels, blood pressure, smoking and drinking habits, practice of physical exercise, etc. Tables that present probabilities of years of survival for each age group have been calculated from epidemiological studies. They give some idea as to which level of potential longevity an individual may belong, but the prediction is valid only in statistical terms, i.e., for groups not for individuals. Moreover, in order to make a good evaluation of a person's potential longevity, the list of factors that should be taken into account is almost endless. Apart from physical variables and a record of life style, personality and psychological status have to be considered because they influence longevity as well. Several of these factors can be identified or quantified through a routine visit to a physician. On the other hand, several other factors are just as important, but their evaluation is technically complicated because elaborate medical and laboratory analyses are required: for example the nutritional status, immunocompetence, antioxidant capacity are extremely important but are very difficult to measure and are subject to statistical variability. Many tests are not yet validated or available in standard practice. In the case of the antioxidant capacity of the body, for example, it is not clear what analyses should be done, and current methods used to assess the antioxidant capacity of a given individual are not very specific or sensitive (Kanter, 1994).

Results of the measures of all the above factors could be analyzed by computer to provide a statistical evaluation of a person's potential longevity. And, of course, studies aimed at slowing senescence or postponing death should take into account the individual's biological rather than chronological age.

Human Studies on Food Restriction

The effects of FR on human longevity are unknown, although very old people often attribute their longevity to their restrained eating habits. No formal FR studies have ever been carried out in humans. However, particular population subgroups with specific traditional diets or religious habits serve as examples of natural human experiments of FR. People from these subgroups have higher longevity. They eat a healthy diet, are often vegetarians, and abstain from coffee, tobacco, and alcohol. However, it is difficult to attribute their higher longevity solely to diet and life style. Other factors such as strict morals and/or religious beliefs or intense interpersonal supports that prevent stress could be relevant also to their longer life span.

There are also surgical techniques that have been applied to decrease the total amount of calories ingested in cases of morbid obesity; these interventions consist of the removal or blocking of parts of the digestive tract. Patients develop a malabsorption syndrome and lose weight, but they may also develop arthritis or other severe medical complications with some interventions. Such surgical interventions are now outdated, as are the high-protein, low-calorie liquid diets. These extreme measures are proposed only in cases of morbid obesity.

Recommendations

Starting a FR diet is a decision that should be taken only after extensive dietary information and medical advice. Walford, a physician and researcher, wrote a book on how to implement a FR diet and explained how well he felt despite his 2000 calories per day intake, but the limits of a one-man experiment are obvious (Walford and Walford, 1994).

Perspectives

It will be very interesting to extend observations made with food restriction in animals to human populations, but studies are difficult to carry out for many reasons, some of them quite obvious. Even if a restricted diet could be imposed on well-defined groups of individuals, it would be impossible to keep all other parameters that influence life span constant. Moreover, the genetic heterogeneity of people would have to be taken into account. It would be difficult therefore to interpret the effect of FR. Despite the impossibility of carrying out such a study, it is reasonable to assume, by inference from animal studies, that FR may be beneficial to humans.

Chapter 2

The Role of Diet in Longevity

The popular saying "tell me what you eat, and I will tell you who you are" emphasizes the role that food plays in our health, our psychological life, and our physical well-being. Moreover, the discovery that a moderate food restriction can increase life span of laboratory animals to a considerable extent is a good illustration of the effects of food on health. Scientists and physicians recognize that diet plays a crucial role at all stages of people's life. Infants, children, and adolescents should receive a diet sufficient to ensure physical and mental growth. Adults should avoid becoming overweight. This is of particular concern in the United States where the occurrence of obesity is stunningly high (30% of the population, particularly among women and minority groups). Obesity is associated with greater morbidity and the incidence of diseases such as diabetes mellitus, hypertension, cardiovascular diseases, stroke; also, the incidence of some forms of cancers rises with increasing weight (Kuczmarski et al, 1995).

Diet is especially important at times of illness or excessive stress or in case of digestive problems. Specific illnesses require special diets. Similarly, it is of paramount importance to make sure that elderly subjects are nourished properly and do not suffer from deficiencies in essential nutrients. Intermittent and minor digestive symptoms are common throughout life, but they become increasingly more frequent with aging.

A balanced diet has a positive impact on health in general and may influence the incidence or the severity of many diseases. As such, it is an important factor in longevity. Epidemiological studies on the negative or the beneficial effects of different diets have provided important conclusions. In the 20-year follow up of the Framingham Heart Study (longitudinal study), for example, it was shown that a higher intake of fruits and vegetables was related to a somewhat lower risk of stroke in men (Gillman et al, 1995).

Nutrition influences gastrointestinal function, blood pressure, immune function, and cognitive abilities. The diseases of middle and old age considered to be related to nutrition include obesity, coronary heart disease, cancer, diabetes, osteoporosis (with its associated enhanced risk of bone fractures), and even infectious diseases, such as tuberculosis and pneumonia.

The consumption of unhealthy foods is promoted by the advertising of food corporations that sell highly processed food. Moreover, labels such as "natural", "no additives", "no white sugar", "no artificial coloring", can be misleading because they do not inform consumers about what the products really contain. The new processed, refined, or convenience foods (hamburgers, French fries, sodas, frozen meals) often outnumber basic, healthy, unprocessed foods on store shelves and family tables. In 1986, dietary objectives were set by the National Cancer Institute (Greenwald et al, 1986) for the year 1990: "By 1990, the per capita consumption of fiber from grains, fruits, and vegetables will increase to 15 grams or more per day. (In 1976–1980, the per capita consumption of fiber from these sources was 8 to12 grams.) By 1990, the per capita consumption of fat will decrease to 30% or less of total calories. (In 1976–1980, the per capita consumption of fat was 40% of total calories.)" Retrospectively, these objectives were not met, and it is unclear if the objectives set for the year 2000 (to decrease the per capita daily consumption of fat from 40% to 25% or less of total calories) are realistic or not.

It is important to recognize that, just as children are not miniature adults, elderly people are not simply adults with additional years. Their total food intake tends to be lower than that of adults and matches their reduced physical activity. Their food requirements are not necessarily that of younger people, and their preferences for food are different. One of the main problems encountered in the elderly population is the prevalence of malnutrition, which is due to such diverse causes as depression, solitude, or dental problems, etc.

In the last few years, physicians have realized that there is a lack of knowledge about the nutritional requirements of old people. Recommendations on diets and dietary supplements specifically addressed to elderly people are being set up. If poor dietary intake is the main cause of vitamin deficiency in elderly people, there is strong evidence that aging itself affects the requirements for specific vitamins (see Chapter 2). For example, the 1989 recommended dietary allowances (RDAs) seem to be too low for elderly people for vitamin D (5 micrograms), riboflavin (vitamin B2, 1.2–1.4 milligrams), vitamin B6 (1.6–2 milligrams) and vitamin B12 (1.6–2 micrograms), and too high for vitamin A (800–1000 micrograms).

Recommendations

Many people do not have a proper diet or have deficient eating patterns. Those eating junk food on a regular basis, for example, need to take vitamin supplements. Eating habits should be changed, and information about what to eat and how to select healthful food wisely should be publicized regularly in local newspapers and in schools. The dietary recommendations issued by the National Research Council, summarized in Table 10, can serve as basic guidelines (National Research Council, 1989).

Table 10. Summary of national research council recommendations

1. Reduce total fat intake to 30% or less of calories.
   Reduce saturated fatty acid intake to less than 10% of calories and the intake of cholesterol to less than 300 milligrams daily.
2. Every day eat five or more servings of a combination of vegetables and fruits, especially green and yellow vegetables and citrus fruits.
   Increase starches and other complex carbohydrates by eating six or more daily servings of a combination of breads, cereals, and legumes.
3. Maintain protein intake at moderate levels.
4. Balance food intake and physical activity to maintain appropriate body weight.
5. Alcohol consumption is not recommended. For those who drink alcohol beverages, limit consumption to the equivalent of 1 ounce of pure alcohol in a single day.
6. Limit total daily intake of salt to 6 grams or less.
7. Maintain adequate calcium intake.
8. Avoid taking dietary supplements in excess of the RDA in any one day.
9. Maintain an optimal intake of fluoride, particularly during the years of primary and secondary tooth formation and growth.

More recently, Willett (1994) reviewed the interrelations between diet and disease, and proposed nutritional guidelines that he drew from comparative studies of the dietary characteristics and life expectancy and disease rates in different countries such as the United States, Greece, and Japan. His recommendations are: (1) Minimize intake of saturated fats and partially hydrogenated vegetable fats, particularly those from dairy products; (2) Eat large amounts of fruits and vegetables, particularly those foods that are rich in carotenoids and vitamin C; (3) Increase dietary carbohydrates, primarily in the form of starches and complex carbohydrates; (4) Reduce animal protein intake.

A general recommendation is to have a balanced healthy diet, to eat three meals a day, eat only the amount of calories needed for the level of physical activity and tissue regeneration, ingest plenty of fiber and small amounts of fat. The role of fiber is essential since fiber tends to lower blood cholesterol and protect from diverticulosis and digestive cancers.

Special lectures on diet and on the composition of foods, i.e., their content in fat, sugar, and protein should be organized especially for elderly people. Information is useful, but it is also necessary to monitor the food intake of the elderly and to supplement their diet with specific vitamins, minerals, and fibers. Such counselling is already carried out in several countries but needs to be developed further. Changing people's dietary habits is not an easy task. Improving the diet of older people can only have beneficial outcomes, not the least of which is that it could help alleviate depression and feelings of loneliness.

Perspectives

A healthy diet in association with regular exercise and weight loss (see Chapter 4, pages 172–177) seems to reduce risk factors for atherosclerosis and coronary artery diseases, not only in young adults but also in older people. Such life style modifications have important socio-economic implications.

Vitamin Supplements

It is vital to include sufficient amounts of vitamins in the diet, but in spite of the progress achieved in vitamin research, there is still confusion as to what people should do. There are no clear answers about the type of vitamins and the doses necessary for each person, but an abundant literature exists on the deleterious effects of vitamin deficiency (for a review, see Somer, 1992). Severe vitamin deficiencies induce recognizable signs and symptoms such as scurvy, beriberi, or xerophthalmia. At present, medical doctors recognize that vitamins are particularly important for elderly people whose needs may be greater than those of young people. On the other hand, no study has shown unambiguously the role of vitamins in curing diseases, other than those secondary to vitamin deficiency.

In a telephone survey, Stewart and his collaborators (1985) estimated that 40% of the United States population took daily supplements of vitamins or minerals. With such widespread use of vitamins, there is opportunity for misuse, defined as the ingestion of vitamins in inappropriate dosages or for purposes that have not been medically approved. Hopes are that vitamin supplements might be helpful in the prevention of diseases, in fighting cancer and heart disease, and in postponing the ravages of aging. In spite of these hopes, vitaminotherapy to promote health is still a speculative area of preventive medicine.

We list the established and possible benefits of the main vitamins. The antioxidant effects of vitamins A, C, E, and β-carotene are described elsewhere (see Chapter 3).

Vitamin A

Vitamin A deficiency causes loss of lacrymal secretion, opacity and ulceration of the cornea which leads to blindness in children from poor countries (xerophthalmia), gingivitis, and dry skin. The established benefits of vitamin A, or β-carotene (**provitamin** A), are the prevention of night blindness and xerophtalmia. Vitamin A and β-carotene have numerous potential benefits. They may retard macular degeneration (a common cause of blindness among the elderly) and reduce the risk of breast, lung, colon, prostate, and cervical cancer (Ziegler, 1989; Barros et al, 1986; Byers et al, 1987; Van Eenwyk et al, 1991). They could affect the incidence of heart disease and stroke (Bruning, 1994). Unlike most carotenoids, vitamin A and most retinoids are toxic when taken in excessive amounts. Adverse effects, including loss of appetite, headache, blurred vision, hair loss, and liver damage, are observed in daily doses over 25,000 IU taken over several months. The body absorbs them efficiently, but cannot eliminate excessive amounts. Retinoic acid (retin-A), a synthetic derivative of vitamin A, is prescribed for the oral treatment of juvenile acne. It helps prevent the formation of the plugs of oil and/or reduces the formation of excess oil. It is teratogenic, however, in large doses. Recently, there has been much interest in the external use of retinoic acid because of its effects on wrinkles and on skin spots. In clinical trials, retin-A was shown to retard or even reverse the aging of skin. It is a potent drug that can cause peeling and reddening of the skin for many weeks. In the USA, retin-A (for external use) is available by prescription, whereas in several other countries, it is an over-the-counter drug. The long-term effects of retin-A are not known; therefore, people using it should be cautious.

Vitamin B1 (thiamine)

Severe deficiency of vitamin B1 causes beriberi. Less severe deficiency causes numbness and paralysis of the legs and arms, gastrointestinal problems, and cardiac and respiratory distress. In adults, severe vitamin B1 depletion due to malnutrition or alcoholism can lead to brain damage. Vitamin B1 is required for the normal functioning of all cells and especially nerve cells. It is involved in cellular processes that break down carbohydrates, fat, and proteins that convert excess carbohydrate to fat. No known toxicity has been reported when the vitamin is taken orally.

Vitamin B2 (riboflavin)

Deficiencies in vitamin B2 cause malformation and retarded growth in infants. Mild deficiencies are more common and are manifested by burning of the eyes, loss of vision, soreness of the mouth and tongue. Vitamin B2 is important for normal development and growth, for the production of certain hormones, and the formation of red blood cells.
No known toxicity is identified.

Niacin (nicotinamide)

Deficiency of niacin causes cracks of the tongue, dry skin, diarrhea, and dementia. Niacin belongs to the group of B vitamins. It plays an important role in the release of energy from carbohydrates. It is involved in the breakdown of proteins and fats, in the synthesis of certain hormones, in the formation of red blood cells, and in the detoxification of drugs. Niacin prevents pellagra. It could act as a possible cancer inhibitor. Too much niacin can lead to jaundice and liver damage.

Vitamin B3 (pantothenic acid)

Deficiencies in pantothenic acid can be induced in laboratory animals fed a refined diet without pantothenic acid. Symptoms are weakness, cardiovascular and digestive problems, infections, dermatitis, muscle cramps, lack of coordination. Pantothenic acid is converted into coenzyme A, an important catalyst in the breakdown of fats, carbohydrates, and proteins for energy. It is necessary for the production of fats, cholesterol, bile, vitamin D, and red blood cells and participates in the synthesis of certain hormones and neurotransmitters. Excess pantothenic acid causes diarrhea.

Vitamin B6 (pyridoxal)

Deficiency causes peripheral neuropathies and dementia. Infants born with a metabolic defect that prevents them from using vitamin B6 develop mental retardation and uncontrollable convulsions. Vitamin B6 has been shown to help prevent anemia, skin lesions, and nerve damage. It could protect against neural-tube defects in fetuses. Adverse neural effects of high doses of vitamin B6 are numbness in the mouth and hands and difficulty in walking.

Vitamin B7 (biotin)

Deficiency in biotin is rare and may result from poor absorption of the vitamin in infants. Symptoms are dermatitis, conjunctivitis, hair loss, anemia, depression. Symptoms can be accompanied by an enlargement of the liver. Biotin is essential (like folic acid, vitamin B12, and pantothenic acid) in cellular reactions that break down fats, carbohydrates, and proteins. Significant amounts of biotin are produced by the intestinal bacterial flora. No toxicity is reported for biotin.

Vitamin B12 (cobalamine)

Deficiency affects the growth and repair of all cells of the body, for example blood cells, leading to pernicious anemia. Poor vitamin B12 intake or absorption results in defects in the formation of nerve cells resulting in nerve damage. Clinical manifestations include disorientation, numbness, moodiness, agitation, hallucinations etc. In aged people, deficiencies of Vitamin B12 are associated with forms of senile dementia, depression, or psychosis (Potter and Orfali, 1993). Vitamin B12 has been shown to prevent pernicious anemia. It is an essential factor for the proper functioning of the brain and nerves. It may prevent neural-tube defects in fetuses during the first six weeks of pregnancy. In adults, it may protect against heart disease and nerve damage.

Vitamin C (ascorbic acid)

Scurvy is a disease resulting from lack of vitamin C in the diet. It was lethal to many early navigators who did not take citrus fruits on board. Vitamin C cannot be synthesized by the body (together with vitamin E and A and β-carotene) and is an essential antioxidant. Vitamin C is a reducing agent that serves as a donor of electrons and has therefore a powerful antioxidant action (see Chapter 3). In addition, it plays an important role in many cell functions such as the synthesis of hormones and neurotransmitters, the detoxification of harmful chemicals, the metabolism of cholesterol, the repair of tissue damage, etc. The RDA is 60 milligrams of vitamin C per day (although many, like Linus Pauling, recommend much higher amounts), and this dose is sufficient to treat and prevent scurvy. Less severe deficiency may cause bleeding of the gums and ecchymosis. Possible adverse effects of high levels of vitamin C (over 5 grams a day) are intestinal gas and loose stools.

Vitamin D

Deficiency causes bone malformation and muscle weakness. The established benefit of vitamin D is the prevention of rickets. It may help prevent osteoporosis (Palmieri et al, 1988). Some scientists believe that vitamin D could be essential in preventing breast cancer and other cancers such as colon (Lointier et al, 1986) and prostate cancer.
An excess of vitamin D causes the buildup of calcium deposits that can interfere with the functioning of muscle or kidney and lead to an increased risk of premature atherosclerosis.

Vitamin E (tocopherol)

Vitamin E deficiency affects the immune system: there is a lower resistance to infection and a reduced antibody and lymphocyte response.

Vitamin E helps prevent retrolental fibroplasia (an eye disorder in premature infants) and anemia. It acts as a powerful antioxidant (vitamin E uptake and metabolism are closely linked to those of selenium) (see below). Adverse effects of high doses of vitamin E (over 1200 IU per day), such as nausea, flatulence, diarrhea, headache, and fatigue, have been reported.

Vitamin K

Vitamin K deficiency causes haemorrhage. It is known to promote blood clotting by participating in the synthesis of coagulation proteins by the liver. It may play a role in prevention of osteoporosis and cancer.

Folic Acid

Folic acid deficiency causes anemia (general pallor) and stomatitis. It is necessary for various metabolic processes, including the synthesis of DNA. Folic acid may protect against cervical dysplasia (precancerous changes in cells of the uterine cervix). Possible benefits of folic acid may be to protect against heart disease, nerve damage, and neural-tube defects.

Recommendations

Guidelines are urgently needed for optimal consumption of vitamins as the amount required may vary, depending on age, sex, genetic predisposition, and life-style. Since 1978, the Federal **Food and Drug Administration (FDA)** in the USA has set the rules and decided on the Recommended Dietary Allowances (RDA), considering teenage boys as the standards. Recently, the FDA has reevaluated these figures to adapt them for the average American adult's nutritional needs. The name is now RDI, or Reference Daily Intake. The doses of vitamins indicated reflect the amount needed from food to maintain satisfactory health, which means the amount necessary to avoid symptoms arising from deficiencies. As discussed in Chapter 3, much higher doses of vitamins are recommended to get protective effects against cancer and cardiovascular diseases. Table 11 lists the amounts recommended for all essential vitamins.

FDA's recommendations for daily requirements of vitamins in the USA are still the subject of discussion and the recommendations given in Table 11 must be regarded as temporary. The 1989 RDI are probably too low for elderly people for vitamin B2, B6, D, and B12, but it is probably appropriate for vitamin C and folate. (Note that many people follow the advice of the late Linus Pauling and take higher amounts of Vitamin C.)

Table 11. Vitamins recommended by the FDA (RDI)

	people under 50		people above 50	
	men	women	men	women
Vitamin A (RE)	1000 μg	800 μg	1000 μg	800 μg
β-carotene	5 mg	5 mg	NA	NA
Vitamin C	60 mg	60 mg	60 mg	60 mg
Vitamin B1	1.5 mg	1.1 mg	1.2 mg	1.0 mg
Vitamin B2	1.7 mg	1.3 mg	1.4 mg	1.2 mg
Pantothenic acid	4–7 mg	4–7 mg	NA	NA
Niacin	19 mg	15 mg	15 mg	13 mg
Vitamin B6	2 mg	1.6 mg	2.0 mg	1.6 mg
Vitamin B12	2 μg	2 μg	2 μg	1.6 μg
Vitamin D	5 μg	5 μg	5 μg	5 μg
Vitamin E	10 mg	8 mg	10 mg	8 mg
Vitamin K	65 μg	65 μg	80 μg	65 μg
Biotin	30–100 μg	30–100μg	30–100μg	30–100 μg
Folic Acid	200 μg	180 μg	200 μg	180 μg

The amount of vitamins indicated are those given by the Food and Nutrition Board, Commission on Life Sciences, National Research council, Recommended dietary allowances. 10th ed., National Academy Press, Washington DC, 1989.
RE: retinol equivalent, 1 RE = 1 μg of retinol or 6 μg of beta carotene; NA: not available; mg: milligrams; μg: micrograms (see also Somer, 1992; Russel and Sutter, 1993).

FDA experts agree that a moderate daily intake of many vitamins will not hurt anybody, but they do not accept the claims that high doses are efficacious in preventing diseases or prolonging life. Moreover, the long-term adverse effects of high-dose vitaminotherapy are not known.

Many nutritionists consider that everyone's basic needs in vitamins are provided by a balanced diet, rich in vegetable and fruit and low in fats. If physicians agree that a proper diet reduces a person's risk of developing many of the common degenerative diseases, such as heart disease, stroke, high blood pressure, cancer, diabetes, osteoporosis, and obesity, they are still debating whether people should take high doses of vitamins to further help prevent these chronic diseases.

Although healthy adults may not necessarily have to take a daily supplement of vitamins, these supplements are highly recommended for people with special conditions, including alcoholics, smokers, and those on restrictive diets, who tend to be poorly nourished. In elderly people, the situation is complex, because they may show biological evidence of deficiency in spite of normal serum concentrations of vitamins. For example, many elderly showed elevated serum concentrations of metabolites that result specifically from a deficiency in vitamin B12, folate, and vitamin B6, while the concentrations of these vitamins were normal. Supplements of all three vitamins reduced the elevated serum levels of the metabolites. Thus, the serum concentration of metabolites was found to be a more sensitive indicator of early vitamin deficiency than serum levels of vitamins (Naurath et al, 1995). However, the consequences of reducing metabolite concentrations to normal values remains to be evaluated in regard to clinical improvement.

Perspectives

In his book, *Prescription for Longevity*, Scala (1992) advises taking vitamin and mineral supplements as a way of ensuring that no essential nutrient is lacking.

Much more information should become available in the next few years, and recommendations on the choice and dosage of each vitamin might be more precise in the future. However, much of this information may continue to reflect an ongoing fight between FDA experts and proponents of large vitamin supplements. The latter are genuinely convinced of the benefits of high dosages, or they may produce or sell these vitamins and so resist, for financial reasons, the attempts of the FDA to control this part of the pharmaceutical market. In many other countries outside the U.S., vitamins are not considered as food supplements, but rather as medications that are subject to the strict regulatory control of health authorities.

Mineral Supplements

Minerals and trace metals play as important a role as vitamins. Many of them act as cofactors necessary for enzymes in the antioxidant systems. The minerals known to have this enzyme-cofactor action are iron, selenium, zinc, manganese, and copper. With the exception of selenium, few therapeutic claims have been made in favor of supplements of minerals at doses above the RDA.

Calcium

Calcium deficiency causes osteoporosis. Calcium is the most abundant mineral in the body and most of it is located in bones and teeth. Sufficient amounts of calcium are needed throughout life but especially during periods of growth, pregnancy, and lactation. Calcium is essential also for blood clotting, muscle contraction, nerve transmission, and for the transport of nutrients across cell membranes. In addition, it is essential for the activity of many enzymes and hormones. The current RDI of 800 milligrams appears to be insufficient to prevent osteoporosis and should be increased to 1000–1200 milligrams. Higher doses are recommended for postmenopausal women who are not taking estrogen. A recent study showed that an amount of calcium equivalent to 1700 milligrams/day and vitamin D (twice the RDI) were necessary to retard mineral loss in women, three to six years after menopause (Aloia et al, 1994). To achieve such an intake, calcium supplements are necessary.

Copper

Copper, like zinc, is required by superoxide dismutases and protects against free radical damage. Its role in the antioxidant system and consequently in inflammation processes explains the popular use of this metal in arthritis. An adequate daily intake of copper is 1.5–3 milligrams for adults. As an excess of zinc seems to interfere with copper absorption, many doctors advise their patients to take copper supplements when zinc is prescribed. A recommended ratio of zinc to copper is ten to one.

Iodine

Deficiency in iodine leads to the endocrine disease known as hypothyroidism and to the development of goiter. More than half of iodine in the body is concentrated in the thyroid gland where it is a component in the synthesis of thyroid hormones. These hormones play an important role in metabolism, growth, and reproduction and in the production of energy from food and, as such, are a regulator of body weight. Iodine is also important for nerve cell functions. The RDI for iodine is 150 micrograms. Foods like raw cabbage, peanuts, and cauliflower are inhibitors of the thyroid gland and must be consumed with caution, if the diet is otherwise low in iodine. Doses higher than 25 times the RDI can produce an enlarged thyroid gland similar to goiter seen in iodine deficiency.

Iron

Iron deficiency is accompanied by impaired immune functions, anemia, fatigue, and weakness. The highest concentration of iron is found in the hemoglobin of red blood cells where it plays an important role in the supply of oxygen to tissues. Iron functions as cofactor for the antioxidant enzyme catalase. It is also needed by the immune system.

Iron is beneficial when bound to a protein structure, but as an unbound molecule (free iron), it is potentially harmful because it can generate free radicals. Iron has no side effects at doses 3 to 4 times the RDI. At

200 milligrams a day, nausea, cramping, constipation, or diarrhea can occur.

Magnesium

Deficiencies in magnesium affect all tissues, notably the heart, the kidneys, and nerves. Magnesium is abundant in all tissues. It participates in the conversion of carbohydrates, fats, and proteins to energy. It plays an important role in muscle contraction and relaxation in association with calcium: cacium stimulates, while magnesium relaxes muscles. It is also necessary for nerve transmission. High doses of magnesium might impair absorption and use of calcium. Toxic levels could build up in patients with kidney failure. The symptoms of magnesium toxicity are weakness and difficulty breathing.

Manganese

Manganese deficiencies are reported to affect antibody response and activity of several immune cells. Manganese is needed by the nervous system for the synthesis of the neurotransmitter, dopamine. It also plays a role in the antioxidant system, which implies that an absence of sufficient manganese could play a role in cancer and other degenerative diseases. Lung diseases and problems of the central nervous system have been reported in workers exposed to industrial dust and manganese intoxication. An adequate daily intake is estimated to be 2 to 5 milligrams for adults.

Phosphorus

Phosphorus is found in all foods of plant or animal origin, and deficiencies are not described in humans. Phosphorus is the second most abundant mineral in the body after calcium and is found mostly in bones and teeth. It is a component of all soft tissues and is essential to growth, reproduction, metabolism, and repair of body tissues. It helps activate the B vitamins. The recommended intake of phosphorus corresponds to the amount recommended for calcium, or approximately 1000 milligrams. Excessive intake of phosphorus occurs in people who consume diets rich in meat, soft drinks, or diets low in calcium. This results in an imbalance in the ratio of calcium to phosphorus and might contribute to osteoporosis. Compounds related to phosphorus have been found to increase bone mass. A recent study reported positive effects of a new biphosphonate on bone mineral density and the incidence of fractures in postmenopausal women (Liberman et al, 1995). Biphosphonates are synthetic analogues of inorganic pyrophosphate, an endogenous regulator of bone turnover that inhibits bone resorption in vitro. It is the unique structure of the side chains of this new molecule that inhibits osteoclast-mediated bone resorption. The advantage of biphosphonates is that they have a high affinity for hydroxylapatite and are resistant to metabolism by endogenous phosphatases.

Selenium

Deficiencies are manifested in animals by weight loss, hepatic problems, cardiac and skeletal muscle degeneration. Selenium is part of the important antioxidant glutathione enzyme system. Glutathione breaks down hydrogen peroxides and scavenges lipid peroxidases, thereby protecting cell membranes and DNA from oxidative damage. Intake of selenium increases the activity of glutathione peroxidase. Selenium might protect against cancer, stimulate the immune system, and protect the heart from oxidative damage. Doses above 4 milligrams of selenium per day can be toxic and cause fingernail thickening, hair loss, and fatigue.

Zinc

Deficiencies in zinc result in slow healing, impaired taste, smell, or vision, loss of appetite, and susceptibility to infection. It can also impair fertility in men. It should be emphasized that several diseases, as well as alcohol, steroids, oral contraceptives, and diuretics interfere with zinc absorption. Many enzymes such as DNA and RNA polymerases as well as superoxide dismutases require zinc to function. Zinc is needed also to maintain the structure and function of cell membranes. The relation of zinc to the antioxidant system could explain its role in controlling immunity, inflammation, and in preventing cancer. Zinc is highly concentrated in the eye, and supplements of zinc could prevent macular degeneration in elderly people. Excess of zinc (2 grams per day and more) causes nausea and vomiting.

Recommendations

Attention should be paid when taking supplements of any mineral, because ingestion of one mineral can affect the balance of the other minerals. It is therefore important to follow the advice of physicians before taking mineral supplements. Table 12 lists the recommended daily intake for each mineral.

Table 12. Recommended daily intake of minerals

	men	women
Calcium	1000–1500 mg	1000–1500 mg
Copper	1.5–3 mg	1.5–3 mg
Iodine	150 μg	150 μg
Iron	10 mg	15 mg
Magnesium	350 mg	280 mg
Manganese	2.5–5 mg	2.5–5 mg
Phosphorus	1000 mg	1000 mg
Selenium	70 μg	55 μg
Zinc	15 mg	12 mg

N.B.: mg: milligrams; μg: micrograms.
The amount of minerals are those indicated by the National Research council. (Recommended dietary allowance. 10th ed., National Academy Press, Washington DC, 1989; see also Somer, 1992.)

Perspectives

As is the case with vitamins, there is no precise information on the long-term effects of a moderate daily supplement of minerals, with the exception of calcium and selenium. Selenium is a popular mineral because of its antioxidant action (see Chapter 3).

Unsaturated Fats, Omega-3 and Omega-6 Fatty Acids

The lower incidence of coronary heart disease in Greenland Eskimos has long puzzled the medical community, and it has been attributed to a high consumption of fish. In the last twenty years, many studies have been carried out to find out what specific compound or compounds in fish have vascular and cardioprotective effects. It turns out that this beneficial effect is due to the large amounts of polyunsaturated fatty acids contained in fish, especially the alpha-**linoleic acid**, called omega-3 (ω-3) and gamma-linoleic acid, or omega-6 (ω-6). A normal diet contains three types of fats: hydrogenated fats (also called saturated oils), polyunsaturated, and monounsaturated oils. The process of hydrogenation is the addition of hydrogen atoms to fat molecules; this alters their structure and makes them dangerous to health. Hydrogenated or saturated fats are easy to recognize because they become hard and opaque at cold temperatures. They have been implicated in cholesterol build-up, heart disease, and cancer. Polyunsaturated oils stay transparent when refrigerated. The problem with polyunsaturated oils is that they are prone to oxidation, which gives rise to free radicals.

Omega-3 and ω-6 are called essential fatty acids, because they are the precursors of other lipid molecules in the body. A deficiency in essential fatty acids is thought to be a cause of chronic degenerative diseases, such as cancer, heart disease, high blood pressure, and stroke. It is also thought to have detrimental repercussions on mental functions, since 60% of the brain is made up of lipids. The respective roles of ω-3 and ω-6 and the role of the balance between ω-3 and ω-6 (Debry and Pelletier, 1991) are more complex than previously thought and have not yet been clearly defined. The equilibrium between the levels of essential fatty acids, the level of enzymes known to desaturate fatty acids and the rate of peroxidation (process leading to the production of free radicals), which is correlated to the degree of unsaturation of fatty acids, is subtle and may change with age.

Intense investigations have been carried out to evaluate the beneficial effects of ω-3 supplementation in the diet. An epidemiological study on aging Nordic populations showed that the plasma levels of arachidonic acid (AA), a cellular membrane lipid with proinflammatory, proaggregatory, and vasoconstrictive properties, were reduced by increasing fish consumption; mortality due to cardiovascular diseases dropped correspondingly (Gudbjarnason et al, 1991).

The effects of a daily low dose of ω-3 fatty acid on cytokine production and lymphocyte proliferation were analyzed in young (23–33 years) and older women (51–68 years) (Meydani et al, 1990b). In both groups of women, inducible production of IL-1β, tumor necrosis factor, IL-6, and IL-2 was suppressed, but the suppression was more dramatic in older women who also showed significantly reduced mitogenic responses to phytoheamaglutinin (PHA). The effects of ω-3 supplementation on IL-1, IL-6, and tumor necrosis factor can have important implications in several diseases: these cytokines play a role in the pathogenesis of inflammatory diseases. IL-1 and IL-6 have been implicated in the pathogenesis of osteoporosis by causing bone resorption (see Part I, Chapter 3; Gowen and Mundi, 1986). Supplementation with ω-3 could also be important in rheumatoid arthritis and psoriasis because ω-3 decreases prostaglandin E_2 (PGE_2) production (Vischer, 1992; Kremer et al, 1987). On the other hand, the decrease by ω-3 of IL-2 production and PHA mitogenesis in older women is disturbing because antigen and mitogen-stimulated IL-2 production has been reported to decline with age and to be a contributory factor to the well-documented decrease of T-cell function with aging (Meydani et al, 1990b). The effect of ω-3 on IL-2, and by extension on T cells, is undesirable and might counteract the other positive effects of ω-3. Although the reduction in cytokine production by ω-3 may have beneficial anti-inflammatory effects, the suppression of IL-2 production and lymphocyte proliferation may compromise cell-mediated immunity in older subjects and lead to an increased incidence of infectious diseases and tumors.

Recommendations

Omega-3 and ω-6 polyunsaturated fatty acids are believed to have more beneficial than harmful effects in humans, including the elderly. It is well recognized now that diets high in cholesterol and saturated fat increase LDL-cholesterol and triglyceride levels (Goldberg and Schonfeld, 1985). Moreover, pathological conditions such as diabetes mellitus, end-stage renal disease, as well as obesity and physical inactivity cause hypertriglyceridemia and low HDL levels (Brunzell, 1984).

To keep HDLs up and LDLs down, some experts suggest switching from saturated fats found in dairy products and in palm and coconut oils, to monounsaturated fats found in oils made from olives, almonds, sunflowers, and rapeseed. Good sources of unsaturated oils that provide essential fatty acids are flaxseed, pumpkin, wheat germ, soybean, and walnut oils. Fish oils in general may play an important role in the prevention of heart diseases, high blood pressure, stroke, and even cancer. It is therefore important to watch one's diet carefully, avoid saturated fats (contained in meat), and include sufficient amounts of fish oils in the diet (Fahrer et al, 1991). Fish, like salmon, sardines, and tuna are good sources of ω-3. They should be kept in the refrigerator and eaten fresh or lightly cooked, because ω-3 and ω-6 are destroyed by heat, light, and exposure to oxygen.

Perspectives

Further research is necessary to understand the precise mechanisms by which diet (together with exercise or weight loss) alter lipoprotein lipid profiles in young and elderly people. Moreover, data on the status of fatty acid distribution in the elderly are lacking, which makes precise recommendations concerning supplementation with polyunsaturated fatty acids more difficult (one study indicates that a diet rich in fish oils and soybean oils has a beneficial influence on plasma lipids in old rats (Suzuki et al, 1985). The roles of lipoprotein lipase, hepatic lipase, and lecithin cholesterol acyl transferase (LCAT) in the regulation of HDL and triglyceride metabolism have not been studied in this age group. Similarly, the role of the liver and intestine on lipoprotein synthesis and clearance in the elderly has not been investigated. Whether aging *per se*, independent of life styles, alters LDL receptor or nonreceptor lipoprotein processing of cholesterol is still an unsolved question.

Chapter 3

Antioxidants and the Chemoprevention of Cancers and Cardiovascular Diseases

Vitamins, Minerals, Coenzyme Q10

There is increasing evidence that oxidative processes, in particular free radical damage, play a significant role in a number of chronic diseases such as cardiovascular or inflammatory diseases, cataract, and cancer. As was discussed in the first part of this book, tissue damage can be induced directly by free radicals from exogenous sources, such as air pollutants and tobacco smoke, irradiation, metabolites of certain solvents, or drugs. Tissue damage can also be induced through increased oxidative stress from endogenous sources. An example is the damage caused by the resorption of blood after an episode of ischemia.

Practical measures can be implemented to lower the sources of exogenous free radicals. However, the fight against endogenous free radicals is more difficult, because endogenous radicals are inherent in the aerobic metabolism process of living organisms, and because the exact nature of the chain of events that amplify the initial damage by free radicals is still somewhat mysterious. For example, oxidation of LDL-cholesterol is directly related to the development of atherosclerosis. In the process of cleaning up the oxidized form of LDL, macrophages are poisoned and killed. Dead macrophages then pile up along the endothelium of blood vessels, forming one of the elements of atherosclerotic plaques. While the general process is known, the molecular mechanisms of these alterations have not been determined in sufficient details to establish a truly preventive pharmacological treatment.

Fortunately, free radicals are eliminated by endogenous and exogenous antioxidants. Antioxidants, or free radical scavengers, combine with free radicals to turn them into harmless chemicals. There are numerous endogenous antioxidants that constitute the body's own natural defenses and limit free radical damage. Some are enzymes like superoxide dismutase and glutathione peroxidase that combine with oxygen radicals to form hydrogen peroxides. Phospholipases clean up free radical damage in cell membranes, and nucleases and glycolases help repair free radical damage to DNA. Others are chemicals such as uric acid, ascorbic acid (vitamin C), glutathione (a three-amino-acid peptide), as well as amino acids containing sulphur in the -SH form, specifically cysteine and methionine. Bilirubin, the waste product of hemoglobin, is a strong free radical scavenger that increases the activity of β-carotene in some tissues. Melatonin is also a free radical scavenger. In addition to these molecules, minerals also play a role in antioxidation reactions, even though they are not themselves direct antioxidants but rather cofactors that attach to and activate the endogenous antioxidant enzymes. The above endogenous antioxidants are certainly highly protective for the organism. An indirect proof of their biological efficiency is given by diseases caused by their deficiencies; for example, the familial form of amyotrophic lateral sclerosis (ALS) is accompanied by mutations in the enzyme superoxide dismutase (SOD1).

The diversity of endogenous antioxidants suggests that there are several ways by which the natural capacity to fight oxidative damage could be potentiated. At present, there is no way to fight oxidation by increasing the levels of endogenous antioxidant enzymes, despite the claim that oral formulation of superoxide dismutase is beneficial (this enzyme is a protein and is destroyed rapidly by digestive enzymes from the gut). The body does not depend solely on its own resources to control oxidative damage; it is helped by antioxidants and minerals from the diet and also by man-made antioxidants that are active after oral administration. The antioxidative actions of β-carotene, vitamin C, vitamin E, quinones, and selenium are discussed below.

Beta-carotene

β-carotene, one of many carotenoids, is a precursor of vitamin A. It is present mostly in unchanged form in the body and acts as a free radical scavenger. As early as 20 years ago, an association between a diet poor in carotenoids and an increased risk of lung cancer was reported in Norway (Kvale et al, 1975). This finding was confirmed by studies on a total of several hundred thousand people in the following years. Apart from these studies based on questionnaires about individuals' diets, low serum levels of β-carotene are also associated statistically with a higher incidence of several cancers, in particular those of the lung (Stähelin et al, 1991), pancreas, and stomach (Comstock et al, 1992).

Whether exogenous β-carotene administration might protect against the occurrence of cancer is a question that has been addressed by many research groups. One of the studies was carried out on about 30,000 healthy Chinese adults from the Linxian area, where the incidence of esophageal and gastric cancer is high. Participants took one of four different combinations of vitamins and minerals (vitamin A and zinc; riboflavin and niacin; vitamin C and molybdenum; β-carotene, vitamin E, and selenium) or a **placebo** for more than five years. Those who took the combination of β-carotene, vitamin E, and selenium had a 13% lower mortality due to cancer (Blot et al, 1993). Unfortunately, these encouraging results were not confirmed in a Finnish study on the prevention of lung cancer in smokers; subjects used β-carotene and vitamin E, either separately or together, for a period of five to eight years (The α-tocopherol, β-carotene cancer prevention study group, 1994).

In the area of precancerous lesion of the mouth (leukoplakia), the results of studies using β-carotene, alone or in combination with vitamin A, have also been interesting but conflicting. While some studies showed protective effects of β-carotene alone and with vitamin A, with some cases of remission of the lesion, negative results were also found. So, the relative roles of β-carotene versus vitamin A need to be evaluated further. Synthetic retinoids such as tretinoin, isotretinoin, or fenretinide seem to have a much greater potential as chemoprevention agents (for a review, see Szarka et al, 1994) in subjects who have leukoplakia or a first cancer of the lungs or of other sites. For example, in patients who had a head or neck cancer, isotretinoin decreased the risk of having a second cancer for five years: more than 90% of the patients were alive with no second cancers under active treatment while this was the case for only 60% of those who had taken placebos (Hong et al, 1990).

At the present time, several large-scale studies are being conducted. In fact, β-carotene and other retinoids are the most promising compounds for chemoprevention of cancers. They are tested either in the general population, in those at risk such as smokers, or in those who already had a primary tumor. These trials are carried out on large populations and for extensive periods of several years. Their complexity and cost are impressive. The β-carotene and retinol efficacy trial (CARET) for chemoprevention of lung cancer in high-risk population, smokers and asbestos-exposed workers, illustrates this. Pilot studies were run from 1985 to 1988, and, as of April 1993, 4000 asbestos-exposed subjects and 11,105 heavy smokers had been recruited. The trial will continue until early 1998, at which point the benefits of β-carotene and retinol treatment will be analyzed. Scientists who launched the study estimate that if β-carotene and retinol can lead to a 23% reduction in lung cancer incidence in CARET, such a result would extrapolate, if supplements with this vitamin were generalized, to saving 34,000 lives annually in the United States (Omenn et al, 1994).

While the only side effect noted with β-carotene, the natural compound, is a yellowing of the skin, vitamin A can have serious side effects, and the synthetic retinoids show a significant toxicity. Clinical trials on synthetic retinoids are being carried out. Fenretinide, for example, might be a promising drug, particularly for breast cancer patients, since it accumulates preferentially in the breast and seems to have side effects that can be tolerated (Costa et al, 1994). Lower doses of synthetic retinoids induce fewer side effects, but the protection against cancer is not as good. When chemoprevention is interrupted, the protective effect is lost. Issues about the clinical usefulness and best tolerated doses of the different carotenoids are still being argued. Their applications in the primary or secondary prevention of cancers of different sites are still controversial, but ongoing trials should bring some important answers in the coming years.

Vitamin C

Vitamin C may help reduce the risk of cancer and heart disease and retard macular degeneration in the elderly. Epidemiological studies show that vitamin C intake (mostly from fruits and vegetables) might protect against throat, mouth, stomach, pancreas, cervix, and breast cancers (Block, 1991). Some studies show that to obtain a measure of protection against cataracts or cancer, as much as 1000 milligrams of vitamin C a day may be required, which is substantially higher than the average daily requirements of 60 milligrams per day and would represent the equivalent of about 20 oranges per day. The late Nobel Prize winner, Linus Pauling, recommended a daily dose of up to 5 grams of vitamin C a day to avoid cancer (he himself was taking even higher doses).

There are many epidemiological studies on the incidence of cancer in relation to the intake of fruits and vegetables, the major source of vitamin C. In their review, Block and his collaborators (1992) reported that 128 out of 156 studies found statistically reduced risks of various cancers in people consuming diets rich in vitamin C. In another meta-analysis, aimed at evaluating the effect of diet on breast cancer, the same inverse relation has been noted in 12 studies (Howe et al, 1990). Other studies show that vitamin C protects against mouth and esophagus cancers. These studies indicate that by increasing the consumption of food rich in vitamin C, important health benefits could be achieved.

Studies in animals have shown some protection against certain cancers using vitamin C, but results are less favorable than with vitamin E or β-carotene.

Vitamin E

Vitamin E, one of the most potent antioxidants, is referred to as a chain-breaking antioxidant because it blocks the chain reaction process that propagates the peroxidation cascade along cell, mitochondrial, or nuclear membranes. Vitamin E appears to be especially important in protecting membrane phospholipids from free radical attack, possibly by interacting with the polyunsaturated fatty acids of the phospholipids (Kelly, 1988). The protective effects of vitamin E have been extensively studied. Some relevant studies are cited below.

Vitamin E helps make oxygen use more efficient in the brain, heart, and other organs (Gaby and Machlin, 1991). Vitamin E has been shown to have stimulatory effects on the immune system: it lowers the production of prostaglandins in immunocompetent cells and increases **cell-mediated immunity** in young and old animals. In a double-blind study, a significant improvement of the delayed-type hypersensitivity reaction was observed in patients over 60 years, who were given vitamin E supplements (Meydani et al, 1990a). Vitamin E has been shown to protect against cancers of the dietary tract, possibly by preventing common food preservatives like nitrates and nitrites from turning into cancer-causing **nitrosamines** (Buiatti et al, 1990). Reports show that 400 IU of vitamin E every day (associated with 300–600 milligrams of vitamin C) can prevent the development of cataracts in up to 70 percent of susceptible patients, compared with people who did not take the vitamins (Jacques et al, 1988; Robertson et al, 1989). 3200 IU of vitamin E a day (associated with 3000 milligrams of vitamin C) given as a preventive measure to parkinsonians early in the disease delayed the need for L-Dopa in these patients for as long as two-and-a-half years (Fahn, 1992).

Vitamin E and β-carotene have beneficial effects on atherosclerotic plaques. The first step in the oxidation of LDL in endothelium membranes is partially prevented by the free radical scavengers, vitamin E and β-carotene (Esterbauer et al, 1991). Similar effects have been observed in vivo. The resistance to oxidation of LDL is increased by vitamin E supplementation, but the efficacy of vitamin E in protecting LDL varies from person to person (Dieber-Rotheneder et al, 1991). In their recent convincing studies, Willett and his collaborators have presented data suggesting that vitamin E can help prevent heart disease (Rimm et al, 1993; Stämpfer et al, 1993). The researchers followed about 120,000 middle-aged men and women for up to eight years and found that those taking daily supplements of at least 100 IU of vitamin E reduced their risk of heart disease by 40 percent. The role of vitamin E in cardiovascular diseases might be much greater than generally acknowledged. According to the results from a multisite longitudinal study (the WHO/MONICA Project) conducted in Europe, countries where men had low median vitamin E concentration showed a mortality rate of 400 to 500 per 100,000 inhabitants, while in countries where vitamin E concentration was high, it was less than 100 (Gey et al, 1993). This would place low vitamin E levels at the top of the list of risk factors for cardiovascular disorders (and resulting deaths), whether myocardial infarction or stroke. Stated more technically, vitamin E accounted for 63% of the difference in ischemic heart disease mortality between countries (stepwise and multiple regression analysis of 16 populations), while total cholesterol accounted for only 17%. Low vitamin A accounted for an extra 4%, i.e., it had a minimal influence.

Selenium

Once thought to be harmful, selenium is now recognized as an essential nutrient that seems to offer protection against cancer and other diseases, to stimulate the immune system, and to have anti-inflammatory properties. Several mechanisms might explain the beneficial effects of selenium on the incidence of cancer. The most frequently quoted is activation of the selenium-dependent glutathione peroxidase, an enzyme that controls damage from free radicals, but effects of selenium on immune responses, apoptosis, or carcinogen metabolism have also been mentioned.

Selenium can be taken as organic or inorganic compounds or salts. They all seem to be effective in protecting against experimental cancers. In laboratory animals, selenium is highly protective against cancers induced by carcinogens that affect different organs. Conversely, animals fed with diets poor in selenium and vitamin E develop early cancers more frequently. In humans, several epidemiological studies have shown that cancer patients have low levels of selenium. In 1969, Shamberger and Frost suggested that cancer was more frequent in areas of the United States where the intake of selenium was low. Other researchers, in the United States and other countries, reached the same conclusion, especially for cancers in men. However, a few studies did not find such a relation. For example, the content of selenium in toenail (taken as a surrogate for tissue content) was analyzed in a cohort of more than 60,000 nurses followed prospectively between 1982 and 1986 (Garland et al, 1995). It was found to be unrelated to the occurrence of cancer (937 cases, including breast cancer) diagnosed during 41 months. In fact, a small positive association between selenium levels and the occurrence of cancer was observed. Unfortunately, the study does not take into account the effects of other variables that might modify the content of selenium in toenails, such as age, smoking, or the consumption of alcohol. These epidemiological studies illustrate the complexity of the research on diet in relation to cancer, and the necessity to rely on more than a single epidemiological study. Large-scale trials to test the role of selenium as the sole agent in cancer prevention have not been published yet, and according to Szarka and collaborators (1994), none were being conducted as of 1994.

Selenium also has been implicated in the epidemiology of cardiovascular disorders. In a prospective study of more that 3000 Danish men, those with low selenium concentration in serum had a 70% increased risk of ischemic heart disease during the next 3 years. Selenium values were lower in smokers, older men, those with little physical activity, or men from lower social classes, but the authors noted that the association was independent of these and other cardiovascular risk factors. It is noteworthy that men from the lowest social class had a relative risk of ischemic heart disease that was three times higher that those from the highest social class, and this difference was not explained by the better selenium status of richer men (Suadicani et al, 1992). One explanation for the inconsistencies concerning selenium levels and human cancers or ischemic diseases might be that the relation between selenium levels in blood and the activity of glutathione peroxidase is linear up to a threshold value at which a further increase in selenium is not accompanied by an improvement in glutathione peroxidase.

With the exception of some northern European countries and areas of China, people of most countries have adequate blood levels of selenium, i.e., around 100 to 150 micrograms/liter. Supplementation with selenium might thus be useful only for subjects with levels below 50 micrograms/liter. This, together with the above epidemiological data, does not justify the large consumption of selenium in the Western world.

Coenzyme Q10

Coenzyme Q10 (CoQ10) or ubiquinone is one of the coenzymes synthesized by all cells. It is essential in enzymatic reactions that produce energy from carbohydrates. Without it, the chain of cellular energy is broken. CoQ10 is an integral part of mitochondrial membranes where it is involved in the synthesis of ATP, the basic energy molecule of cells. CoQ10 has a structure similar to that of vitamin E, but it is not a vitamin and no deficiencies have been identified. This compound, synthesized by all cells in the body, is a potent antioxidant useful in preventing free radical damage.

CoQ10 may be a promising drug in the prevention of several diseases and may have a place in the treatment of congestive heart failure and angina. At the recommended doses (10 to 30 milligrams per day), it is safe, with no reported toxicity. It has been prescribed in higher doses (up to 100 milligrams per day), with apparent success, to patients with declining **cardiac outputs** (Bliznakov and Hunt, 1986). CoQ10 has been reported to have positive effects in patients with periodontal diseases or gingivitis. It could have a beneficial role in diseases involving the immune system, such as allergies and asthma, lupus erythematosus, multiple sclerosis, diabetes, and possibly neurodegenerative diseases. CoQ10 has been reported to have significant life-enhancing effects in mice (Bliznakov, 1981). However, its role in extending life in humans and postponing or treating diseases remains to be shown.

Melatonin

The antioxidant properties of melatonin have been described in a number of scientific publications, and the interest in this molecule is increasing in the medical community. The roles and possible benefits of melatonin are discussed further in Chapter 5.

Recommendations

Null and Feldman (1993) introduced the concept of antioxidant fitness, which they refer to as the ability of the body to neutralize free radicals. By understanding the role played by antioxidant defenses and by learning how to support the body in its defense against free radicals, people might increase their health potential, even though there is little evidence at present that exogenous antioxidants taken as diet supplements increase life span.

Table 13 proposes a list of antioxidant supplements that may be beneficial particularly for people with certain clinical disorders and people who are physically very active.

In spite of the theoretical importance of antioxidant supplementation, there are still few experimental studies that report their protective effects for healthy adults. Warning about using megadoses of antioxidants should be given. For example, in the presence of iron, vitamin C acts as a potent prooxidant. Similarly, β-carotene can serve as prooxidant at high oxygen partial pressures and can therefore be dangerous for exercising subjects if ingested in high doses (Kanter, 1994).

Before taking any drug or vitamins and minerals mentioned in this (and other sections), it is strongly recommended to consult with a physician.

Perspectives

The reported relation between maximum life span and the level of DNA repair (Hall et al, 1984), and between life span and plasma urate levels in

Table 13. Suggested daily intake of antioxidants

	Men and women
Beta carotene	30–180 mg
Vitamin C	150–1000 mg
Vitamin E	50 mg–1.5 g
Selenium	50–200 µg
N-acetyl-L-Cystein	300 mg
Coenzyme Q10	10–300 mg

N.B.: the range of doses indicated corresponds to those recommended by different authors. Many researchers and clinicians suggest using the higher doses to preserve optimum health (see also Helzlsouer et al, 1994).

primates (Cutler, 1985; Ames et al, 1981) suggests that antioxidant mechanisms are important factors in longevity. The hypothesis is that if the protective system against oxidative damage breaks down, cellular metabolism is disrupted and processes leading to the destruction of the cell may start. This means that if the process of free radical generation is controlled by appropriate protective mechanisms, organisms should survive longer. A first step in this direction would be to set up means to evaluate the antioxidant status of an individual. This would give a better assessment of the role of antioxidants as protective agents.

The burden of proof that exogenous antioxidant compounds have a protective effect against clinical consequences of free radical damage is still ahead, although recent studies bring support to some beneficial medical effects of antioxidants. There is suggestive evidence that low levels of plasma antioxidant vitamins are associated with an increased risk of subsequent cancer mortality (Stähelin et al, 1991) and that supplementary antioxidant vitamin intake reduces the progression of coronary artery atherosclerosis (Hodis et al, 1995). As most antioxidants are well-known natural substances, they cannot be patented. Thus, it is difficult to get the pharmaceutical industry interested in financing the expensive trials that might demonstrate the efficacy of antioxidants in preventing the occurrence of diseases and delaying death.

Aspirin

Acetylsalycilic acid, or aspirin, is a nonsteroidal anti-inflammatory drug (NSAID) that decreases the tendency of blood platelets to aggregate within vessels. This explains why it affords protection against the recurrence of cardiovascular and cerebrovascular accidents, when these are due to atherosclerosis. This beneficial effect is observed for myocardial infarction, transient ischemic attacks, unstable angina, and stroke (Solomon and Hart, 1994). However, the protection is not complete: in an analysis of 145 randomized trials on 100,000 subjects, comparing antiplatelet therapy to placebo, the reduction in vascular events was in the order of one quarter in high-risk subjects (Antiplatelet Trialists' Collaboration, 1994). In the primary prevention of vascular events, aspirin was also beneficial for men who had never suffered from a myocardial infarction, and the US Preventive Services Task Force now recommends low doses of aspirin every other day for asymptomatic men over 50 years of age for the prevention of myocardial infarction (Couch, 1993). Coronary heart disease is an equally important cause of death in women, but the data on aspirin for primary prevention in women are limited. Woods (1994) suggested that only women with high risk of coronary artery disease should take aspirin. It is critical to carry out more randomized trials of aspirin in women, because the risk-benefit ratio may be different in women and in men; in particular, the ratio of the incidence of stroke to that of myocardial infarction is higher in women than in men (Women's health study research group, 1992; Rich-Edwards et al, 1995). Aspirin does not seem to have efficacy in primary prevention of stroke, although it does prevent recurrence of such events (Solomon and Hart, 1994).

Antiplatelet primary or secondary prevention of vascular events has been studied formally in probably more than 200,000 subjects. Despite these impressive research efforts, the issues of aspirin dose, of associated side effects, and of efficacy of alternative medications are still to be clarified. The usual dose of aspirin of 250 to 750 milligrams per day, and also low doses of 30 to 100 milligrams per day administered even on alternate days, have a protective effect and selectively inhibit thromboxane A2 synthesis. In fact, it might be that low doses have a more consistent effect on platelet adhesiveness. Unfortunately, severe hemorrhagic side effects do occur even at low doses. Concerning the issue of alternative medications, the following questions have been explored. The addition of dipyridamole to aspirin potentiates the efficacy of aspirin with a 30% to 50% decrease in the risk of recurrence of vascular events in particular strokes. Another antiplatelet drug, ticlopidine, might be more efficacious than aspirin in the secondary prevention of strokes, but with a risk of neutropenia, i.e., a decrease in the number of white blood cells. When strokes occur in subjects who have auricular fibrillation, a frequent form of cardiac rhythm irregularity, the prescription of oral anticoagulant drugs, such as warfarin, seems more efficacious than aspirin, with up to a 70% reduction in the risk of future strokes, and only moderate increase of severe bleeding over aspirin (Couch, 1993; Solomon and Hart, 1994).

Aspirin might have other protective effects. There are epidemiological indications that NSAIDS protect against colorectal cancer. There are also suggestions that the incidence of Alzheimer's disease is less in subjects who have consumed aspirin or other NSAIDS throughout their life. Further trials on large numbers of subjects, specifically on people over 70 years of age, followed over many years, are still needed.

The above results have led many people to start taking aspirin regularly; the risk is digestive bleeding, in particular gastric ulcers. Some consider that the risk is such that aspirin should not be used in the primary prevention of vascular events in subjects who have a low risk of suffering from cardiovascular diseases. Smokers over 50 years with familial history of hypercholesterolemia should take aspirin. Before deciding on taking aspirin, it is wise to ask the advice of a physician.

Chapter 4
Life Styles

Several modifications in our life style turn out to be efficacious in improving health and, consequently, prolonging life: eating a healthy diet; counteracting free radicals; keeping physically active; avoiding stress, cigarette smoking and excessive alcohol consumption; and perhaps, above all, keeping a positive attitude in life. These various measures are discussed below.

Social Network

Social interactions have a strong influence on health and longevity (see Part I, Chapter 2). Unfortunately, aging is often accompanied by a decrease is social interactions. This may be due to our own behavioral changes with aging, the loss of friends and relatives, retirement, etc, but it is also the fault of society as a whole. In 1972, Simone de Beauvoir wrote in her book, *The Coming of Age*: "We are told that retirement is the time of freedom and leisure: poets have sung the 'delights of reaching port'. These are shameless lies. Society inflicts so wretched a standard of living upon the vast majority of old people that it is almost tautological to say 'old and poor': again, most exceedingly poor people are old. Leisure does not open up new possibilities for the retired man; just when he is at last set free from compulsion and restraint, the means of making use of this liberty are taken from him. He is condemned to stagnate in boredom and loneliness, a mere throw-out. The fact that for the last fifteen or twenty years of his life a man should be no more than a reject, a piece of scrap, reveals the failure of our civilization; if we were to look upon the old as human beings, with a human life behind them, and not as so many walking corpses, this obvious truth would move us profoundly. Those who condemn the maiming, crippling system in which we live should expose this scandal. It is by concentrating one's efforts upon the fate of the most unfortunate, the worst-used of all, that one can successfully shake a society to its foundations. In order to destroy the caste system, Ghandi tackled the status of the pariahs; in order to destroy the feudal family, Communist China liberated the women. Insisting that men should remain men during the last years of their life would imply a total upheaval of our society . . . And it is this old age that makes it clear that everything has to be reconsidered, recast from the very beginning. That is why the whole problem is so carefully passed over in silence; and that is why this silence has to be shattered. I call upon my readers to help me in doing so."

Simone de Beauvoir's statements are a harsh description of the fate of old people, a revolt against the progressive social isolation into which they often fall. The situation she describes is still prevalent in many western countries, and negative attitudes toward the elderly are alarming. Robert Butler, a recognized authority on aging, cited the major concerns expressed in many countries: "Can we afford the growing numbers of older persons? Must these growing numbers lead to intergenerational conflict? Will there be stagnation in the productivity of society as a result of the growing numbers of older persons?" (Marwick, 1995). In spite of these discouraging statements, there are hints of changes in public awareness. Statesmen and the medical community are more and more concerned with the fate of elderly people, and they are looking for practical solutions to keep them active and more involved in public affairs. In some countries, proposals to create central agencies have been made. The purpose of these agencies would be to gather information on all kinds of tasks that could be executed by elderly people and to establish contacts between companies, offices, and elderly people who wish to participate.

Recommendations

There is no doubt that elderly people have to be socially involved. Retired people should be encouraged to indulge in their hobbies, whether these include keeping busy in their former professional activities or finding new areas of interest. They should keep active, see their friends, travel in groups, etc. Discussions with young volunteers who visit elderly people on a regular basis emphasize the fact that many are courageous, generous, and loving. They are curious about what happens in the world, they want to keep active, and, most importantly, they have a good self-image.

It is important that the practical and intellectual experience that people have acquired during their professional life be useful to society. Many men and women who occupied leading positions in state affairs and in companies or universities do not wish to stop working entirely at the time of their retirement. Special measures should be implemented so that they could act as consultants or keep some responsibilities in their areas of expertise, and thus continue their activities on a part-time basis. Elderly people, should strive to keep their independence and should consider themselves as part of society. This is not an attitude they should adopt only when they reach their sixties or seventies. It should be the continuation of behaviors adopted earlier. A prerequisite for living to old age is to live in a society that is structured, a society that recognizes its history and traditions and respects its elderly people. Religions, the history of nations, stories of the past told by parents, help maintain memories and keep rituals alive. Family gatherings, feasts that unite family members and friends are social events that allow elderly people to keep their role in the society.

Recommendations should not be addressed only to elderly people, but also to society as a whole, and to those who will be the elderly of tomorrow. Younger people should change their preconceived and possibly negative perception of elderly people. Often, they are so concerned with their own lives that they neglect giving elderly people an opportunity to express themselves. They should be aware of the need to offer aged persons the chance to volunteer and help in community affairs.

Perspectives

The increase in the elderly population has been termed a longevity revolution. This revolution will be even more dramatic when the millions of baby-boomers born between 1946 and 1964 turn 65 years of age. The urgent need to implement drastic changes in social policy is obvious. There is progress: an association of international leadership centers on longevity and society has been created with the goal of developing humane and effective policies for government and private organizations to answer the needs of an expanding aging population.

Are Aging and Happiness Compatible?

In a remarkable analysis of the factors that determine happiness in old age, Cyrulnick (1993) emphasizes that the development and adaptation of mental functions are intimately linked to the cultural and social environment in which people live. The way people perceive the world and the way they can express their emotions will determine how they will feel in their old age, and these mental mechanisms already take root during youth. Cyrulnick notes that a stable environment, together with a certain level of stress, are necessary early in life to allow the optimal expression of an individual's psychobiological potential. A happy and equilibrated upbringing contributes to happiness later in life. It is likely that there is an innate aptitude for happiness, but it is also possible to create an environment in which one can grow and age happily. Elderly people who have an active intellectual life, who live, read, move, fight, meet other people, travel, love, or hate tend to have a better life.

Being able to talk about one's past and present experiences is a vital process. By speaking about their lives, about painful memories deeply imprinted (and never talked about), elderly people express their emotions, order their memories, and maintain contact with people. Psychologists can help elderly people to reexamine their lives and be conscious of their past. Talking about past unresolved conflicts can be a very soothing experience. It also helps elderly people keep their place in society. When alone, they tend to think about their misfortunes and the unhappiness in their lives; social ties help them cope with their emotions.

Sexuality

Sexuality encompasses many levels of expression and is not restricted to the ability to have intercourse or to its frequency. Sexual health and normality are notoriously difficult to define; sexual intercourse, masturbation, and affectionate interactions with others have more influence on vitality, and perhaps longevity, than previously anticipated. An active sex life has many positive physiological effects; for example, it is said to regularize menstrual periods in women and lower the incidence of hot flushes in menopausal women (one more illustration that behavior has feedback effects on physiological functions). Recent studies have suggested that educated middle-class and upper-class elderly men are able to continue sexual activity, despite erectile dysfunction, by enjoying alternative practices such as mutual masturbation or oral sex. In contrast, men from a lower socioeconomic background report complete loss of erectile function and tend to cease all heterosexual activities: only 29% practised mutual masturbation and 16% oral sex. Their attitudes toward these practices were negative (Cogen and Steinman, 1990).

Maintaining an active sex life probably has beneficial repercussions on life in general. The desire to stay young and sexually attractive implies a search for physical fitness and potency. A person who is attentive to his or her physical appearance at 20 or 30 years, is likely to continue to do so at 70 or 80 years. Comfort and Dial (1991) remind us that sexuality is lifelong and extends well beyond the reproductive years. It is an important factor in self-esteem and satisfaction and may well influence the way one ages. Sexual activity need not diminish with age, but can even improve. Later in life, there is less urgency, less performance pressure for men; for women, a greater self-assurance and less exhaustion due to child bearing and rearing are factors that can improve their sexual life.

Recommendations

Sexuality in elderly people should be more publicized, with the information that sex does not stop at age 70, or even 90. Society would thus be aware that older men and women do have active sex lives. This could eliminate preconceived ideas that the elderly do not need or desire affection and love. Nowadays, a lot of information is available to assist senior citizens in dealing with emotional and physical problems relating to sexuality. Physicians should therefore show sensitivity to these problems and should not conceal them or attribute them to the devil of age. The idea of an accepted decline with aging into a sexless state should be reconsidered and rejected (Comfort and Dial, 1991). Men and women of all social classes, but perhaps more specifically from lower socio-economic classes, need to be educated about sexual practices. The problems of sexuality that elderly men and women face are frequently amenable to a sex therapy approach that emphasizes the improvement in the couple's intimacy, and the expansion of their sexual flexibility (Kaplan, 1990). Apart from psychological help, several compounds or hormones have been proposed to increase sexual arousal and desires. At present, there is little information on these therapies. For example, the effect of exogenous testosterone in the maintenance of potency in older men is yet to be determined, but there are indications that testosterone therapy for women and men enhances libido, as suggested by Morley (1991).

Perspectives

All studies reported so far examined the effects of aging on sexuality in elderly individuals. None discussed the issue of whether sexuality could retard the aging process. To carry out such studies would obviously be difficult in view of the numerous other factors that influence aging, but the results might well be very interesting.

Physical Exercise

Forty years ago, most elderly people moved slowly, and their bodies looked worn out. The mere idea of an eighty year old person running a marathon was in the realm of fantasy. Times have changed. Review the chart depicting the time required to run the New York marathon in relation to runners' age (Figure 21). A fair proportion of men and women over 60 take on this strenuous challenge and, what is even more remarkable, finish the run in not much more than twice the time of the fastest young runner. According to physiologists, the upper limits of the physical fitness and capacities that they observe today in old people will certainly be exceeded tomorrow.

With the exception of people having physically strenuous jobs, bodies do not wear out and deteriorate from overuse; instead they decline from underuse. It is well established now that physical inactivity, often accompanied by overeating, results in pronounced decline in cardiovascular performance, muscle mass, glucose tolerance, high-density lipoprotein (HDL) levels, and triglyceride metabolism. It is possible that the higher prevalence of coronary artery diseases, generalized vascular disease, myocardial infarction, and stroke in older populations is, in part, the

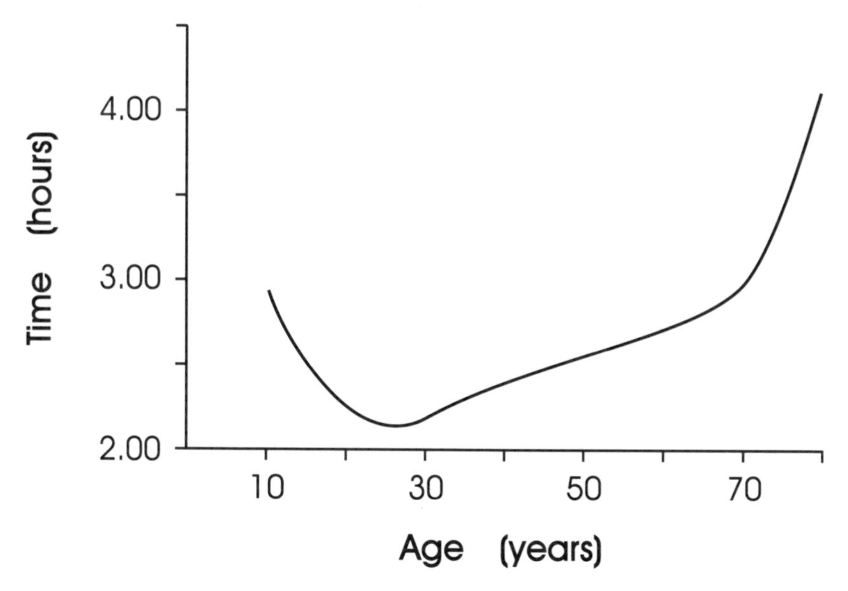

Figure 21. The fastest times in the New York marathon according to the runners' age.
(Redrawn from Walford and Walford, 1994.)

consequence of a lack of physical activity. Even short periods of inactivity, for example, confinement to bed-rest, induce these detrimental changes.

How does physical training modify physiological variables and disease progression? Does physical exercise influence longevity? Studies conclude unanimously that exercise is a kind of panacea for many physical ailments. It is the trigger for a cascade of physiological changes that are beneficial. Even though exercise alone is not the primary cause of all these positive effects, better physical condition often brings about weight loss; and this may have a synergistic effect on cardiovascular performances, glucose tolerance, high-density lipoprotein (HDL) levels, and triglyceride metabolism (Seals et al, 1984a).

Below we summarize the effect of physical exercise on physiological systems and organs.

Effect of Exercise on Cardiovascular Diseases

Aging brings about a loss of physical endurance and the decline of cardiovascular fitness. This, of course, has an important impact on the functional performance of older individuals, and usually has a profound effect on the delivery of blood and oxygen to muscles. The Baltimore Longitudinal Study of Aging (Shock, 1984) confirmed this and indicated that there is a substantial decline in muscle mass with aging. The main questions are:

does exercise directly or indirectly (through reduction in lipids, improvement of glucose tolerance, etc.) retard atherogenesis, lower peripheral vascular resistance, and reduce blood pressure? Medical doctors compared the maximum oxygen uptake or VO_2 max (an important physiological index of aging of the cardiovascular system) of athletically trained and untrained (physically inactive) elderly individuals. As is the case with other physiological variables, there was a large variability in VO_2 max among the elderly, and some had a VO_2 max comparable to that of young untrained individuals. Interestingly, the decline in VO_2 max in highly conditioned athletes, aged 50 to 82 years, who remained competitive, was almost negligible, less than 2% per decade (Pollock et al, 1987). This suggests that the observed decline in VO_2 max with age can be attributed in great part to physical inactivity. With few exceptions, most older individuals usually give up their habits of physical activity throughout their adult years. Recently, Kash and his collaborators (1993) conducted a longitudinal study over a 25-year period with 15 men who were 33–54 years old at the beginning of the study. These men agreed to follow a well-defined training program that varied over the years but always included one hour of strong physical exercise three times a week. Determinations of VO_2 max, blood pressure, heart rate, body fat, and careful recording of the training regimen were performed on a regular basis. Kash and his collaborators confirmed previous findings that aging causes a decrement of VO_2 max, due to a decline in heart rate and stroke volume. However, the decline of these subjects' cardiovascular function was only one half the average decline quoted in the literature. Sustained exercise in these men had retarded the age-associated changes in physical capacity over the 25-year period. Exercise therefore maintains muscle mass and the cardiovascular responses of older individuals. This positive effect suggests that exercise could also prevent vascular deterioration due to atherosclerotic coronary artery disease.

Effect of Exercise on Hypertension

Hypertension is frequent in the age group over 65 years, and it is one of the major risk factors for cardiovascular and cerebrovascular events at all ages, getting more critical with advancing age. There are good reasons to believe that endurance exercise training may be a physiological way to lower blood pressure in the case of moderate hypertension in the elderly (Cade et al, 1984).

Effect of Exercise on Glucose Tolerance

As many as 40% of people over 65 have diabetes and need to be treated. The decrease in glucose tolerance with age, even in healthy individuals, is often associated with the development of a hyperinsulinemia resulting from the stimulation of the pancreatic beta-cells by elevated glucose levels. Regular aerobic physical activity (and a higher VO_2 max) has been associated with enhanced sensitivity to insulin and better glucose tolerance in older individuals (Schneider et al, 1984). These studies suggest that vigorous exercise for at least 30 to 45 minutes should be performed at least every other day to maintain normal glucose and insulin metabolism.

Effect of Exercise on Hyperlipidemia

Lipid metabolism is complex and is affected by many factors, such as hereditary diseases, diet, medication, alcohol comsumption, and cigarette smoking. Although the effects of these factors have not been studied in a systematic way in the elderly, they probably all have a profound negative influence (Brunzell, 1984). Many experts now think that the overall cholesterol level in the general population is less significant than the ratio of good cholesterol (high density lipoprotein (HDL)-C) to bad cholesterol (low density lipoprotein (LDL)-C). With age, there is a trend towards a rise in triglyceride levels and a fall in the ratio of HDL to total cholesterol and LDL-C by 20% to 30% (even though there is heterogeneity in the lipoprotein lipid profiles of older people). Increased levels of plasma cholesterol, triglycerides, and LDL-C and reduced levels of HDL-C are major risk factors for the development of coronary artery disease (Bierman, 1985). Indeed most heart disease victims have high levels of LDL-C and low levels of HDL-C. The reasons for this are not clear, but one suggestion is that HDL-C may absorb artery-damaging toxins that are released when fats are broken down in the blood. Physical training can definitely improve the regulation of lipid metabolism: older athletes have higher HDL-C and lower triglyceride and LDL-C levels than their sedentary peers, and they tend to be less obese. Their levels are comparable to those of younger athletes. It is important to stress that the favorable modifications of blood lipids are observed in subjects who exercised and lost weight at the same time (Seals et al, 1984b), indicating that both physical fitness and body fat are determinants of changes in lipoprotein lipids in older people.

Effect of Exercise on Osteoporosis

Osteoporosis, a generalized decrease in bone mass affects 25% of older men and 50% of older women. It predisposes to bone fracture and thus represents an enormous health problem. In men and in women, osteoporosis is also increased by many negative life-style habits such as cigarette smoking, excess alcohol consumption, caffeine, and most of all by lack of exercise. Clinicians distinguish two types of osteoporosis: type I, or postmenopausal osteoporosis, which occurs at the time of menopause in women and is linked to estrogen deficiency; and type II osteoporosis, or senile osteoporosis, which occurs in men and women over 70 years of age. Treatment of postmenopausal osteoporosis with vitamin D_3, which regulates calcium uptake from the intestine, has not given clear results in humans, unlike the results obtained in studies in dogs which show bone mass increase with vitamin D_3 (see Part I, Chapter 3). One explanation could be the existence of a genetically determined variability among humans to regulate calcium absorption and renewal of bone. Researchers discovered recently that there are two versions of the gene encoding vitamin D_3 receptors. One of these alleles is linked with stronger skeleton development and the other with a weaker one (Morrison et al, 1994). A person who inherits two copies of the gene that confers low bone density will run a higher risk of having osteoporosis early. Fortunately, bone density is determined not only by heredity but also by environmental factors. A good way to counteract the genetic effects for people identified as vulnerable to osteoporosis is to build up bone density by increasing calcium intake and by exercising.

A number of cross-sectional studies suggest that bone loss can be attenuated by increased physical activity. Prospective studies demonstrate that postmenopausal women who exercise regularly will gain bone mass (Dalsky et al, 1988). Furthermore, it is not high-intensity aerobic physical activity but rather weight-bearing activity that increases bone mass. Weight-bearing activity includes exercises like walking, jogging, stair climbing, and treadmill walking. Results from a detailed analysis of the role of strenuous exercise in postmenopausal women receiving no estrogen replacement were intriguing: women, who ran 10 miles or more per week exceeded sedentary women of the same age and height in bone density of the spinal column, but also of the radius, an arm bone (Nelson et al, 1988). This means that a systemic factor is acting and not simply mechanical influences. Of particular interest, the level of plasma growth hormone was three to seven times higher in the trained women. This is a striking example showing that exercise can reverse an age-related decrease of an endogenous substance. These women also had higher levels of plasma 1,25 dihydroxyvitamin D_3 which should favor increased retention of calcium and a higher level of somatomedin C. Parathyroid hormone and estrone were lower in the exercisers, while calcitonin and estradiol were unchanged. Clearly, exercise, together with an adequate calcium intake, is one of the most important factors that may protect older people from osteoporosis. Type II osteoporosis is an excellent example of an age-associated disease involving selective changes in gene regulation that is reversible through hormonal changes.

Effect of Exercise on Mood

Last but not least, exercise is known to affect people's moods. It is common knowledge that after one hour of running or strenuous exercising, most people feel physically exhausted but mentally relaxed, if not slightly euphoric. Physical exercise is known to stimulate the synthesis of endorphins, which might explain this feeling of well-being. For many subjects, physical exercise fills the role of a mild antidepressant.

Recommendations

Often, what is regarded as the result of aging is, in fact, a consequence of disuse of bodily functions. Physical activity is by no means a prevention or a cure for the diseases of old age, but it could possibly delay the onset or clinical evolution of some degenerative diseases. Exercise is important at all ages. Strength, flexibility, and balance are reduced with aging, but they wane less quickly if people continue exercising throughout their life. Ideally, men and women should maintain physical fitness as they age, but this is sometimes difficult because of professional or family responsibilities. The exact type and amount of exercise that is beneficial for a given individual is difficult to establish. Repeated physical exhaustion, as in marathons, carries its own risks: competitive women athletes become amenorrheic and have a higher risk of developing osteoporosis. Exercise must be adapted to age and/or physical capacities. This is especially true for older people. Exercising helps them lead an active and independent life, and it is therefore advisable to encourage older people to exercise, whatever their age. People, young or old, will exercise only if they are motivated, and, quite often, organized gymnastic or stretching classes are the best way to instigate elderly people to exercise regularly. Jogging and stretching-type exercises are more appropriate than hard competitive running and aerobic exercises for the majority of elderly individuals. For reasons that have not yet been elucidated, hard and strenuous exercising does not have the same beneficial physiological effects as regular weight-bearing exercises. People 50 years and over should walk regularly, at least two or three times a week at a fast pace, for an hour or so.

Two recent studies, on close to 27,000 men show unambiguously that vigorous physical activity is associated with longevity (Blair et al, 1995; Lee et al, 1995). Physicians should therefore encourage their unfit patients to start exercising.

Smoking

There is no doubt that tobacco has positive effects that justify its popularity. Tobacco increases vigilance, reduces boredom, and improves transitory feelings of fatigue. It lowers the level of anxiety; it is an excellent anxiolytic, which explains the favorable impact of tobacco on so-called negative emotions such as tension, sadness, or irritation (O'Connor and Stravynski, 1982). Nicotine is a stimulus barrier drug, and smokers say that tobacco helps them cope with overstimulation. Because of these strong effects of tobacco on behaviors and psychological states, it is not easy to quit smoking.

Tobacco smoking is considered a major health hazard and is definitely a life-shortening drug. According to North American official reports, cigarette smoking is the single most important preventable cause of death in our society. Smoking accounts for more that 80% of all lung cancers (Figure 22), 25% of heart attacks, 80% of bronchitis and emphysema cases. There are 2000 deaths annually from cigarette-sparked fires. Smoking is said to kill almost 400,000 American and 60,000 French people annually, i.e., more than 1000 persons a day in these countries alone. The noxious effects of tobacco on longevity are such that, according to statisticians, if people did not smoke tobacco, hundreds of thousands of men and women would experience a life expectancy gain of 15 years. An increase in life span of that magnitude would be comparable to the increase in longevity predicted from the complete elimination of all cancers not caused by tobacco use. The epidemic of chronic diseases and deaths caused by smoking has now extended to many Asian countries, in particular, China. Until recently, few Chinese suffered from heart disease. Unfortunately, smoking has recently become widespread in that country, and, as a consequence, a rapid increase in the incidence of tobacco-associated diseases is observed. At current rates of tobacco smoking, about 50 million Chinese now under the age of 20 will eventually be killed by tobacco.

The results of a huge epidemiological study carried out at the University of Oxford were published recently (Peto, 1994; Doll et al, 1994). About 34,000 British physicians were followed for a period of 40 years, from 1951 on. This study showed unambiguously the dangers of long-term use of tobacco. About 50% of regular cigarette smokers were killed by their habit. Those killed by tobacco in middle age (35 to 69 years) lost an average of 20 to 25 years of life expectancy. The excess mortality was from diseases that are caused by smoking: cancers of the lung, mouth, esophagus, pharynx, larynx, pancreas, and bladder, and from chronic obstructive pulmonary diseases, and from vascular diseases and peptic ulcers. Kinlen and Rogot (1988) present evidence that smoking can cause leukemia also.

The noxious effects of cigarette smoking are numerous: aside from inducing lung cancer and increasing the rate of other cancers and cardiovascular disease, tobacco is thought to induce disturbances in metabolism, especially that of lipoproteins. It also plays a role in the loss of skeletal bone. The diseases caused by tobacco are due mainly to the tar of tobacco smoke, containing polynuclear aromatic hydrocarbons, a class of

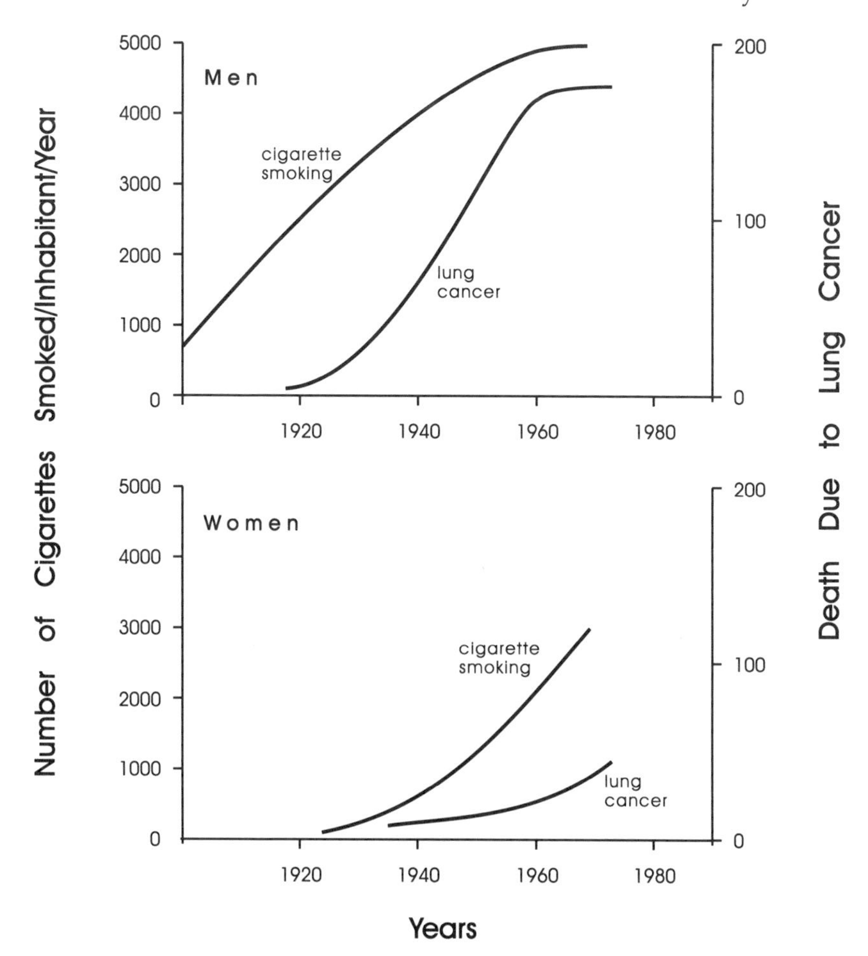

Figure 22. Mortality due to lung cancer in comparison with cigarette smoking per US inhabitant, from 1900 to 1980.
The two diagrams present the increase in cigarette smoking and the resulting increase in deaths due to lung cancer per 100,000 inhabitants per year, for men (upper diagram) and for women (lower diagram). (Redrawn from Cairns, 1985.)

chemicals that is believed to cause between 40% and 80% of all human cancers. These substances can cause mutations in the DNA, by binding to and chemically damaging DNA molecules. Tobacco tar also contains aldehydes, toxic chemicals that cross-link molecules (and thereby inactivate them), and heavy metals, such as lead, radioactive polonium, and arsenic. These heavy metals poison several enzymes, particularly those with sulphydryl groups, which are often protective antioxidants.

Recently, another mechanism by which tobacco is noxious was discovered. Imaizumi (1989) reported that compounds from cigarette smoke increase the blood concentration of a phospholipid called platelet activating factor (PAF), released by circulating blood cells such as neutrophils, basophils, monocytes, macrophages, platelets, as well as by vascular endothelial cells. Later, Johnston and his colleagues showed that the increase in PAF is due to the inhibition of PAF acetylhydrolase (PAF-AH), an enzyme that breaks down PAF (Miyaura, 1992). PAF acts in exceptionally small concentrations and across short distances. Not all its physiological actions have been clarified yet, but it is known to enhance platelet activation and formation of blood clots and to be a causal factor of inflammatory damage within the body. Abnormally high concentrations of PAF contribute to a variety of disorders, such as diabetes, cardiovascular disease, arterial thrombosis, anaphylaxis, renal disease, systemic lupus erythematosus, and liver cirrhosis. In addition to causing numerous ailments, PAF has profound cardiovascular effects. When given intravenously, it causes systemic hypotension in several animal species. These observations explain in part why smokers have a higher risk of vascular and lung damage. It may also explain why women who smoke cigarettes and take oral contraceptives face an increased risk of stroke—estrogens themselves increase PAF levels in the blood.

The effects of PAF are complex, and there is controversy concerning the exact mechanism of the hypotensive effect of PAF. Some researchers postulate that it is due to dilation of resistance vessels; others have found that the systemic hypotension is due to pulmonary hypertension or decreased blood volume. There is also controversy over which molecules induce the depressor effect: it could be mediated by vasodilatory prostaglandins, or thromboxane A2, or leukotrienes. Yamanaka and his colleagues (1992) studied the different physiological effects of PAF and suggested that PAF causes a transient hypotension due to a prostaglandin-independent dilation of resistance vessels, followed by a prolonged hypotension, apparently the result of venodilation induced by vasodilatory prostaglandins.

A recent study has shown that tobacco can prevent vasodilation of coronary arterial walls by affecting directly their endothelium (Nitenberg et al, 1993). This could explain partly the incidence of myocardial infarctions in young people (30–40 years old) who are chronic smokers, and have normal coronaries, no hypertension and no hyperlipidemia. Whereas vasodilation is observed in nonsmokers after injections of acetylcholine, smokers suffer an important vasoconstriction of their coronaries. One of the factors synthesized by endothelial cells that influence vasomotricity, is nitric oxide, which could play an important role in this process.

Nicotine, although not as much a health problem as the polynuclear aromatic hydrocarbons and aldehydes from tobacco smoke, has risks of its own. It causes blood vessels to constrict, thus increasing blood pressure. It also causes an increase in blood lipids, particularly cholesterol.

Recommendations

The study on British medical doctors reported above (Doll et al, 1994) underlines the fact that the effects of smoking are reversible and that stopping smoking is beneficial. If people stop in middle age, before getting cancer or other diseases, they avoid some, perhaps most, of the later risks of death from tobacco. The best recommendation is to quit smoking, but this advice is easier to give than to follow. Although many smokers are able to quit smoking initially, the majority will resume their habit and live through several successive withdrawal attempts before they are able to quit smoking definitively. The problem is to find the most suitable withdrawal method. Some are based on behavioral and relaxation techniques to limit withdrawal symptoms and induce a state of mind that helps to maintain abstinence. Another method is to use nicotine-containing chewing gums. These chewing gums have minor side effects, such as throat or gastric irritation and excessive salivation. There are also nicotine transcutaneous devices that, for some, are more efficacious than chewing gums. In a meta-analysis of 17 double-blind placebo-controlled studies on smoking cessation using nicotine patch, subjects receiving the active patch had abstinence rates more than twice as high as subjects who received the placebo patch. On the other hand, adding intensive counseling did not increase abstinence rates further after six months (Fiore et al, 1994).

A few nonnicotine drugs may help quit smoking. Clonidine, a **presynaptic** $\alpha 2$-adrenergic receptor agonist, as well as beta-blocking drugs have been tested in clinical trials with limited success. Diphenylhydantoin, a drug used to control epileptic seizures, improves learning in animals and increases their life span. It is also a stimulus-barrier drug, like nicotine, and has therefore been tried in smoking-withdrawal. However, none of these strategies can substitute for the desire and determination of smokers to interrupt their toxicomania (substance abuse, and in severe cases, substance dependence).

Smokers who cannot help but continue smoking should at least reduce smoking risks: the advice is to use cigarettes that are relatively higher in nicotine compared to tars; unfortunately, the amount of nicotine is generally low in the low-tar cigarettes, and smokers smoke more of them to get a satisfactory level of nicotine intoxication. Smokers should also take supplements of vitamins and minerals to counteract some of the negative effects of cigarettes (see Table 14). Inhalation of tobacco smoke depletes the tissues of vitamin C, and a normal dietary intake of vitamin C may not be sufficient to maintain an adequate level of this vitamin in the body. Vitamin E, an antioxidant, may reduce damage to lung tissue from tobacco smoke. Other vitamins and minerals that protect against the development and progression of lung cancer include selenium, vitamin A, and β-carotene. Several epidemiological studies have provided evidence that

Table 14. Daily vitamin and mineral supplements suggested for smokers

Vitamin A	5–10 mg
Vitamin B1	10–30 mg
Vitamin B2	10–20 mg
Niacin	50 mg–1 g
Vitamin C*	0.5–2 g
Vitamin E*	300 mg
Choline	1–3 g
Beta Carotene*	30–180 mg
Selenium*	250 µg
Zinc	50 mg
Folic Acid	1000 µg
Cysteine	0.3–1 g

(g: gram; mg: milligram; µg: microgram; *: most recommended compounds).

vitamin and mineral supplements are associated with a reduced risk of lung cancer. However, in a recent study on nearly 30,000 male smokers of 50 to 69 years of age from Finland, no reduction in the incidence of lung cancer was found after five to eight years of dietary supplementation with vitamin E or β-carotene or both. This is somewhat of a surprise, and one might wonder whether the lack of positive effects could not be attributed to the low doses of antioxidants prescribed in this study (The α-tocopherol, β-carotene cancer prevention study group, 1994).

Perspectives

Research is needed to better understand the toxic as well as the pleasant effects of tobacco. This will help find drugs that could mimic the effects of nicotine and be used as substitutes for tobacco, either as a less dangerous toxicomania or in a phase of withdrawal from smoking. Another direction is the synthesis of nicotine receptor antagonists; one such drug, mecamylamine already exists, but its toxicity prohibits its clinical use. Long-term studies on the efficacy of daily vitamin and mineral supplements for smokers (as indicated in Table 14) are also needed so that firm recommendations, instead of suggestions, can be made.

Whereas most research has focused on smoking cessation, there is also concern about preventing young people from starting to smoke. Indeed, advertisements by tobacco companies are subtle and aim to attract future long-term consumers. It is important to set up more information campaigns and provide youngsters with explanations about the deleterious effects of tobacco and on why people smoke. Effective strategies for coping with smoking situations and resisting social pressure should be discussed by educators with young people. Social pressure from peers, the media, and family members who do not smoke can encourage young people not to smoke.

In many public places, smoking is now forbidden, and even though such prohibitions limit the freedom of individuals, it does discourage smokers and may lower their consumption of tobacco. Making smoking tobacco illegal would protect a lot of people, but would also add an another product to the list of highly profitable illegal drugs.

Ultimately, we must deal with the fact that humans as well as laboratory animals can become addicted to tobacco, alcohol, or other drugs; the issue is to understand why living beings stimulate themselves with such compounds. Interestingly, laboratory animals self-administer nicotine at a much lower rate than they do cocaine or opiates. The question is to find out why humans consume a drug such as nicotine that has a low addictive potential in animals but induces severe dependency in humans.

Apart from the scientific problem, there are societal issues that need to be debated in order to win the war against tobacco. Health insurance companies should not invest in tobacco industries any more (Boyd et al, 1995), and the American Medical Association should make it clear that it will no longer accept grants from the tobacco industry (Blum and Wolinsky, 1995). Recently, Brown and Williamson company documents have revealed confidential reports on tobacco industry activities regarding research on nicotine addiction and manipulation of internal and external scientific research. These documents show that, in spite of their vehement public denials, tobacco companies knew all along that smoking is addictive and that tobacco smoke is carcinogenic (Glantz et al, 1995). The concerted avoidance of liability on the part of tobacco industries gives much food for thought. Unfortunately, the political and economic pressures to maintain a powerful tobacco industry are enormous, and sadly, the recent decisions of the United States government to impose regulatory measures to limit teenage smoking will have only a small impact. The tobacco industry continues to work to develop new markets. In particular, women are a major group that is increasingly influenced by cigarette advertising; women are smoking more and more, nearly as much as men, and mortality figures reflect this sad increase.

Alcohol Consumption

Epidemiological studies in humans have shown that apart from causing specific diseases in humans, such as cirrhosis of the liver, myocarditis, and the slow and tragic deterioration of the central nervous system (Arendt, 1989), excessive alcohol consumption leads in general to premature death.

This is the consequence of alcohol itself as much as self-neglect, poor hygiene, and poor nutrition. Sleep problems, loss of libido, depression, and anxiety may also be caused or aggravated by excessive consumption of alcohol. In later life, an excess of alcohol is especially damaging, because there is a reduction in body water and lean body mass, which leads to higher levels of alcohol in the blood after a given dose. This may explain in part why older adults consume less and less alcohol as they age (Adams et al, 1990). Furthermore, hypnotics, tranquillizers, antidepressants, and certain other drugs used by elderly people can heighten the risk of acute alcohol intoxication.

The effects of alcohol on longevity have received less attention than have those of tobacco. One reason for this may be the difficulty in determining at which point the amount of alcohol ingested starts to be detrimental. Many studies have assessed the effects of alcohol on longevity in combination with the effects of other life habits. Breslow and Breslow (1993) listed seven health practices as risk factors for higher mortality; these are excessive alcohol consumption, smoking cigarettes, being obese, sleeping fewer or more than 7 to 8 hours, having very little physical activity, eating between meals, and not eating breakfast. Excessive alcohol consumption ranks high on the list, but as most people who drink large amounts of alcohol are also heavy smokers, conclusions of studies on alcohol often include the effect of tobacco smoking as well. In a study on nearly 7000 men and women aged 65 years and over living in the USA, LaCroix and his collaborators (1993) investigated how unhealthy behaviors, in particular alcohol consumption, smoking, and physical inactivity were related to the duration of active life. They have concluded that all three factors have a strong negative impact on the elderly. Excessive alcohol consumption is considered a key factor in increasing the risk of falls and bone fractures. However, in contrast to cigarette smoking, alcohol does not seem to have only detrimental effects. Results show that moderate drinkers—those who drink 1 ounce of alcoholic beverage (28.35 grams) or less per day—are more likely to maintain mobility than are nondrinkers (LaCroix et al, 1993). This supports the conclusion of a previous study, in which moderate alcohol consumption of less than 1 ounce per day was related to a higher level of physical function in older adults (Guralnik and Kaplan, 1989).

Few studies on long-term alcohol consumption have been carried out in animals. In a recent study, the influence of chronic exposure to ethanol on senescence was assessed in a line of rats genetically selected for their voluntary consumption of high amounts of ethanol (Hervonen et al, 1992). Unexpectedly, in this particular strain of rats, there were no significant differences in the life span of ethanol-fed rats compared with controls. The authors postulated that these rats had specific capacities to compensate for the impact of ethanol. Further studies are needed to determine whether rats selected for a low level of voluntary consumption of ethanol and an inadequate defense against its effects, would show changes in life span when fed large amounts of ethanol.

The idea that a moderate alcohol intake is better than abstinence was questioned by Shaper and his collaborators (1988) who argued that the conclusions of many studies, namely that moderate drinkers have the lowest rate of mortality due to cardiovascular diseases as opposed to nondrinkers and heavy-drinkers, may not be correct, because these studies did not take into account the past histories of patients. Many nondrinkers may have been exdrinkers who gave up drinking for health reasons and therefore bias the results of the studies. This view is however not supported by all scientists who think that a moderate amount of alcohol (about one glass of wine per day) is not necessarily detrimental. It may increase HDL-C levels, and thus may have favorable effects on fibrinolysis and coagulation, thereby decreasing the risk of coronary heart disease. As stated by Marmot and Brunner (1991): "Against the possible protective effect of moderate alcohol consumption on the risk of coronary heart disease has to be set the direct effect on blood pressure and the possible effect on risk of stroke." The issue is probably more complex than it might seem at first sight. In a recent study, Fuchs and his collaborators (1995) show that among women, light to moderate alcohol consumption is associated with a reduced mortality rate, but this survival benefit would be confined largely to women at greater risk for coronary heart disease. Clearly, large prospective studies on the effects of alcohol are very useful and have shown a consistency of findings in population as diverse as Puerto Ricans and Japanese physicians, but these studies have their flaws. It will be necessary to carry out further studies on defined populations, such as women or men in specific age groups, or ethnic groups, or women and men with genetic susceptibility to diseases, etc, to assess the effects of small amounts of alcohol on the incidence cardiovascular diseases and other diseases.

Recommendations

Limiting alcohol consumption to improve health cannot be questioned. Everybody, and particularly the elderly, ought to watch carefully their consumption of alcohol and should not exceed one to two glasses of wine a day or their equivalent. Unfortunately, the pattern of a person's health practices tends to persist over the years. For example, people who have consumed large amounts of alcohol during their early life are likely to continue doing so in their old age (Breslow and Enstrom, 1980). The good news is that there is growing evidence that "it is never too late to change, for example, to stop drinking an excessive amount of alcohol, or quit smoking, or increase physical activity" (Kaplan and Haan, 1989).

Several drugs (naltrexone, acamprosate, serotoninergic antidepressants) are being proposed to help induce or maintain abstinence. In particular, naltrexone, an opioid receptor antagonist has produced promising results in reducing the risk of relapse in alcohol-dependent patients. Its exact mode of action is not known, but it is reported to reduce the pleasure associated with drinking (Volpicelli, 1995). Acamprosate, which acts on excitatory amino acid receptors, is also very promising. As is the case with other pharmacological agents, psychosocial support improves significantly the outcome of the treatment.

Perpectives

As with other toxicomanias, the understanding of the biological bases of the predisposition to alcoholism would be a great advance for public health, but it is a complex issue. The area in the brain called the ventral tegmental area (VTA), with its surrounding and projection areas, represents the anatomical system that governs the function of self-stimulation, i.e., the urge to repeat consumption of compounds that bring satisfaction. The VTA certainly did not evolve through millenia just for our appreciation of alcohol, nicotine, or morphine: it subserves important other functions, tied mostly to two monoaminergic neurotransmitters, dopamine and noradrenaline. In theory, one would like to find a way to specifically influence the VTA (through medication or genetic engineering techniques) in order to control toxicomanias without affecting negatively the other functions of the VTA. Although it is difficult to say whether such an approach will be feasible in the near future, research in neuroscience is progressing at a fast pace, and there is some hope that solutions to this problem will be found.

In parallel with these biological approaches, other efforts are needed at the societal level. In all cultures, subgroups of people defined by their educational and social levels, show a higher rate of drug addiction. When the conditions of life bring no pleasure, no self-fulfillment, and no hope for a constructive future, resorting to the chemical stimulation of the VTA is understandable. It fulfills the vital function of having pleasure. Societal changes that prevent young adults from becoming drug addicts will be difficult to implement without a more favorable evolution of society as a whole. Working toward such a change in society would seem as reasonable and logical an investment for society as research programs on the biochemical mode of action of drugs in the brain. Effective legislation in this area could provide work for politicians for a lifetime.

In the near future, there are two directions of research that should be fruitful. The first is the search for a less toxic alcohol deterrent than disulphiram or calcium carbimide. If alcohol is taken when either of these drugs is consumed, a severe physical reaction is induced, which therefore curtails further drinking; however, they have adverse effects on hepatic and central nervous systems. A second direction would be to develop ways of assessing individual predisposition to alcoholism. This would include a genetic analysis (inherited tendency to consume alcohol) as well as the identification of given personality traits that may be related to alcoholism. Preliminary findings point to the role of a subgroup of dopamine receptors and to thyroid gland function as candidates in a predisposition to alcoholism. Developing screening tests that can be used in the clinical setting might help to define subgroups of the population and possible prophylactic measures.

Controlling Stress

Excessive stress is one of the major obstacles to longevity. Indeed, as indicated in Part I, Chapter 2, life stressors have deleterious consequences on endocrine and immunological systems, in addition to inducing depression or drug addiction. It is perhaps more appropriate to speak of strain rather than stress, because strain describes the subjective evaluation of a given stressor for a given individual. Indeed, when faced with the same painful event, people differ considerably in their strain level; the tendency to succumb to or to cope with identical life events depends on the inherited biological and psychological sensitivity to strain, on life experiences, and on the values or philosophy of each person. Reducing the level of strain is a priority for everyone. In fact, major corporations recognize the necessity to focus on diminishing employees' level of strain in order to increase productivity. Such measures have medical and financial advantages. Today, numerous books are written on this issue, and professional counselling aimed at strain reduction in our private and professional lives is gaining popularity.

Recommendations

There are many methods to reduce strain (see Table 15).

Psychological approaches to strain reduction can be in the form of either psychotherapies, self-help books, or formalized programs of stress reduction. Relaxation techniques are classified under physical methods since, apart from their psychological benefits, they also act directly on neurophysiological and biochemical variables. Deep relaxation induces physiological alterations such as decreased oxygen consumption, heart rate, arterial blood pressure, respiratory rate, and arterial blood lactate. Increases in skeletal muscle blood flow and in the intensity of slow alpha waves of the electroencephalogram are also observed. These changes are different from those reported during simple relaxation, that is, sitting quietly. The practice of the relaxation response technique may be useful in counteracting the physiological and pathophysiological states associated with increased mental and physical arousal. Benson (1983) set up a list of

Table 15. A few methods to reduce the strain level

Psychological approaches
— revise the values that govern your life
— decrease exposure to stressful events
— seek support from professionals (psychotherapists, psychiatrists), family members or friends
— avoid solitude and increase contacts with others

Physical methods
— practice physical exercise
— learn relaxation techniques

Medical methods
— antistress medications

instructions that elicit what he calls the relaxation response. Subjects assume a relaxed position in a quiet environment, with eyes closed, and are asked to repeat a word or phrase and to keep distracting thoughts away.

Other techniques such as direct suggestion and mental imagery using hypnosis are reported to modulate delayed-type immunoreactivities (Zachariae et al, 1989). Finally, there are medical approaches to stress reduction that are useful when other methods fail. For example, many people now turn to the new antidepressants like fluoxetine (Prozac), or other serotonin reuptake blockers, to reorganize their lives and get more enjoyment out of life. Unfortunately, these drugs do not produce beneficial effects in all subjects, and their side effects can be unpleasant and render them unacceptable. Other medications that decrease the biological response to stress are the beta-blockers and the anxiolytics. Beta-blockers are prescribed in some cardiovascular diseases and for high blood pressure. They antagonize the action of noradrenaline on some of its cell-membrane receptors. Thus, they decrease the cellular consequences of an excessive release of noradrenaline, such as occurs in stress. In cases of acute psychological stress, such as during university qualifying examinations, beta blockers control tremor, accelerated heart rate, and anxiety. The antistress action of beta-blockers is limited to the period following the taking of the medication. In cases of chronic anxiety or panic attacks they are less effective. They should not be prescribed to people with asthma or cardiac insufficiency. There is no indication that beta-blockers can have sufficient antistress effects to prolong life, and this is also true of antianxiety drugs such as the benzodiazepines, e.g., diazepam or alprazolam. The latter drugs have side effects such as excessive sedation, amnesia, and difficulty in concentration. Relying on benzodiazepines for the control of chronic strain is not recommended, although these anxiolytics serve as a last resort for many stressed people.

Perspectives

Scientists and medical doctors are actively involved in finding new ways to counteract the effects of stress and hypersecretion of glucorticoids on the brain. As was discussed earlier (see Part I, Chapter 2), an excess of glucocorticoids can lead to neuronal deterioration, in particular in the hippocampus. In rodents, this hippocampal senescence may be slowed down by artificially reducing the glucocorticoid exposure via adrenalectomy and replacement with low doses of glucocorticoids. Finding equivalent treatments in humans is obviously difficult, but behavioral techniques that lower strain can also lower glucocorticoid levels. Finding a pure anxiolytic drug, devoid of other effect, is a goal of many research departments in pharmaceutical industries. Whether such an anxiolytic would afford sufficient protection against the biological consequences of stress in predisposed persons will have to be evaluated.

Chapter 5

Substitution of Endogenous Compounds and Hormones

The question that we discuss in this chapter is whether elderly subjects should receive replacement therapy of hormones and/or growth factors, such as human growth hormone, melatonin, sex steroids, and thymus hormones. The reasons to schedule such replacements is that the synthesis and secretion of these compounds decrease significantly over the years, and these changes are thought to play a role in the manifestations of senescence. As a matter of fact, several syndromes secondary to deficiency of these compounds in young subjects show changes that resemble those of senescence. The benefits and the risks of these replacement therapies are being evaluated through in vitro studies, animal studies, short-term neuroendocrine studies in small numbers of subjects, and in large field trials.

Melatonin

The gradual involution of the pineal gland accompanied by a decline in melatonin production with age is considered by many researchers as a key factor in the programming of senescence (see Part I, Chapter 3). The changes in plasma melatonin level or in its major urinary metabolite, 6-hydroxymelatonin, and the modifications of melatonin circadian rhythm are used as indicators of aging in humans (Nair et al, 1986; Sack et al, 1986). Because of the important properties of melatonin, both as a regulator of circadian rhythmicity and as an antioxidant, scientists have proposed that the diminished output of melatonin could be one of the mechanisms of aging. Other scientists postulate that it is not only the loss of melatonin but the persistence of serotonin (melatonin is a serotonin derivative) that induces a chain of events accelerating aging processes. They suggest that while melatonin has anti-aging effects and is regenerative during the night, the actions of serotonin would promote aging processes (Grad and Rozencwaig, 1993; Rozencwaig et al, 1987). The protective effects of melatonin against free radicals, especially the highly reactive hydroxyl radicals, is postulated to have important implications, particularly in neurodegenerative disorders. The progressive damage of neurons by hydroxyl radicals may even be regarded as one of the major irreversible processes of aging. Recently, Lissoni and his collaborators reported that the administration of melatonin (40 milligrams per day) in association wih interleukin IL-2, could amplify the antitumor activity of IL-2 in patients with metastatic colorectal cancer, breast cancer, nonsmall cell lung cancer, and endocrine cancer (Barni et al, 1995; Lissoni et al, 1994; Lissoni et al, 1995). The postulated mechanisms are that melatonin acts indirectly via the immune system or that it inhibits directly tumor growth factor production. These studies emphasize the role of melatonin as an antitumor agent.

The various physiological properties of melatonin have led scientists and medical doctors to test its potential life-enhancing effect in rodents and humans. Pierpaoli and his colleagues (1991) reported that melatonin had rejuvenating effects in old mice and substantially extended their life span. Moreover, pineal glands grafted into the thymus of old recipient mice (genetically identical) induced a 12% increase in their survival, in addition to maintaining the structure and the number of cells of the thymus (Pierpaoli and Regelson, 1994). A life-enhancing effect of endogenous melatonin was also observed indirectly. One of the positive effects of food restriction (FR, see Chapter 1) is to reduce the generation of reactive oxygen species and to preserve antioxidative defense responses (including melatonin production). Reiter (1992) showed that FR increased life span in experimental animals as expected, and, more importantly, it preserved pineal structure and function in aging animals. The rhythm of melatonin secretion in these animals was similar to that of younger animals. Since exogenous administration of melatonin to nonfood restricted animals increased their survival, scientists postulated that one of the mechanisms by which FR increases life span is by increasing extrapineal melatonin production, such as in the gastrointestinal tract where melatonin is synthesized. This could explain the increase in serum melatonin levels and suggests that pineal deficiency of melatonin could be compensated for by peripheral sites of melatonin production, a fact that remains to be established.

In humans, melatonin is reported to have a number of physiological effects. Melatonin (given orally at doses as high as 6 grams) affects brain activity, induces sleep, and sometimes relaxation (Lerner and Nordlund, 1978; Anton-Tay et al, 1971). In women, melatonin is believed to play a role in postmenopausal osteoporosis (Sandyk et al, 1992).

There is some epidemiological evidence also that the pineal gland may be involved in the determination of human longevity. Patients afflicted with retrolental fibroplasia (these patients can not perceive any light, and blindness occurs within the first few weeks of life) were found to have a higher life expectancy than age-matched controls (Lehrer, 1979). Their higher longevity was postulated to be due to melatonin because their pineal gland is stimulated chronically and maximally, producing high melatonin levels since there is no physiological light-mediated melatonin suppression during daytime.

Recommendations

The efficacy of melatonin is recognized in alleviating the manifestations of jet-lag. It may also have some effect in regularizing sleep. In a recent study, the quality of sleep (sleep efficiency) was found to be greater in elderly people suffering from insomnia, who were treated with controlled-release melatonin than with a placebo. Sleep latency decreased only moderately, and total sleep was not affected (Garfinkel et al, 1995). People who wish to take melatonin should probably take 2 to 4 milligrams regularly around 9 to 10 PM each evening. In order to mimic the effects of endogenous melatonin, a slow-release formulation that maintains high concentration during the night is ideal. There is no official recommendation for melatonin as a substitutive treatment in elderly subjects.

A report on the therapeutic effects of melatonin in a case of chronic sarcoidosis suggests that melatonin might be useful as an immunoregulatory drug in specific disease cases.

Physicians must be cautious when administering melatonin to patients because it is not without risks. When given to young animals, it delays the onset of puberty and decreases fertility. Reports indicate that chronic administration of melatonin has antigonadal effects in some animals that are seasonal breeders and that it could possibly have similar effects in humans. We emphasize that there is no strong data indicating that melatonin is indeed a life-extending compound for humans.

Perspectives

There is a justifiable interest in the physiological actions of melatonin and in its direct and indirect beneficial effects. It is still open to question whether melatonin can serve as a potent pharmacological agent, with protective effects against neurodegeneration and against the mutagenic and carcinogenic actions of hydroxyl radicals.

There is evidence in animal models that melatonin may retard the development of processes that contribute to reduced life span, in particular through its antioxidant actions. Melatonin can therefore be considered a life-extending hormone in rodents. Reiter suggests that, considering the drop in melatonin with age, the analysis of the amplitude and duration of the nocturnal peak can be indicative of the biological age of the organism (Reiter, 1995). The evidence that melatonin plays a role in human longevity is weak, even though there is a general belief that consistently reduced or altered melatonin production could be associated with situations such as aging, cancer, or depression. It is however too soon to claim that administration of melatonin is going to work wonders as a life-extension drug (Reiter and Robinson, 1995). More studies need to be carried out to specify the role of melatonin on various physiological functions and to determine the effective dose and mode of administration of this hormone.

The role of melatonin as an adjuvant to anticancer medications seems promising (Lissoni et al, 1994; Lissoni et al,1995).

Melatonin is the source of several biologically active metabolites (Hardeland et al, 1993). In future work, these metabolites will be the focus of attention, and it is possible that they might serve as future pharmacological tools. Some of these metabolites are candidates for regulators of GABA-dependent chloride channels while others act as modulators of serotoninergic receptors.

Thymus Hormones

The role of the thymus in longevity is seen with scepticism. Indeed, experiments designed to restore normal immunological functions by grafting thymus glands from young animals to old animals resulted in T-cell restoration to normal levels, but no life span increase of these animals was observed. Similarly, attempts to prolong life have been carried out with little success by injecting synthetically prepared thymic peptide hormones that have different biological properties, like thymosin-α1, thymopentin, thymulin, and the octapeptide THF-γ2 (Doria and Frasca, 1987). Whether such treatments will increase life span remains to be demonstrated.

These hormones, particularly THF-γ2, are very attractive for clinical applications, for example in human diseases such as primary congenital immune defects, T-cell defects secondary to cancer chemo- or radiotherapy, and in systemic viral infections. THF-γ2, injected in immunodeficient aging mice, enhances the frequency of mitogen-responsive T cells and T-helper-cell activity. Administration of this or other immunological peptides, looks particularly attractive for treatment of T-cell deficiencies in the elderly to increase immune responsiveness to exogenous antigens (Goso et al, 1992).

Sex Hormones

A huge literature describes the changes in endogenous sex hormone levels with senescence, and a few effects on life span are documented. The most spectacular findings are the physiological effects of estrogen therapy in menopausal women. This replacement therapy is probably the most striking demonstration that a natural aging process can be slowed down in humans. Menopause is the time when sexual steroid production in females ceases. Estrogen deficiency has many negative effects, not the least of which are the acceleration of osteoporosis, and the incidence of hot flushes (a transient vasodilation on the face and thorax, due to the response of hypothalamic thermoregulatory centers). It is now well established that estrogen replacement in women will prevent osteoporosis, suppress hot flashes, and retard cell atrophy in many parts of the reproductive tract. However, there has been a lot of debate about the positive and negative effects of estrogen replacement therapy. Estrogen therapy has been reported to have undesirable side effects, such as weight gain and breast tenderness. The major concern has been that women treated for many years might have increased risk of breast cancer. In a recent and comprehensive analysis, Belchetz (1994) summarized all published studies on the proven and postulated benefits or risks of this treatment. Treatment with estrogen alone decreases mortality overall, principally through its effect on heart disease, but it is not without danger. An increased incidence of endometrial carcinoma has been observed after treatment with estrogen alone. On the other hand, combined treatment with estrogen and progesterone seems to give more favorable results. Belchetz (1994) reports also that life expectancy is increased by two years on average among women who use replacement hormones. In the last year, more studies have reported the advantages and disadvantages of hormonal therapy for postmenopausal women. The Postmenopausal Estrogen/Progestin Interventions trial (PEPI trial) was conducted on 875 postmenopausal women aged 45 to 64, to assess the influence of estrogen, with or without progestin, on heart disease risk factors. High-density lipoprotein cholesterol, fibrinogen, insulin, and blood pressure were monitored for a period of 3 years. The conclusion was that estrogen alone or in combination with a progesterone, improves lipoproteins and lowers fibrinogen levels (The writing group for the PEPI trial, 1995). This may contribute to the decreased risk for cardiovascular disorders such as myocardial infarction. This study provided the opportunity to evaluate the effect of the various hormonal treatments on the endometrium. The results showed that women who received estrogen alone were more likely to develop hyperplasia than those given either estrogen and progesterone, or a placebo. The authors of the study advise physicians to prescribe progesterone cyclically or continuously for endometrium protection. The administration of progesterone has the additional advantage that endometrial hyperplasia reverts to normal in women treated with estrogen alone (The writing group for the PEPI trial, 1996). The protective effect of estrogen/progesterone therapy in reducing the risk of cardiovascular disease may be explained in part by the action of estrogen on lipids, and also by the direct action of estrogen on arteries. Treatment with estrogen/progesterone is reported to reduce the level of a factor synthesized by endothelial cells, endothelin-1 (ET-1), which causes vasoconstriction and has been implicated in myocardial infarction (Ylikorkala et al, 1995). In a recent study, the effects of estrogen on weight gain were analyzed in women who had been on hormonal replacement therapy either intermittently or continuously for 15 or more years. No significant effects were found (Kritz-Silverstein and Barrett-Connor, 1996).

Whether men undergo a phase of andropause is still debated. However, since many older men complain of a lack of energy, mild depression, and lowered libido, some clinicians prescribe testosterone and claim that this improves the well-being and sexuality of their patients. Specialized institutes even prescribe testosterone on the basis of the patient's endogenous hormone levels. This might not be correct, for even at the age of 20, the range in sex hormone levels among individuals is quite large. Also, concerning testosterone replacement, it is startling to recall the peculiar practices that were carried out sixty years ago. At that time, thousands of transplants of monkey testicles were performed on older men by numerous surgeons in France and other countries, who claimed that transplanted men experienced a substantial improvement in their mental capacities (i.e., better memory), improved aptitudes for intellectual work, and, above all, a complete rejuvenation—growth of hair, younger outlook, lower blood pressure, and many other positive physiological and psychological changes (Voronoff and Alexandrescu, 1930). These medical practices did not convince the medical community at that time. Today, the implementation of ethical rules has made such medical practices unacceptable. That male sexual hormone and neuropeptide replacement could have rejuvenating physiological effects on older men remains an interesting possibility.

Recommendations

The variety of effects of a hormonal therapy in menopausal women makes it difficult to draw a generalized conclusion. The decisions about who should be treated and for how long, whether estrogen alone, or estrogen and progesterone together should be prescribed, or neither, depend on many factors such as a woman's age, her medical history, the indication for treatment, and the potential risks of side effects. Postmenopausal hormonal replacement can only be scheduled after consulting a physician. In spite of these warnings, the conclusion of recent multicenter trials indicate that the benefits of a hormonal replacement therapy seem to outweigh the negative effects.

In men, the benefits of testosterone replacement have not been well studied, but it is known that such treatment increases the risk of prostatic cancer.

Perspectives

In recent years, scientists claimed that male sexual hormones were implicated in the observed 10% to 20% shorter life span of males compared to females in many mammalian species, including the human. The longer life span of women has been attributed to their extra X chromosome, while men's Y chromosome has been held responsible for their shorter life span (Hamilton et al, 1969). Because of these and other observations, scientists proposed that by inactivating the genes encoding these hormones, men might live an extra 12 to 14 years. At present, nothing is clearly demonstrated, and more studies need to be done in this area, namely on the effect of male sexual hormones on life span, and on the identification of sex-determining genes, their role during development, and their influence on life span.

Dihydroepiandrosterone

Dihydroepiandrosterone, or DHEA, and its sulphonated metabolite DHEA-S, are steroid hormones produced by the adrenal glands and released in the circulation. DHEA is a precursor to several corticosteroid hormones. The level of DHEA (and DHEA-S) that is secreted varies in humans throughout life. It is detected in children, increases during late childhood and adolescence, reaches its maximum value at about 25 years, and then decreases steadily to 10% of its maximum value at age 70. Interestingly, the level of DHEA decreases in all people, but those who had high levels in their twenties, still have relatively high levels in their old age, whereas those with low levels in their youth end up with very low levels in their old age. DHEA may therefore indicate individual differences in aging or lifespan. As the levels of DHEA in the brain are about 6.5 times higher than in other tissues, the substantial drop in DHEA synthesis with age has been implicated in the loss of mental functions.

In human populations, prospective studies indicate that DHEA levels are inversely correlated with breast cancer incidence (Bulbrook et al, 1971), and with death from cardiovascular disease after 50 years of age (Barrett-Conner et al, 1986). These observations triggered research on the effects of exogenous DHEA in rodents. Studies showed that DHEA is involved in converting the body's excess glucose to energy. When given in the diet, it improved the diabetes condition in mice carrying the *obese* mutation (Coleman, 1978; Coleman et al, 1984). DHEA also has been found to play an important role in many metabolic pathways. DHEA was able to suppress breast tumors in a short-lived, moderately obese strain of mice (Schwartz, 1979). The chemical induction of tumors in other organs was also inhibited by DHEA (Schwartz and Tannen, 1981). DHEA has shown remarkable effects in inhibiting other cancers and kidney and vascular diseases. The effect of DHEA on vascular diseases has been quite impressive: fed to rabbits it reduced the size of atherogenic plaques after aortic endothelial injury (Gordon et al, 1988). Furthermore, Steinberg and his collaborators (1989) showed that DHEA acts on the peroxidation of LDL particles, a key factor in atherogenesis. Lopes and Rene (1973) proposed that the effects of DHEA on various metabolic pathways could originate from a common mechanism, i.e., the direct inhibition of glucose-6 phosphate dehydrogenase, a major source for the cofactor NADPH and a contributor to lipogenesis. Inhibition of lipogenesis by DHEA may in fact mimic the action of food restriction (FR). Other scientists postulated that the mechanism of action of DHEA involved sex steroids, because DHEA is converted to estrogen or testosterone, and these hormones are known to reduce fat mass. These experimental results support a protective role for DHEA in various pathological processes.

A few studies were carried out in humans. DHEA given to postmenopausal women induced a pronounced stimulation of natural killer lymphocyte activity (Casson et al, 1993). In a six month study on 30 human subjects, 40 to 70 years of age, DHEA given at a daily dose of 50 milligrams restored DHEA levels to those of young adults within 2 weeks. A twofold increase in serum levels of androgens was noted in women, while there was only a small rise in androstenedione in men. IGF-I levels increased. Two thirds of the subjects noticed a remarkable increase in physical and psychological well-being (Morales et al, 1994).

As Birkenhager-Gillesse and his collaborators (1994) wrote: "Many uncertainties concerning the role of DHEA-S in the neuro-immuno-endocrinological network have yet to be unravelled, and the question remains whether the age-related decrease of DHEA-S is related to organ-specific failure on the level of the adrenals or the gonads, or whether it is a result of changes in feedback or regulatory mechanisms".

Recommendations

Several authors recommended daily supplements of DHEA for elderly subjects. This recommendation cannot yet be endorsed because of the lack of large-scale trials. DHEA has some androgenic effects (masculinization), which are undesirable to women, and are dangerous for men in cases of predisposition to prostatic cancer.

Perspectives

DHEA has been tested in rodents and rabbits, and it is not yet known if it has the same beneficial effects in humans as it does in rodents. Because of its interesting physiological actions on fat metabolism in rodents, DHEA could represent a potentially useful drug in the treatment of obesity in humans. It is likely that DHEA is not the wonder drug it was recently advertized to be, but its positive effects on cellular metabolism could be useful in disease prevention and, possibly, life extension. At the present time, DHEA has not been approved by the FDA and is not available by prescription, but trials are being carried out with AIDS and cancer patients.

Human Growth Hormone

One of the striking alterations that all people experience in middle and late adulthood is the shrinking of lean body mass and the expansion of adipose tissue. The loss of skeletal muscle is the most visible sign of these changes, but other organs such as liver, kidney, spleen, skin, and bone also lose weight with age. This is due in part to the slow but continuous decline in growth hormone secretion (hGH) after the third decade of life (Rudman et al, 1981). Physical exercise or tests involving insulin or arginine injection that stimulate hGH secretion in young subjects fail to do so in old subjects. Analysis of hGH in elderly people during the first phase of slow-wave sleep does not show the pulses of secretion of hGH in blood that is seen in younger persons. As discussed previously (Part I, Chapter 3), the decrease in total sleep time and changes of sleep pattern in elderly people perturb many physiological parameters, and the reduction of hGH release could be linked to sleep disturbances. The diminished secretion of hGH is accompanied by a fall in insulin-like growth factor I (IGF-I), a hormone with a growth-promoting action, produced by the liver in response to hGH.

These observations led scientists to propose that the structural changes of the aging body could be the result of the reduced availability of hGH in late adulthood and could be corrected in part by the administration of hGH, which is now available as a biosynthetic product, produced by genetic engineering techniques. A study reports the effects of synthetic hGH administered to a group of men from 61–81 years old (Rudman et al, 1990). These men had a low plasma IGF-I concentration (350 U/I compared with a range of 500–1500 U/I in healthy men of 20 to 40 years). They received 0.03 milligrams of hGH per kilogram three times a week during six months. Their plasma IGF-I responses were then similar in magnitude to those of younger people, indicating that the hepatic response of older people to hGH was not impaired, but that low plasmatic IGF-I concentration in older men resulted from this hormone deficiency and was not due to a resistance to hGH. An increase in lean body mass and a decrease in adipose tissue mass was observed in these men; these changes were similar to those seen in children and young adults treated with the same doses of hGH. This early trial has triggered several studies on the role of hGH in elderly people. Taaffe and his collaborators (1994) evaluated whether hGH administration is capable of maintaining the improvement in muscle strength that was achieved after 15 weeks of resistance exercise training in older men. They found that exercise training was highly useful in increasing muscle strength but that addition of hGH after the training period did not further improve the results obtained by training alone. On the other hand, lean body mass increased in the group of subjects receiving hGH rather than placebo, with no changes in body weight. This study indicates that hGH adds little to a program of exercise and should not be administered to those elderly subjects who are capable of exercising regularly. Aside from the beneficial effects of hGH on lean body mass, and also to some extent on renal functions, this hormone has a negative influence on bone, where it accelerates bone turnover (Holloway et al, 1994). Moreover, gynaecomastia and carpal tunnel syndrome have been observed.

Recommendations

The hormone hGH plays a central role in many physiological functions, and its lowered secretion with aging may accelerate aging processes. There is no indication at present, that exogenous hGH prolongs life, but, on the other hand, it could be an interesting candidate hormone to control some of the ill effects of aging. For example, treatment of elderly men and women with hGH to prevent atrophy of muscle and thinning of the skin could be beneficial. More studies have to be carried out, however, before hGH can be prescribed in these circumstances. It is of course also necessary to check whether prolonged treatment with hGH may have adverse effects.

Perspectives

The hGH endocrine axis involves the hypothalamus, the hypophysis, and the liver; its regulation is as complex as that of other endocrine axes, with time-related changes (ultradian and circadian rhythms), and with serum hormone-binding proteins and specific cell receptors. As a consequence, there are many potential sites for the pharmacological manipulation of this axis. Most current studies report the results of administration of recombinant hGH, but since hGH cannot be administered orally, other molecules are being developed. The release of hGH is under the influence of the hypothalamic growth hormone-releasing hormone (GHRH), a peptide of 42 amino acids. GHRH and shorter versions of this peptide are efficacious when administered orally. This is the case of the hexapeptide GHRP-6 or nonpeptide compounds such as L-692,429. Another hGH secretagogue is the amino acid, arginine, or the sedative and hypotensive drug, clonidine. Finally, the direct administration of IGF-1 has not been tried to our knowledge. In conclusion, the optimal way to administer hGH or similar acting compounds is still unresolved, as is the true extent of its beneficial effect.

Chapter 6
Medications and Alternative Medicine

Medications that might prolong life fall in two main categories: medications with specific effects aimed at the treatment of disorders, and medications that might improve longevity of the general population rather than of subjects suffering from given disorders.

Anti-aging Medications

The possible anti-aging effects of a few drugs have been selected for discussion. It is unlikely that these drugs will be useful in human life extension.

Phosphatidylserine, arginine, acetylcarnitine and **choline** are all natural substances. They are produced by the body and were shown to decline with age. Ingestion of these substances does not prevent the gradual destruction of neurons in many age-related dementias.

L-deprenyl

L-deprenyl (an inhibitor of monoaminoxidase, MAO-I), a drug discovered in the sixties, extended life span in rats and increased their sexual activity (Knoll, 1988). However, Ingram and his collaborators (1993) failed to find any effect of chronic treatment with oral L-deprenyl on survival or motor performance in mice. Their conclusion is that the issues of dosage, route of administration, and different responses between species have to be carefully taken into consideration to demonstrate effects of L-deprenyl on aging processes. MAO inhibitors induce higher levels of dopamine in the brain, associated with a feeling of well-being and energy (Knoll et al, 1965). L-deprenyl also stimulates the enzymes superoxide dismutase (SOD) and catalase, two enzymes known to block oxidation processes and free radical generation. There is a suggestion that L-deprenyl is efficacious against memory impairment in Alzheimer's disease patients, with some improvement in cognition, anxiety, and depression (Piccinin et al, 1990). L-deprenyl has been approved by the FDA in the USA for the treatment of Parkinson's disease. Its indications as an antidepressant and as a drug supposed to increase sex drive have not been approved by the FDA.

Codergocrine

This drug has been used for a long time and is approved by the FDA for the treatment of cognitive dysfunction in the elderly. It is said to improve memory, social behavior, and sleep in older individuals. It acts by enhancing brain metabolism of glucose and interacts with neurotransmitters involved in brain function. Its beneficial effects are marginal at best.

Phenformin, Metformin, and Buformin

These antidiabetic drugs are used to treat adult-onset diabetes to help normalize glucose metabolism. Metformin may also stimulate the immune system in older people, and there is some evidence that it could be useful in treating atherosclerosis.

Centrophenoxine

Centrophenoxine has antioxidant actions. This compound has been tested in rats and mice and has been claimed to be effective in increasing their learning capacity and in reversing the accumulation of age-related lipofuscin in the central nervous system. Scientists do not agree on the exact mechanisms of action of centrophenoxine. Whether it slows down the normal process of aging in the brain is not known. BCE-100 is derived from centrophenoxine and is supposed to be more efficient in its antioxidant activity. It is also supposed to be more efficient in improving central nervous system higher functions. Clinical trials are underway for BCE-100 as an Alzheimer's disease treatment.

Tetrahydroaminoacridine

Tetrahydroaminoacridine (THA) or tacrine has been shown to improve short-term memory in Alzheimer's patients. This drug acts by helping to preserve acetylcholine in the brain and has modest positive effects on daily-living activities in these patients. No reports exist on the potential effects of THA on life extension in animals. The drug is hepatotoxic, and other compounds with a similar mode of action are being studied clinically.

RU 486

RU 486 (mifepristone) is a steroid compound that blocks progesterone receptors (antiprogestin drug) and causes abortion. The use of RU 486 to abort unwanted pregnancies is legal in France but banned in the United States and other countries. RU 486 also blocks cortisol receptors. Many scientists think that the use of RU 486 should not be restricted to its abortive action. RU 486 is believed to have an enormous potential as an anticancer drug (progesterone has been implicated in the spreading of breast cancer cells in the body). It could have a wide range of other possible applications: in the treatment of Cushing's syndrome, endometriosis, hypertension, obesity, osteoporosis, wound-healing, and benign brain tumors (meningiomas). It might also block some of the biological effects of stress. None of these uses are approved by state regulatory agencies such as the FDA.

Cancer Drugs

Drugs against cancer have no anti-aging effects per se, but inasmuch as they suppress or delay cancer growth, they indirectly increase life span.

Scientists are exploring a new way of controlling cancerous growth. Their goal is to accelerate the natural programmed cell death mechanism that all cells, including cancerous ones, possess, and to make the programmed death come about in a shorter time than that required for cancer cells to multiply and spread. A drug derived from turnips, called thapsigargin, has been used to initiate the death mechanism in human prostate cancer cells, essentially driving them to commit suicide. Prostate cancer cells are a good target for the new technique because they grow slowly and thus tend to resist chemotherapies that preferentially attack fast-growing cells. The drug has been successful in the most virulent of prostate cancer cells from both humans and rats. It triggers a buildup of calcium in the cancer cells. The actual mechanism of action of thapsigargin is thought actually to block calcium channels in an open conformation that results in excess calcium entry from outside the cell. Excess calcium activates DNAses and precipitates the natural death sequence of the cells. Scientists are now thinking of modifying thapsigargin in such a way that it will kill prostate cells preferentially and will not penetrate healthy cells elsewhere in the body. At the same time, they are determining whether this drug can also induce cell death in breast cancer cells.

Scientists recently reported the synthesis of a chemical compound called dynemicin A that was first isolated in 1987 from soil samples. Originally an antibacterial agent, dynemicin A has been found to have the remarkable ability to attack cancer cells by preventing them from dividing. It is supposedly more potent than any known chemotherapeutic agents in killing cancer cells while sparing normal cells. It also gives fewer side effects than existing cancer drugs. As dynemicin A has been found to be toxic in animal studies, synthetic versions of dynemicin A have been made. The new molecules show almost no side effects and are effective against leukemia as well as breast and lung cancers in mice. If human trials prove to be as successful, this agent could be one of the most effective chemotherapeutic drugs.

The reason why normal cells escape the killing process by the drug remains a mystery. Scientists hypothesize that normal cells, unlike cancer cells, retain their capacity to repair DNA damage caused by the drug.

Recommendations

None of the above drugs are recommended for long-term consumption and for life extension, despite enthusiastic recommendations from a few authors. All of them have side effects, and what their benefit-to-risk ratio will be after decades of consumption is unknown, but it is likely to be unfavorable.

Perspectives

Several drugs have been shown to prolong the life of laboratory animals. For example, chronic administration of flumazenil (an antagonist of the benzodiazepine site of the central GABA-A receptor) to rats, beginning at the age of 13 months, was reported to increase their life span (rats exposed to flumazenil lived 1.7 months longer than aged-matched control rats, on the average) and to protect them from the age-related loss of cognitive functions (Marczynski et al, 1994). The anti-aging effects of flumazenil is thought to be due to its inhibitory actions on the GABAergic system. According to the benzodiazepine/GABAergic hypothesis of brain aging, there are abnormally strong benzodiazepine/GABAergic influences in aged individuals which would promote neurodegeneration. In a broader sense, it is possible that the inhibition of binding of hypothetical endogenous anxiogenic substances on the benzodiazepine receptor might play a role in the effect of flumazenil on aging.

Even though flumazenil and other analogous drugs are potentially of interest to humans, extensive research must still be done in animals before human trials can be carried out. Even then, large scale studies (megatrials) over long periods will be necessary to demonstrate the efficacy of these drugs in life extension.

Alternative Medicine

Alternative medicine is popular at present, and bookstores abound in books devoted to these treatments. Clinics and private institutes advertise such treatments, and medical insurance companies have recently discovered that, since these therapies are often less expensive than traditional treatments, it is worthwhile to include them in the coverage they offer.

Alternative medicine is a term that includes a large variety of different therapeutic remedies and techniques: chiropractic, acupuncture, massage therapy, homeopathy, etc. **Chelation** therapy, high dose vitaminotherapy, health foods, herbal medicines and so-called smart pills are also frequently mentioned as alternative therapies. People use alternative medicine for many reasons. Some have a preference for so-called natural therapies. Others turn to these unconventional therapies because they are disappointed with the results of conventional therapies. Several studies reported that between 10% and 54% of patients who suffer from diseases for which there is no treatment, such as cancer, human immunodeficiency virus (HIV) infection, or Alzheimer's disease had used alternative medicine.

Herbal Medicines

Herbal medicines and other natural compounds contain multiple components, and their pharmacology is difficult to study. Despite extensive research, few of these medicines have a recognized role in curing disorders or influencing aging. Moreover, this field of therapy is somewhat conflicting. Those who promote herbal medicines claim that these medicines are endowed with qualities of efficiency and gentle actions, with no side effects. They forget that substances like nicotine, morphine, aspirin, digoxin, vitamin C, and several others are powerful compounds derived from herbs. Traditions and beliefs about herbal medicines and other natural compounds are such that scientific facts are difficult to disentangle from folklore. For example, the continuing demand for medicines derived from rhinoceros horns or from tiger organs, used to achieve increased sexual performance or be endowed with the power of these animals, is a catastrophe for the survival of these endangered species.

Acupuncture

Acupuncture may be considered one of the most popular forms of complementary medicine worldwide. It involves a substantial number of different techniques. It is a traditional therapy in Chinese medicine and is used frequently for its analgesic effects (Cheng and Pomeranz, 1987; Pomeranz and Chiu, 1976). In Europe and the United States, acupuncture is used for numerous ailments and diseases. An increasing number of physicians are trained in these techniques and apply them with reported success in many cases, particularly in the treatment of pain. Unfortunately few controlled clinical trials on the efficacy of acupunture treatments have been reported. This makes it difficult to evaluate the efficacy of acupuncture techniques for the time being.

Some studies report the stimulation of immune responses by acupuncture. It might affect the immune response by bringing about the release of opioid peptides from the pituitary gland (Kato et al, 1983). Recently, Fujiwara and his collaborators described the effects of acupuncture point stimulation on the induction of plaque-forming cells (PFC) in spleen cells of BALB/c mice (Fujiwara et al, 1991). They proposed that helper T cells, derived from the bone marrow, are activated via neurotransmitters, e.g., epinephrine and/or enkephalin, released from the adrenal gland under the influence of acupuncture.

Acupuncture might play a role in postponing aging by boosting the immune system, but these effects are not well documented in the scientific and medical literature. It would be very useful to carry out more in-depth studies designed to find out the effective mechanism of acupuncture.

Recommendations

People who resort to alternative medicine therapists should be aware that most of these therapists do not pretend that they can treat severe and life-threatening diseases such as bacterial infections, cardiac insufficiency, myocardial or cerebral infarction, or cancer. Choosing alternative medicine as the exclusive means of cure may be dangerous when one suffers from a severe disease for which classical medicine and hospital-based techniques have proven efficacy. Such a choice might well hamper longevity.

Although some of the unconventional remedies may be harmless, others may give undesirable side effects or induce adverse reactions.

Perspectives

Alternative therapies may prove to be effective for a number of problems, not the least of which is stress reduction. It is therefore important to test the long-term effects of these therapeutic approaches by carrying out longitudinal follow-up studies.

Chapter 7
Surgical Strategies

Plastic Surgery

The speciality of plastic surgery has blossomed in the last decade. Rejuvenating and reshaping parts of the body that look either worn out or ungraceful can have important positive psychological effects and give patients a new lease on life. Just about any part of the body can be reshaped: face, nose, breast, hips, thighs, knees, arms, abdomen, and so on. Surgeons are using new approaches and the classical scalpel is replaced frequently with new technologies, such as the laser and liposuction techniques. Nowadays, an increasing number of men and women resort to the many face-lifting techniques, for personal and even professional reasons. Face lifting or other plastic surgery interventions, tightens their skin and therefore modifies their features in a way that makes them look less tired and more fit and competitive. More than ever before, people change jobs, and men and women in their forties or fifties think that they can increase their chances of getting a good professional position if they look younger. About ten years ago, women and men undergoing face lifts were made to look many years younger. Today, such drastic interventions and changes are no longer fashionable. Candidates for face lifts want more subtle changes, for example, getting rid of wrinkles or looseness of the skin.

Recommendations

Regardless of the type of body transformations or techniques used, plastic surgery remains a surgical intervention, and the trauma involved should not be undervalued. Candidates for plastic surgery should discuss fully their motives and the ensuing risks of the surgery with their surgeon. They should schedule enough free time for the hospitalization and healing process and should make their decision with the support of their spouses and other family members. Surgical interventions impose acute stress for which one should be well prepared.

Replacement of Body Parts

Aging is a matter of wear-and-tear and, like any machine, the human body has a tendency to give in to entropy and run down. Replacement of body parts as they get damaged or diseased can prolong life and keep people active much longer. Thousands are alive thanks to transplanted kidney, heart, lung, or liver. Each year, millions of people have some parts of their body replaced. Teams of scientists, physicians, and surgeons now work in close collaboration to replace defective vital organs, including pancreas, some endocrine glands (the thyroid gland, for example), and even neurons in the brain (see below and Chapter 8).

The major problem encountered in transplantation medicine is immune system reactions. To date, no pharmacological compound better than cyclosporin has been found to prevent rejection. Cyclosporin, a fungus-derived drug discovered in the seventies, is very efficient in repressing the immune system and preventing immune cells from attacking transplanted organs. Unfortunately, it has its drawbacks since it makes the body vulnerable to bacteria and virus infections. Many body parts can be replaced without triggering rejection by the immune system. The first body parts that were replaced were teeth. Today, dentists have new sophisticated techniques to replace teeth, using a titanium socket for the prosthetic tooth anchored in the jawbone. Artificial plastic hips have been in existence for more than twenty years. Surgeons are now capable of replacing the knee, one of the most complex joints. Manufactured bioceramic bone made of synthetic calcium phosphate hydroxylapatite, which is one of the natural substances contained in real bone, does not induce immunological graft rejection. Preliminary experiments with bioceramic bone transplanted into the thigh bones of rabbits were successful; the artificial bone formed a foundation on which natural bone grew. Four years after transplantation, the bioceramic material had been entirely converted to real bone. Scientists are synthesizing substitute cartilage. Their approach consists in isolating the stem cells lining a rabbit rib, placing those cells in softened bone, and then transferring the mixture of stem cells and bone into holes drilled in the rabbit's own cartilage. New cartilage is synthesized in the holes in almost 90% of the rabbits tested. These results have prompted surgeons to use a similar approach in humans. They have reported successful treatment of knee injuries with laboratory-grown cartilage cells (Brittberg et al, 1994).

Several groups working in biotechnology companies are actively involved in making artificial skin, with the primary goal of healing patients with burns or wounds. But artificial skin will also probably be used in the future by cosmetic surgeons to replace old skin with a fresh substitute. One way is to make a synthetic mesh lattice in which human skin cells are imbedded. Another is to produce a replacement epidermis made from the patient's own cells.

In recent years, the development of laser photocoagulation therapy as well as other new surgical procedures and new drugs have greatly improved the treatments of vision disorders like glaucoma and cataract. The transplantation of the cornea has now become a standard surgical procedure, and over a million people a year all over the world undergo surgery to replace their damaged corneas. Intensive research is now being carried out on the retina: here the challenge is tremendous since nerve cells and connections have to be reestablished.

Scientists are also searching for ways to synthesize artificial blood, although attempts at making artificial red blood cells (RBC) have failed so far. Synthetic chemicals known as perfluorocarbons which can carry gases like oxygen and carbon dioxide might be useful (the major problem with these compounds is their strong immunogenicity). Scientists have tried to cross-link hemoglobins to form polyhemoglobins, or microencapsulate hemoglobin in the hope that these molecules would remain in the bloodstream longer than free hemoglobin alone.

Implant technology provides new and exciting prospects. This includes, for example, the direct implantation of altered neurons that could synthesize and secrete brain transmitters or express receptors where they are needed in the brain. Implants containing insulinsecreting cells are also engineered to treat diabetes.

Chapter 8

Perspectives in the Detection and Treatment of Diseases

In the last two decades, considerable progress has been made in the detection and treatment of human diseases. The advent of molecular biology and the development of recombinant techniques have led to considerable progress in biomedical research. In particular, it has allowed the isolation and characterization of the gene defects in a number of diseases. This has opened new directions in clinical medicine, in particular in diagnosis of diseases and genetic counselling, and in prospects for prevention and cure. Physicians can now detect predispositions to some diseases long before clinical manifestations: genetic abnormalities that predispose to metabolic diseases such as diabetes, cardiovascular, or neurodegenerative diseases can be detected in young adults. It is also possible to detect prenatally those fetuses that carry inherited diseases, such as mucoviscidosis or thalassaemia. The combined knowledge from the fields of molecular biology and microbiology has opened new avenues in pharmacology and therapeutics and has given rise to new approaches for synthesizing drugs, namely the large-scale synthesis of proteins, peptides, factors, that have been found to have therapeutic potential.

New technologies have also been devised to treat certain diseases considered incurable until now. These new approaches are grouped under the name of gene therapy. Gene therapy includes a variety of techniques; most of them are designed to replace a defective gene with its normal counterpart (Kahn, 1993).

These discoveries are likely to influence people's life span. A brief description of biotechnological approaches and examples of gene therapy are given below.

Biotechnology Contributions in Medicine and Gene Therapy

Genetic engineering techniques include a variety of biochemical and enzymatic procedures that allow the recombination of DNA sequences from various microorganisms or viruses and the isolation and characterization of specific DNA sequences. These techniques proceed from the discovery of, among others, restriction enzymes and other DNA-modification enzymes, and, the working out of methods of sequencing DNA molecules.

Using these biotechnological tools, scientists succeeded first in identifying genes encoding regulatory proteins, enzymes, factors, peptides, and receptors and then proceeded to study their functions. Many of these proteins and peptides turned out to have important physiological actions and were found to restore missing functions. This then stimulated a new area of research using microorganisms and cells in culture, and involving the large-scale production of peptides, proteins, and factors as therapeutic medicine.

Production of Recombinant Proteins

By the end of 1993, over 150 drugs or vaccines, produced by genetic engineering, had been tested in clinical trials in the USA; some of them were ready to be accepted by the Food and Drug Administration (FDA) for clinical use. A large number of these are modulators of the immune system: interferons, interleukins, colony stimulating factors, growth factors, etc. Many newly isolated factors are of neuronal origin.

The production of gram quantities of recombinant proteins (proteins produced by gene cloning) is a multistep process. This process is summarized in Figure 23.

Gene Therapy

Most inherited or acquired diseases resulting from genetic errors cannot yet be cured; their symptoms are treated by classical drug therapies. With the advent of biotechnology and genetic engineering, scientists formulated the idea of using specific DNA sequences or genes either to modulate the expression of a specific cellular or viral gene or to correct the structural anomaly of a mutated gene. This general approach to treating human diseases, gene therapy, has now moved from fiction to reality. The challenge is to introduce the gene of interest isolated from a healthy donor into the cells containing the defective gene of the affected patient. The first attempts to treat a disease by gene therapy were carried out on patients suffering from a severe immunodeficiency caused by mutations in the gene encoding adenosine deaminase (ADA; Anderson, 1992), a disease usually treated by bone marrow transplantation or by repeated injections of enzyme preparations. The general approach, successfully used to treat ADA patients, was applied to patients suffering from other diseases. Figure 24 describes the successive steps involved in the treatment of ADA patients using genetic therapy.

Different strategies are used to introduce genes into patients. A gene of interest, isolated from a healthy donor, can be either: (1) inserted in vitro into the patients own cells and the cells are then reintroduced back into

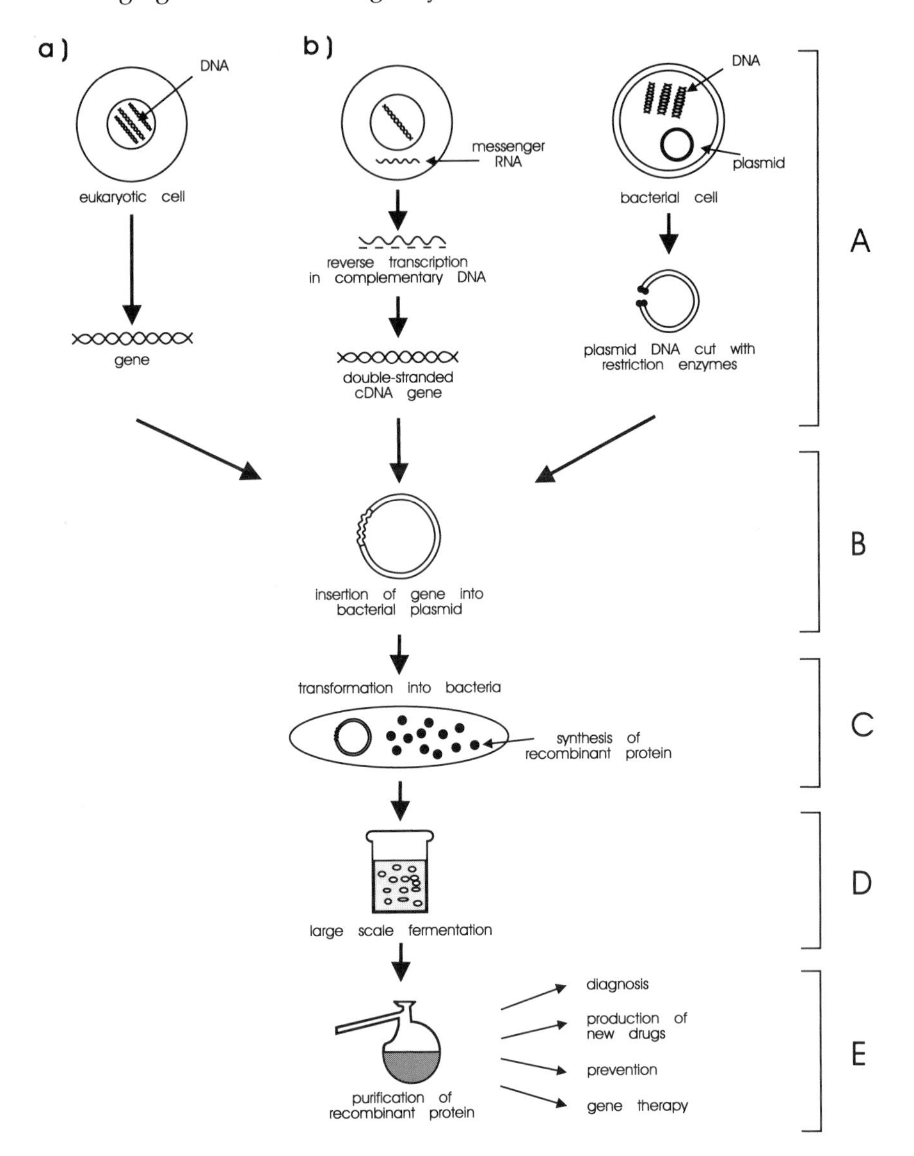

Figure 23. Diagram representing the different steps involved in the large scale production of a recombinant protein in bacteria.

A. Isolation of a gene from eukaryotic cells (a: gene isolated from nuclear DNA; b: gene isolated from the messenger RNA in cytoplasm.)
B. Insertion of the gene into a bacterial plasmid
C. Transformation into bacteria or human cells in culture
D. Large scale bacterial amplification
E. Purification of the recombinant protein

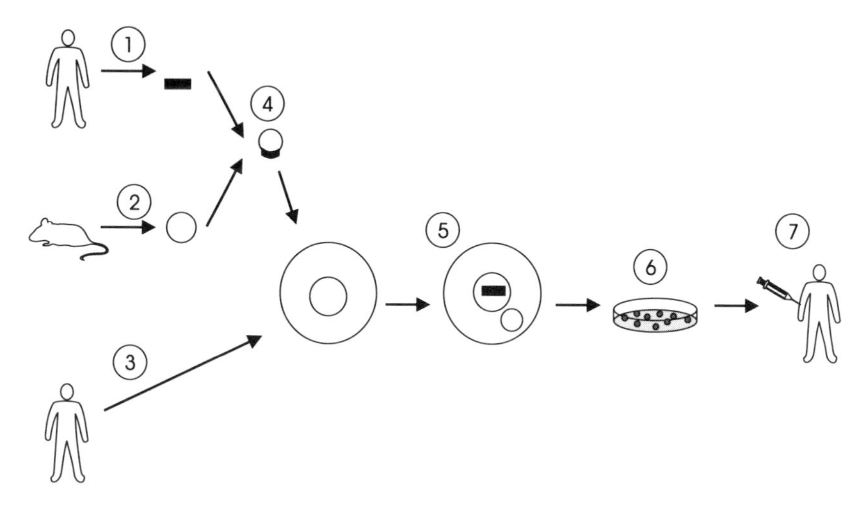

Figure 24. Schematic representation of the transfer of a normal adenosine deaminase (ADA) gene in a patient suffering from a fatal immune disease.

1. The normal ADA gene is isolated from the cells of a healthy individual.
2. **Retroviruses** are isolated from a leukemic mouse and are inactivated.
3. Blood cells from the patient are adapted to tissue culture conditions.
4. The normal ADA gene is inserted in the genome of the retrovirus that serves as a carrier to transfer the gene to appropriate cells.
5. The ADA gene reaches the cell nucleus and triggers the expression of a normal ADA protein.
6. Cultured cells multiply in vitro. Each cell is a factory producing the normal protein.
7. When there is a sufficient number of transformed cells, they are reinjected in the patient. Transformed blood cells little by little supercede diseased cells and deliver the healthy protein in the circulation.

the patient. This is possible with blood cells, for example, which can be retrieved from the patient, cured, and injected back into the patient; (2) introduced into the genome of an attenuated virus (*herpes simplex*, retroviruses, or adenoviruses). The virus is then injected at the diseased site of the patient; (3) inserted into a plasmid which is itself encapsulated into a polylysin core or liposomes. These vesicles are then injected into the patient's blood; (4) injected directly into a patient's tissue, for example, DNA injected directly into skeletal muscle. These techniques have their specific advantages and disadvantages.

It has proved feasible to correct defective gene expression in blood cells or cells from tissues that are easily accessible. Researchers have recently reported a successful attempt to introduce the LDL receptor gene in human liver cells (Grossman et al, 1994). It will be more difficult to correct defective genes in less accessible organs, in particular, the brain.

Contribution of Genetic Engineering Techniques to Medicine

Genetic engeneering techniques are used in many areas of medicine: production of new drugs, diagnosis, disease prevention, gene therapy, and medical research. These techniques are crucial for the detection and identification of genetic defects that cause diseases. The number of diseases found to be the result of a single defective gene is estimated to be around 4000. Out of these, 350 have been studied, and the genes responsible have been characterized in several cases. Some are transmitted, like the hemoglobinopathies, the myopathies, mucoviscidosis, Crigler-Najjar disease (a liver disease), another severe liver disease characterized by the absence of ornithine transcarbamylase, or acquired conditions like some cancers or viral diseases. The genes of many of these diseases have been mapped to specific chromosomes. In many cases, a DNA test is already available or under development. In others, genes have been localized on chromosomes, but have not been isolated, or diagnosis is only available through family-linkage studies. Progress in gene therapy moves fast, and almost every week there are reports of medical breakthroughs. Several examples are given below to illustrate these new therapies.

Adenosine Deaminase Gene

Anderson and his colleagues (1992) attempted treatment of ADA by gene therapy. They first isolated the gene for adenosine deaminase then introduced it into a retrovirus. The virus was then used as a vehicle to introduce the gene into the nucleus of the patient's lymphocytes. White cells modified by this method were injected back into the patient's blood (see Figure 24). As lymphocytes have a limited life span, this treatment must be periodically repeated. Two girls (4 and 9 years old) have been treated with recombinant lymphocytes for two years. There has been a clear clinical and biological improvement of their immune functions with no side effects.

Multidrug-Resistance Gene

Today, chemotherapy of metastatic diseases has proven effective in the treatment of only a few human cancers. For example, cures have been obtained in testicular cancer, some leukemias and lymphomas, Hodgkin's disease, and some childhood sarcomas. However, in other tumors, the treatment is palliative and remissions are transient. One problem in cancer therapy is the failure of many human tumors to respond to chemotherapy because of the development of a broad-spectrum chemoresistance, termed multidrug resistance (MDR): cancer cells that are not quickly destroyed become insensitive to the drugs. While researchers look for new and better chemotherapeutic agents, it is also essential to enhance the efficacy of classical chemotherapy. This involves a close analysis of the biochemical substrate that confers MDR. Gottesman and Ling identified a gene, called MDR1, which is expressed in increasing amounts upon continued chemotherapy (Endicott and Ling, 1989; Pastan and Gottesman, 1991). The protein encoded by MDR1, called P-glycoprotein, a 170-kD transmembrane protein, acts like a chemical pump and extrudes many types of compounds from the cell, including chemotherapeutic agents, thereby conferring a drug-resistant phenotype to cells expressing the protein. About 50% of all cancer cells and many normal cell types express P-glycoprotein in significant amounts; it is found on the surface of epithelia of the kidney, intestine, liver, and pancreas, in the adrenal cortex, the placenta, and in capillary endothelial cells in testis and brain. The reason why some cells, such as mature bone marrow cells and most types of mature blood cells, do not express it, is not known.

Scientists are now introducing MDR1 gene into bone-marrow cells, the cells most vulnerable to chemotherapeutic agents. The first attempts by Sorrentino and his colleagues (1992) yielded encouraging results. Working with mice, they extracted healthy bone marrow cells and injected the animals with cancer cells. A harmless, nonreproducing virus carrying the activated MDR1 gene was then inserted into the extracted bone marrow stem cells. The modified bone marrow cells were reinjected into the mice that were subjected to an intensive chemotherapy. There was no drop in the white cell count, indicating that the treated bone marrow not only survived higher chemotherapy dosages but appeared to reproduce and replace the original weaker marrow. This type of therapy might soon be applied to cancer patients.

Tumor Necrosis Factor Gene

Rosenberg and his colleagues (1988; 1990) treated patients in the advanced stages of melanoma. Their approach was to boost the immune system against the disease. They used lymphocytes targeted to the tumor (TIL) as a way to deliver large amounts of substances with antitumor activity, such as cytokines (Interleukin-2) and other gene products (for example Tumor Necrosis Factor) to the site of tumor. In this way they circumvented the problem of toxicity associated with the systemic delivery of these gene products.

Dystrophin Gene

A major problem with the classical genetic techniques is the mandatory passage to tissue culture. Protocols exist that can circumvent the culture stage, and Ascadi and colleagues (1991) injected the human dystrophin gene directly into mice suffering from myopathy; they reported some success. Their work may point the way to dealing with human diseases such as muscular dystrophy.

Cystic Fibrosis Gene

Recently, a group of scientists reported success in correcting the chloride-ion channel defect in transgenic mice expressing a disrupted channel gene. They transferred the normal gene to airway epithelial cells, using liposomes containing a functional cyclic-AMP-regulated chloride channel. The ion conductance defect in the trachea of transgenic mice was corrected by this gene therapy (Hyde et al, 1993). These studies open the way to applications of gene therapy to the pulmonary defects of human cystic fibrosis.

Protein p53 Gene

The properties of p53 (see Part I, Chapter 5) give hope for future cancer therapies. At present, the majority of cancers are treated with nonspecific drugs that attack healthy cells at the same time as cancerous cells. The discovery of p53 has uncovered one of the fundamental mechanisms of the genesis of many cancers and may be used to devise new and better targeted drugs. Current efforts are directed towards developing diagnostic tools to detect altered forms of p53, which will serve as disease markers. Also, there is interest in finding drugs that repair the altered form of p53 or in modifying directly the p53 gene itself.

LDL Receptor Gene

Human gene therapy took a great step forward when Grossman and her collaborators (1994) announced the success of an ex vivo approach to treating a 29-year-old woman with familial hypercholesterolemia. Their strategy is summarized in Figure 25.

First, about fifteen percent of the woman's liver was removed. Then, the cultured hepatocytes were treated with recombinant retroviruses containing the LDL receptor gene. The transfected hepatocytes were analyzed for uptake of the gene and returned to the patient via injection into the portal vein. Grossman and collaborators demonstrated expression of the LDL receptor in grafted cells after 18 months, and they reported a 17 percent decrease in the woman's LDL level and a 40 percent decrease in her LDL to HDL ratio.

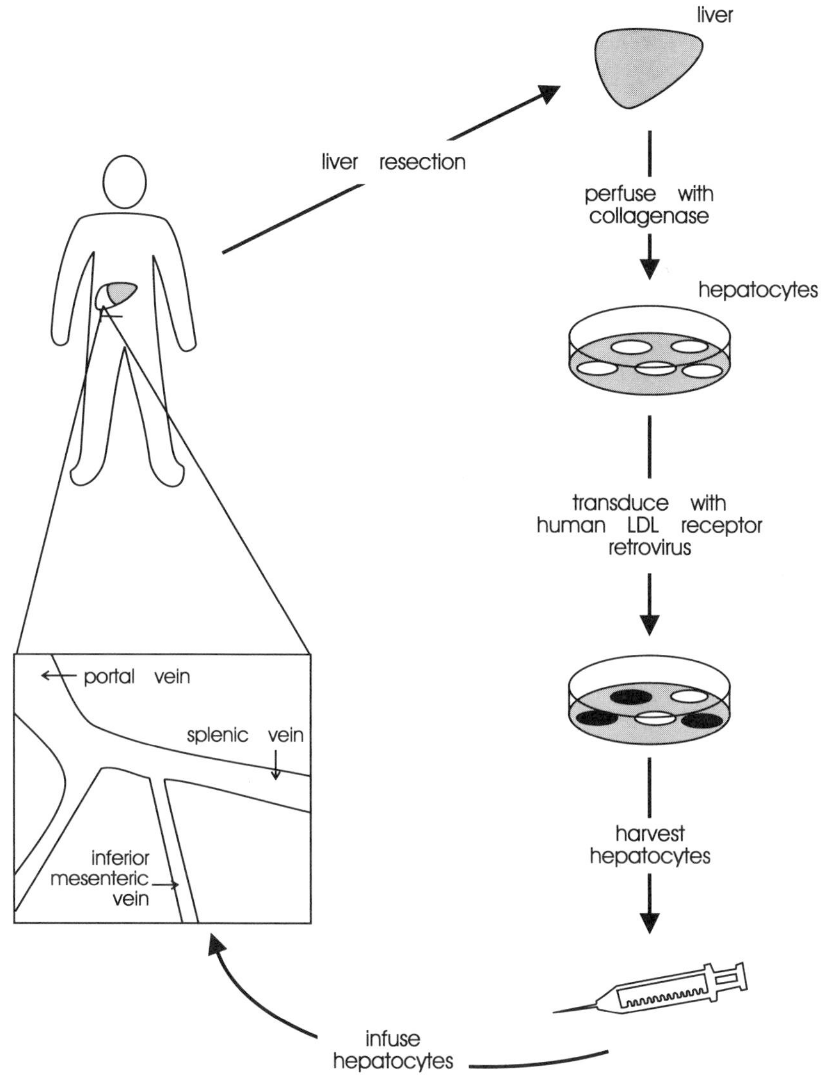

Figure 25. Strategy for ex vivo gene therapy for familial hypercholesterolaemia.

Perspectives

The application of molecular genetics and biotechnology in medicine and medical research has had profound effects on the way researchers and physicians understand, diagnose, and treat human diseases. Thanks to these powerful new tools, it is now possible to have a better understanding of the pathophysiology of many disorders and to identify susceptibility to a large number of diseases. This revolution in molecular medicine has had and will have an important impact on medical practice. Genetic diagnosis will be improved, and this may lead to better counselling of patients and families with inherited diseases. There is hope that better treatments will exist for genetic diseases, and for cancer, AIDS, hepatitis, parasitic diseases, and many others. There is also much enthusiasm for gene therapy, and this enthusiasm is understandable and legitimate in view of the recent spectacular advances made in this field. However, the real clinical efficacy of gene therapy has not yet been demonstrated. Gene therapy is still in its infancy, and investigators in the field are facing serious difficulties in the applications of gene transfer. Sadly, negative results have been reported by different investigators after gene therapy in human patients (Knowles et al, 1995; Mendell et al, 1995). The transfer of human cultured myoblasts in muscles of patients with Duchenne's muscular dystrophy produced no improvement in muscle strength; the myoblasts did not survive. Similarly, treatment of patients with cystic fibrosis with a recombinant adenovirus containing the CFTR gene was not successful because of a low expression of CFTR protein. If the results of these early trials are disappointing, they should not discourage further research and trials in gene therapy. In fact, an important message drawn from the failures is that there is a need for new delivery systems. Scientists are now working on new systems and new approaches to deliver genes in vivo. For example, it has been found that the intramuscular injection of recombinant plasmid can program the expression of specific proteins in myocytes. This has led to the development of new vaccines for infectious diseases. Alternative delivery systems for proteins and peptides are being developed rapidly. Scientists are setting up a variety of methods for oral delivery. For example, proteins or peptides are being packaged in various hydrophobic and hydrophilic polymer formulations or in liposomes to protect them from being inactivated by the gastrointestinal tract. Some delivery systems employ semipermeable membranes to control release of the drug and some take advantage of the specific physiological properties of the tissues involved, such as respiratory tract delivery systems or transdermal patches.

Application of Gene Therapy to Neurological Diseases

Diseases of the nervous system represent a major public health problem. The list of disorders of the nervous system is extensive and includes diseases ranging from irreversible damage of neuronal pathways, such as results from spinal cord injuries, to slow degenerative diseases. About a third of neurological disorders are inherited, the other two thirds occur sporadically at adult age. In most cases, the basis of the pathology cannot be determined, and available treatments are often of little effect on the rapid evolution of the disease. The challenge of neurobiologists and neurologists is to find ways to stop the irreversible neuronal damage after injury and to provide the organism with the ability to reestablish axonal connections, or, in the case of neurodegenerative diseases, to remodel synaptic circuitry in order to restore functional recovery.

Neurotrophic and Neuroinhibitory Factors

In the last ten years, a large number of neurotrophic factors have been discovered and characterized. The possibility that these factors could be of potential use in slowing neuronal loss in a number of neurological diseases has led to the creation of many biotechnology companies that produce large quantities of factors using recombinant biotechnology. Most of these factors have been found to promote neuronal survival when they were tested in experimental animal models of neuronal death. Some of them induce cell migration, others, neurite outgrowth and new synapse formation. Scientists have concentrated their efforts first on degenerative diseases of motoneurons, particularly on amyothrophic lateral sclerosis, a disease characterized by a progressive loss of spinal and cortical motoneurons (ALS; see Part 1, Chapter 5, pages 118–119). The evolution of this disease is inescapable and very rapid, and there is currently no treatment.

Scientists have discovered numerous neurotrophic factors that are capable of preventing death of spinal and cortical motoneurons in vitro and in vivo. These factors belong to several family of molecules: neurotrophic factors such as nerve-growth factor (NGF) and brain-derived neurotrophic factor (BDNF), neurotrophins such as NT-3 (see below) and NT-4/5, glial cell-line-derived neurotrophic factor (GDNF), ciliary neurotrophic factor (CNTF), leukemia inhibitory factor (LIF), as well as growth factors like fibroblast growth factors 2 and 5 (FGF-2,-5), insulin-like growth factor-1 (IGF-1) (Henderson, 1995a). Spinal motoneurons, unlike central motoneurons, project into the periphery and are capable of taking up and transporting neurotrophic factors in a retrograde fashion. Because of this feature, treatment with neurotrophic factors was thought to be and promises to be relatively direct and uncomplicated.

The discovery of these factors has become a major driving force in basic and applied research and an increasing number of studies are being carried out, not only on neurons and glial cells but also on surrounding neuronal tissue, for example on myelin, the white matter sheathing on nerve fibers.

It has been shown that after partial lesions, uninjured axons can respond to the denervation of neighboring areas by expanding their connections within the central nervous system (CNS) of adult mammals. This phenomenon is called collateral sprouting and occurs both in the peripheral and central nervous system. Scientists noticed that in the mature CNS, regions that show a high degree of collateral sprouting are poorly myelinated. On the other hand, in highly myelinated regions, lesion-induced sprouting could be observed at fetal ages but not postnatally. It appeared therefore that when myelin forms, it inhibits further growth of the axons. This is compatible with the observation that, during axonal growth in the developing animal, there is no myelin formation until axons have found their destiny, the target muscle, organ, or tissue they will innervate. Myelin thus has been shown to contain molecules that actively inhibit the growth of nerve fibers. These neurite growth inhibitory molecules have been shown to be expressed on the surface of oligodendrocytes and are contained in CNS myelin; they are expressed only after long-distance fibers have reached their target and stopped growing. These findings suggest that the ability to establish new connections is negatively correlated with the degree of myelination in the CNS. The final proof that myelin is responsible for the inhibition of collateral sprouting has been given recently by Schwegler and his collaborators (1995) who showed that, by experimentally suppressing myelin formation, lesion-induced sprouting is strongly enhanced in the spinal cord. Thus, the failure of axons to regenerate in the central nervous system can be attributed in part to inhibition by surrounding tissues. The role of myelin neurite growth inhibitors is important because they seem to be necessary to maintain the stability of neuronal connections in the normal CNS.

However, in the case of lesions, it would be desirable to prevent the action of neurite growth inhibitors. In 1990, Schnell and Schwab used neutralizing antibodies to block inhibitory myelin-associated proteins and were able to demonstrate regenerating axons in some cases. Since axonal growth depends on the balance between growth-promoting and growth-inhibiting factors, they then searched for neurotrophic factors that would increase the magnitude of regeneration. They found a factor that they called neurotrophin-3 (NT-3), a member of the neurotrophin family, that is expressed in the spinal cord during fetal and early postnatal development (Schnell et al, 1994). NT-3, in combination with inhibition-blocking antibodies, promotes axonal regeneration over remarkably long distances down the spinal cord (provided part of the cord has been left intact). Their reports are the first to show unambiguously that extensive functional spinal cord regeneration is possible.

However, treatment with neurotrophic factors may not always be beneficial. While increasing collateral sprouting with some of these factors may contribute to functional recovery, it could also destabilize the equilibrium between neurotrophic and inhibitory factors and lead to changes in neuronal connectivity after lesions or to changes resulting from altered functional activity. Furthermore, these factors have been used in embryonic motoneurons in culture or in mouse models in which motoneuron death occurs at neonatal or juvenile stages. It is therefore conceivable that these factors may not be active on dying adult motoneurons, or may not protect upper motoneurons, or may even show undesirable side effects. The results of the first clinical trials with CNTF in the treatment of ALS were, in fact, disappointing and emphasize that much more needs to be discovered about the pharmacological and the pharmacodynamic effects of these factors before they can be used as therapeutic tools in ALS (Henderson, 1995b).

The discovery of neurotrophic factors inspired neuroscientists to use them in the treatment of neurodegenerative diseases. Many drugs to treat Alzheimer's disease have been tested in clinical trials but without success, and there is currently no effective method of prevention, treatment or cure for this disease. The hope in using neurotrophic factors has been that, due to their growth-enhancing activity, these factors might stop the pathological process leading to neuron loss and might promote neuronal growth and regeneration. In principle, these peptides would be promising candidate drugs to treat neurodegenerative diseases. However, results of peripheral injection of peptides in animals show that they have a short half-life, and often induce serious side effects, which can lead to death when given at high doses. Furthermore, because of their size and nature, proteins and peptides are not able to cross the blood-brain barrier, and some of them diffuse little in nervous tissue. This inability to enter the brain considerably hampers their use as therapeutic agents to treat neurodegenerative disease. To circumvent this problem, scientists are thinking of synthesizing nonpeptidergic drugs that would be as efficacious and specific as natural peptides. These molecules could be agonists or antagonists of peptide receptors. Several pharmaceutical companies are actively searching for such molecules that could enhance neuronal growth or, conversely, block pathological processes.

Alternative methods are being devised to increase the level of neurotrophic factors in the brain. Recently, two groups of scientists reported the successful transfer of genes in cells of the central nervous system (Akli et al, 1993; Le Gal La Salle et al, 1993). Because adenovirus can infect neuronal cells in vitro and in vivo, these scientists injected recombinant adenovirus stereotactically, either intraventricularly or into various brain areas such as the hippocampus, the substantia nigra, and the striatum. The goal was to use adenovirus as a transfer tool to introduce genes encoding various neurotrophic factors. Reports indicate that the technique is promising and might ultimately be useful in treating neurodegenerative diseases such as Parkinson's or Alzheimer's diseases. However, many technical difficulties have to be solved before this type of treatment can be applied to humans.

Other approaches to deliver peptides in the brain include implants of encapsulated cells that would synthesize peptides in neuronal tissue, or the installation of micropumps. These techniques might allow the slow and steady delivery of neurotrophic factors.

Neuronal Grafts

Considerable work has been done to promote regeneration after spinal cord injury and neuroscientists have used different approaches. One approach was to use grafts of fetal tissue to regenerate axons (Iwashita et al, 1994). Scientists excised a 1.5–2 mm piece of the lower thoracic spinal cord in 1-to 2-day-old rat pups and replaced it with a piece of fetal spinal cord of similar length. The fetal transplants fused with both transected ends to form an almost seamless spinal cord with well-organized white and grey matter and visible spinal roots. An extensive and long-lasting functional recovery was reported in the most successful cases. Another approach was to use fibroblasts that had been genetically modified to secrete nerve growth factor (NGF; Tuszynski et al, 1994). These cells, grafted in the central canal of the spinal cord, induced a robust sprouting response from sensory neurites. This type of response had not been observed in previous grafting experiments using nerve bridges or fetal cells and was attributed to the high level of NGF released locally. In addition, many neurites within the NGF-secreted grafts were enveloped in the processes of cells that had the ultrastructural characteristics of Schwann cells, and these neurites became myelinated. The presence of Schwann cells contrasts with results obtained in the brain. When genetically-modified fibroblasts were grafted in the striatum, for example, sprouting neurites were often associated with astrocytes, and no Schwann cells were observed. This raises the possibility that neurites from the peripheral nervous system may have intrinsic characteristics that promote myelination by Schwann cells rather than by oligodendroglial cells. The challenge now will be to apply these new therapeutic tools to treat patients with spinal cord lesions.

Grafts of fetal neurons, genetically modified cells, or neuronal cell lines have also been used in the CNS in the hope that they will restore damaged neuronal circuits and consequently will reverse functional deficits. There have been encouraging results from studies involving neurons from rat fetuses transplanted into damaged areas of an adult rat brain (Widner, 1993); embryonic cholinergic neurons transplanted into rat brains, in regions previously injured with a cholinotoxin (Emerich et al, 1992); and fetal dopaminergic neurons transplanted into rat brains (Chung et al, 1993). In the study on the transplanted cholinergic neurons, the transplanted cells caused a generalized increase in cholinergic activity, which was attributed to the release of trophic factors from the transplant; some recovery of performance on some memory tasks was seen. In almost all cases, the transplanted fetal cells survived and promoted significant functional recovery in the animal models. This same approach has been used to treat human patients suffering from Parkinson's disease: neural tissue from embryonic substantia nigra was transplanted into these patients and led to partial remission (Fisher and Gage, 1994). In view of the results obtained with trophic factors on the one hand and neural grafts on the other, Tuszynski and Gage (1995) decided to combine transient NGF infusions with grafts of substrates that promote axonal growth regeneration. This combined approach—both preventing host neuronal loss and also promoting axonal growth—induced increased functional recovery compared to earlier approaches.

Scientists are now testing other types of cells for grafting. Multipotential neural precursor cells (stem cells) are interesting candidate cells, because they show a remarkable plasticity and can differentiate into diverse neuronal and glial cell types. These cells could be manipulated toward specific lineages that would synthesize factors of interest and could be grafted in regions that have suffered injury or neurodegeneration.

Biotechnological Approach to Kill Brain Tumor Cells

Some brain tumors are inoperable because they are hidden beyond reach of scalpel or laser. A strategy used to kill brain tumor cells selectively without destroying neighboring tissue was devised by Culver and his colleagues (1992). The rationale of their experimental approach was to induce tumor cells to express a protein that would be the target for a drug. The protein of choice was thymidine kinase, because cells producing it could be turned into a target for antiviral drugs. This so-called suicide gene approach consisted of inserting into mouse fibroblast cells a retroviral vector carrying a gene coding for the enzyme thymidine kinase from *herpes simplex* virus. The transformed fibroblasts were injected precisely into the brain tumor of a rat (Figure 26a). Copies of the recombinant retroviral vector were synthesized and infected the nearby dividing tumor cells, but not the nondividing adult neurons (Figure 26b). Infected tumor cells produced thymidine kinase, and converted ganciclovir, a commonly used antibiotic and an analogue of thymidine, into a cytotoxic product (Figure 26c). The use of mouse cells containing a gene-bearing virus represents one of the most efficient way of introducing genes into a large number of cells. In rats, this form of gene therapy proved to be very toxic to cancer cells. Surprisingly, even cells that were not genetically altered could be killed. Exactly how this bystander effect operates is not known.

Scientists are now improving the technical aspects of this therapeutic approach and are trying to use nonretroviral vectors. They will apply similar strategies to other types of tumors.

Encapsulated Cell Therapy

There is growing interest in using the encapsulated cell transplant method for treating a variety of conditions and, particularly, neurological disor-

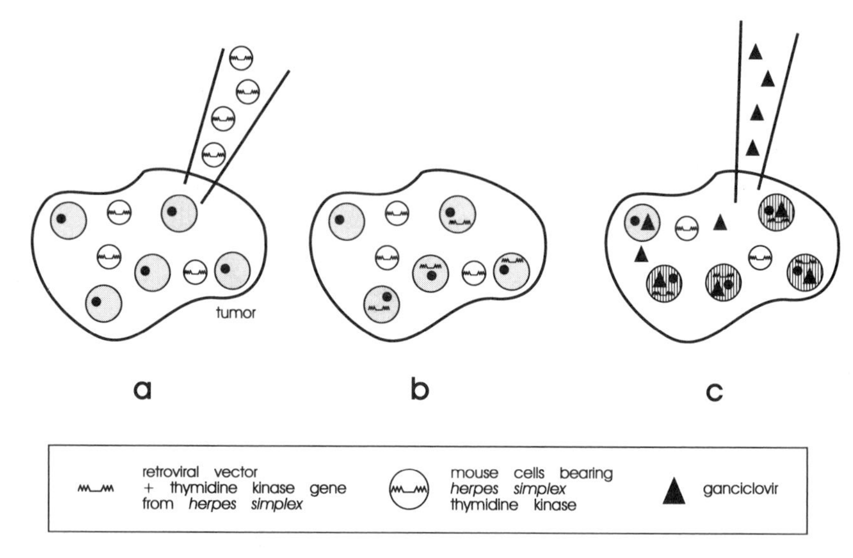

Figure 26. Schematic diagram illustrating the suicide gene therapy. Brain tumor cells are specifically destroyed by transplantation of murine fibroblasts producing retroviruses containing *herpes simplex*-thymidine kinase gene (Culver et al, 1992).

ders. The rationale for using encapsulation in cell therapy is to protect allogeneic and xenogeneic cell transplants from being destroyed by the host's immune system and to protect the host from transplanted cells. An advantage of this technique is that since there is no immune response against the encapsulated transplants, there is no need to use immunosuppressive drugs. Another advantage is that, being encapsulated, the transplanted cells can be removed from the patient. This technique was used successfully for the treatment of chronic pain using chromaffin cells, which were isolated from the adrenal glands of calves and placed in a semipermeable membrane. These cells secrete catecholamines and opioids (natural analgesics). The encapsulated cells were implanted either at the base of the spinal cord or in the lateral ventricle of the brain and they released the analgesics, which bound to receptors locally. Many patients treated with this type of therapy showed a reduction in their chronic pain. This technology is particularly suited for CNS applications, where the major problem is to get drugs to cross the blood-brain barrier. The encapsulated cell method is particularly suited for the treatment of amyotrophic lateral sclerosis. Hamster kidney cells, engineered to express human ciliary neurotrophic factor (CNTF), might be implanted in areas where motoneurons die. Encapsulated cells can secrete CNTF in small doses, and slow release may be a way of avoiding the problems encountered with the administration of large doses of CNTF.

New Targets for Drug Therapy

Pharmaceutical companies have successfully synthesized various compounds that bind and disable specific proteins, and there now are a large number of drugs that block receptors, block ion-channels, or inhibit H^+/K^+-ATPase pumps. Other drugs block enzymes or proteins that act on intracellular signalling pathways. Researchers are pursuing a number of new ways of creating drugs that are targeted at different cellular sites. For example, a new way to block protein function is to block protein synthesis at its very first stage, the transcription of DNA into messenger RNA. Scientists are devising ways either to interfere with the transcription of genes that are known to cause specific diseases or to block or mimic specific transcription factors, without influencing those factors that are essential to keeping the cell alive. Genes that suppress or stimulate tumor cell growth are a good target. Another way to block protein function is to prevent the translation of messenger RNA into proteins.

To block the genetic code, biotechnology companies are working not only to engineer biologically active protein molecules but to make novel drugs made of nucleic acids, lipids, and carbohydrate molecules.

Antisense Oligonucleotides

Protein synthesis can be blocked at the translation level using antisense **oligonucleotides** which consist of short, single-strand DNA fragments that recognize, bind to, and block specific messenger RNAs. In order for the oligonucleotide to bind to the RNA, it must contain a sequence complementary (or antisense) to the target sequence on the RNA. Therapeutic antisense oligonucleotides can be made in two forms: oligonucleotides or expressed oligonucleotides. Oligonucleotides are made by an automated DNA synthesizer. Expressed oligonucleotides are made by expression vectors such as adenoviruses or retroviruses (the same vectors as those used in gene therapy). The vector is administered to the patient, and produces antisense RNA in infected cells. In theory, both approaches should block the translation of endogenous messenger RNA molecules into proteins (Stein and Cheng, 1995).

Oligonucleotides used in vitro, in cells infected by human immunodeficiency virus (HIV), were found to inhibit viral replication. The use of oligonucleotides as a therapeutic agent, antisense therapy, was tested first in animal models for the treatment of hypertension. The idea was to prevent the synthesis of high levels of angiotensin II. Antisense oligonucleotides designed to block the messenger RNA encoding the angiotensinogen, the precursor of angiotensin I and II, were injected into the liver and the brain of hypertensive rats. The treatment was successful. It caused a decrease in blood pressure.

There are serious practical problems encountered with antisense therapy: the destruction of the oligonucleotides by DNAse, which is ubiquitous in the body; the difficulty of delivering adequate concentrations of antisense oligonucleotides into target cells; and the difficulty of keeping them bound to the messenger RNA long enough to inhibit protein synthesis. Scientists are now trying to attach oligonucleotides to liposomes (lipid particles) in which specific ligands can be incorporated so that they can be directed to specific cells.

In spite of initial scepticism concerning the antisense oligonucleotide approach, pharmaceutical companies have now decided to put more effort into this area. For example, antisense oligonucleotides have been designed to inhibit the growth of papillomavirus, which causes genital warts, a condition that is both painful and a risk factor for genital cancer. As another example, human trials have been initiated to test an antisense drug to fight leukemia. The first step of such a therapy consists of removing the bone marrow containing leukemic cells from patients who undergo massive (bone marrow-destroying) chemotherapy, and to incubate it in tissue culture with antisense drugs to eliminate specifically leukemic cells. The treated marrow is then transplanted back into patients to restore their immune systems. These tests are at a preliminary stage (Askari and McDonnell, 1996).

Ribozymes

An approach complementary to the use of oligonucleotides, is to use pieces of catalytic RNA, called ribozymes, that cleave messenger RNA molecules, and thus prevent them from being substrates for protein synthesis (Kiehntopf et al, 1995). The goals are to attack RNA molecules that encode mutated or deleted proteins, which cause uncontrolled cellular proliferation, or to inhibit the expression of genes encoding foreign viral proteins after insertion of this genome into host cells, as is the case in HIV infection or other viral infections. This approach is gaining popularity over the oligonucleotide approach, because ribozymes have true enzymatic activity and cleave RNA, thus making the RNA no longer available for proper protein synthesis. Furthermore, ribozymes could potentially be used to repair mutated RNA. Repair of mutated tumor suppressor genes, such as p53, could be one application. The challenge now is to be able to use this technology in vivo.

Synthesis of Lipids as Drugs

Lipid molecules have long been considered as merely structural building blocks of cell membranes, but their important role in cell function is now being recognized. Not only do they control the movements of materials in and out of the cell, but they also act as mediators of important intracellular processes. For example, the activation of cell-surface receptors causes the release of the lipid **diacylglycerol** (DAG), which turns on protein kinase C (PKC), a key enzyme in the regulation of cell growth and maturation (see Part I, Chapter 3). Uncontrolled cell division and formation of cancer have been attributed in part to an over-expression of PKC. The regulation of PKC is therefore an important target in cancer drug development, and biotechnology companies are synthesizing lipid-based drugs that might regulate the expression of PKC.

Phosphatidic acids represent another class of important bioactive lipids that are the subject of intense investigations. They are not present in cells normally, but appear after cellular injury and may stimulate a series of intracellular events. Ceramide, a lipid produced in response to tumor necrosis factor (TNF), is believed to play a role in stress responses and tumor suppression. It could also be active in apoptotic processes by triggering the initial steps of the cell death program.

Scientists acknowledge now that lipids are important modulators of intracellular signals and that they might play an important role in the regulation of gene expression. As such they represent attractive targets for drug companies. These findings will add a completely new dimension to the research of molecular biologists, genetic engineers, and protein chemists.

The Use of Peptides and Carbohydrates to Block Receptors

Drugs targeted at cell surface receptors have been very successful. Today, small organic molecules are designed with computers, and it is possible to find molecules that bind to receptors with exquisite precision. Peptides and carbohydrates are used to block receptors. Some new applications of these receptor blockers are in the treatment of immune disorders, such as rheumatoid arthritis and multiple sclerosis, and the prevention of immune responses that cause inflammation and organ transplant rejection. Drugs made with carbohydrates have the advantage over protein drugs that they can be packaged in a pill and withstand the effects of digestive tract enzymes. The disadvantage, however, is the difficulty of controlling the structure of complex carbohydrates during large-scale production.

Conclusion

We all know that life is a fatal disease and that aging is inescapable. Instead of lamenting our inevitable doom, we propose taking an objective view of measures that might delay the onset of diseases and postpone aging, and this was the focus of the second part of this book. Several of the answers come directly from popular wisdom. Whereas it is difficult at present to modify genetic defects, it is possible to change one's environment and, to some extent, avoid factors that are detrimental to health. Healthy life habits are necessary, i.e., avoiding stress at work and at home, eating a proper diet without excessive calories, fats, or alcohol; practicing regular exercise and a positive mental attitude, maintaining intellectual and physical activity; and of course avoiding cancer-promoting agents, particularly tobacco. These common-sense habits can delay the onset of disease and definitely prolong life. Unfortunately, there is incredible resistance to applying them in our daily lives, and few individuals assiduously follow these health guidelines. No matter how powerful antioxidants and other nutrients may turn out to be, they will never compensate for the risks associated with deleterious behaviors, nor can they substitute for sensible habits. Life styles depend on upbringing and on the philosophy prevailing in the environment. For example, Europeans tend to have a different attitude from North Americans, many of whom seem to want to extend their lives at all costs. The European diet is notoriously rich in fats, and tobacco and alcohol consumption is quite high. Europeans seem to consider that making sacrifices to add extra years to their lives is not worthwhile, compared to living a self-indulgent and enjoyable, even if somewhat unhealthy, life.

The health of older people is inextricably linked to the society in which they are aging. Many men and women 80 years old and older are still alert and retain the capacity to lead an independent life. Brain deterioration is not inescapable—advanced age is not necessarily accompanied by major declines in cognitive abilities and senility. Readers are encouraged to read Marantz-Henig's book entitled, *The Myth of Senility* (1987). She emphasizes the fact that the term Alzheimer's disease has become a popular diagnosis applied to all people who are losing their memory; she shows that even today, many educated and informed people still believe that a significant decline in mental functioning is inevitable in old age and have a biased attitude toward old people. However, the vast majority of elderly people are still capable of learning, of interacting, of thinking, remembering, and loving. The important message of Henig's book is that: "... the biological changes encountered over time are rather benign and easily transcended, but the social and cultural constraints under which those changes occur can turn the picture from one of a gentle downward slope into an avalanche.... For many years to come, the best prescription for a mentally healthy old age will probably remain what it is today: patience, love and an enlightened awarness that most folks need never go senile" (Marantz-Henig, 1987).

Younger people still disregard the elderly. It is time to change this. Older people need to be valued as a vital resource; the cost effectiveness of this awareness can be enormous. Retirement or the loss of the opportunity to work is often felt as a constricting and debilitating event. It is therefore important to set up more incentives to keep older persons active. Most professional occupations do not provide the opportunity for continued productivity, and society needs to address the problems of integrating retired people. Jobs or volunteer work could be created in areas such as environmental protection, city council activities, library research work, the arts, or community affairs. In addition, formal education should be encouraged for older people and should not be restricted to the first two decades of life. In this respect, many universities have successfully opened up programs for seniors. If society emphasizes the potential for self-development throughout life rather than insist on the overall decline with age, older people will be more likely to lead a productive and fulfilling life. Social contacts are essential to longevity.

Perhaps the most important factor of all is happiness and psychological well-being, the essence of human life. Happiness of the human species has always been the focus of attention of philosophers, psychologists, and social scientists. Obviously, health is an important determinant of overall well-being, with healthier people reporting greater happiness, but the role of happiness itself in affecting life span may be just as relevant. One of the major conditions for aging successfully and in good health is the presence of friends and family. The role of the family perhaps needs to be redefined to emphasize the benefit of possible interactions with older people. Each generation has a role to play and each person should be a resource for the entire family over his or her entire life span. Older people should no longer be a burden for the younger generations but should be included in a network of family ties and mutual support. Older people should learn to make themselves necessary.

The development of research on aging in the United States was closely connected to the establishment of the National Institute on Aging in 1974. Largely through the rapid growth of this Institute and the explosive growth of modern biology, there is now a substantial and rapidly expanding research effort in the biology of aging. Many research centers and institutes on aging have been created in the USA during the last several years. These centers are concerned with all aspects of the problems of aging. Research on aging is ongoing in Europe as well, although not with the same emphasis and at the same pace as in the USA. We hope that current development of the European Community will give the impetus for the foundation of European institutes on aging. The aim of research in biological gerontology is not to prevent aging, a goal certainly beyond our human capacities, but to increase knowledge on the mechanisms of the normal or the accelerated aging process.

Research on aging should include comparative studies of different systems and species. Priority should also be given to research on the interactions between the various biological systems, with emphasis on the nervous system in relation to the immune and endocrine systems. Today, neurodegenerative diseases are the focus of extensive research because, in addition to affecting the life span, their severity and high incidence have an important social and economic impact. Researchers must still identify the underlying causes of these disorders and the secondary changes that occur in the brain. More basic knowledge of the underlying mechanisms will make possible the development of more accurate diagnostic tools, and, hopefully, more effective treatments.

Research on aging involves not only research on fundamental and medical aspects of aging, but also research on the socioeconomic and psychological problems of older people. In 1982, the National Institute of Health published a document called *Toward an independent old age: A National Plan for Research in Aging* (National Research on Aging Planning Panel). The four identified priorities were: "1. Better understanding of the basic processes of aging, 2. Better understanding, preventing, and control of the clinical manifestations of aging and aging-related disorders, 3. Better understanding of the interactions between older people and a dynamic society, and 4. Increased opportunity, motivation, and support for older people to contribute productively to society." What a great program! Scientists and physicians must continue their efforts in research on aging and look for new avenues in the prevention of diseases. More emphasis should be placed on ways to keep healthy throughout life, instead of concentrating primarily on ways to cure illness during the last years of life. More importantly, a new awareness is needed of the aging population and a change in the perception of the problems associated with aging, if people want to improve the length and quality of their lives, as they themselves get older. The young and active generation of today will be the elderly population of tomorrow, and they will benefit from the implementation of these new ideas. By taking care of problems of the elderly now, the younger generation will in effect assure the success of their own old age; they can be altruistically selfish.

The priorities for age-related research for the next 20 years also have been recently described by a committee from the Institute of Medicine in the United States. Here, we quote from the executive summary: "Science offers the best hope to improve the older person's quality of life. Research that is directed and supported properly can provide the means to reduce disability and dependence in old age, and can decrease the burdens on a health system strained to its limits. This is a time for celebration but a time as well for deep concern. It is a time for celebration because gains in life expectancy have resulted in an estimated 33 million Americans 65-years old and over, the majority of whom are vigorous and active well into advanced old age. It is a time for concern because, for a growing number of older people and their families, these added years of life often are burdened by disability, dementia, and the loss of independence" (Lonergan, 1991).

Aging research is quite unique in the life sciences in that it requires involvement of a wide array of scientific disciplines. Research in the basic disciplines of genetics, biochemistry, cell biology, neurobiology, developmental biology, and others are necessary. Many of these research domains presently stand apart from each other because of their focus on specific organs and diseases. It is only by combining the information in all these areas that one will be able to have a global understanding of aging. The knowledge gained in gerontological research will be important for other fields, especially in cell biology, including developmental biology and cancer.

The recent explosion of technological advances in the biological sciences is a definite asset for future research on aging. The use of transgenic animals is important and will provide answers about the role of specific gene products in the etiology of aging. With the new technique of computer-based image analysis, it is possible to analyze the development of the structural complexity of neurons and their modification by environmental factors, such as chemical and physical factors and biological agents such as HIV-1. These new technologies will give further insight into the mechanisms of synaptic transmission and circuit development and will ultimately allow the identification of most molecules that promote neuronal plasticity.

Predictive medicine, which allows early detection of predisposition to diseases, is expanding as a result of recent molecular biology discoveries, and will enable the early identification of organs or systems that might become defective in a given individual.

At the present time, medical interventions to postpone aging or death consist essentially of counteracting the negative influence of our chromosomes at the phenotype level, i.e., to treat symptoms of diseases or prevent their occurrence. These interventions have an obvious and well-recognized effect on longevity, for example, through the control of complications from diabetes, high blood pressure, or through the prevention of myocardial infarction or recurrent psychiatric disorders, but in the majority of cases they do not treat the original cause of disease. The last few years have witnessed fantastic discoveries in the area of the genetic basis of disorders, genetic engineering techniques, and gene therapy. Molecular genetics is becoming a powerful tool to identify individuals predisposed to given disorders. The early detection of disorders, at a time when the future patient is asymptomatic will open new avenues for early prevention. Repairing defective genes can also be envisaged, and there is much hope in the medical community that several diseases causing handicaps may soon find a cure.

We are entering a golden age when measures for the primary prevention of most of the degenerative disorders that are prevalent in advanced societies will soon be available as well as more precise knowledge about nutrition and the use of supplements of endogenous compounds and of vitamins and drugs. The ability to prolong life is indeed within our grasp.

Glossary

Adrenal glands: Endocrine glands located above the kidneys, that synthesize and secrete glucocorticoids and sex steroids as well as catecholamines.

Adrenocorticotropic hormone (ACTH): A peptidergic hormone synthesized in cells of the anterior pituitary. ACTH triggers adrenocortical synthesis and release of glucocorticoids.

Aflatoxin: A carcinogenic mould found in stale nuts and other foods.

Aldosterone: A steroid hormone secreted by the adrenal cortex that stimulates sodium retention by the kidneys.

Allele: One of the possible forms of a gene found at a particular locus on a chromosome.

Amplification (gene amplification): The reiteration of a DNA sequence beyond the level found normally in somatic cells.

Amyloid deposition: Abnormal deposition of protease-resistant protein aggregates in tissue. In the central nervous system of patients with Alzheimer's disease, amyloid deposition is due to accumulation of a fragment of a transmembrane glycoprotein of as yet unknown function.

Antioxidant: A substance that prevents oxidation. At the molecular level, antioxidants prevent the loss of one or several negative charges (loss of electrons by an atom).

Atherosclerosis: Degeneration of chronic lesions located in arteries causing a hardening of arterial walls (arteriosclerosis). These modifications of arterial walls result from the calcification of yellow plaques made of lipids, located on the inner surface of arteries.

Autoimmunity: A condition in which the immune system attacks organs or tissues of the body as if they were foreign substances or cells.

Autonomic nervous system: Part of the nervous system that regulates smooth muscle, cardiac muscle, and glands. It is composed of two divisions, the sympathetic and parasympathetic nervous system. The autonomic nervous system operates without conscious control.

Autosomal gene: A gene located on any chromosomes other than sex chromosomes.

B lymphocytes (B cells): White cell producing antibodies and derived from bone marrow.

Blood-Brain Barrier: A semipermeable series of cell barriers that allow the selective passage of substances and nutrients to and from the brain.

Bradykinin: Bradykinin is one of a group of small peptides, the kinins, released locally in cases of tissue damage or infection. It is among the substances that promote the inflammatory response. Bradykinin has local vasodilator effects; it induces pain through the release of neuropeptides such as substance P.

Calcitonin gene-related peptide (CGRP): A peptidergic hormone encoded by the so-called calcitonin gene and obtained by alternative splicing of the calcitonin messenger RNA. These two peptidergic hormones, calcitonin and CGRP, are derived from the same gene, but have distinct structures and biological activities.

Cardiac output: The volume of blood per minute pumped by the right or left ventricle.

Catecholamines: A group of molecules derived from tyrosine including epinephrine (also known as adrenaline), norepinephrine and L-Dopa, having similar effects to those produced by stimulation of the sympathetic nervous system. Epinephrine and norepinephrine increase blood pressure and pulse rate.

Cell-mediated immunity: The immune response that occurs in the tissues and concerns T lymphocytes and interferon and other cytokines.

Chelation: Binding of metal to a molecule. Chelating agents are used to remove metal from blood or tissues.

Cholesterol: A fat in food from animal origin and also produced in the liver. A precursor for other steroids, i.e., cortisol, DHEA, and sex steroids.

Chromatin: Molecular complex made of DNA and associated specific proteins.

Chromosome jumping and walking: Application of molecular biological techniques including gene cloning and DNA sequencing, to localize a gene on a chromosome.

Circadian rhythm: Cycles that occur approximately every 24 hours in the body.

Clonal senescence: The decreased capacity of cells to proliferate in vitro.

Codon: Three contiguous nucleotides encoding an amino acid.

Collagen: A glycoprotein forming the major component of connective tissues, teeth, and bone.

Colony forming ability: Multiplication of isolated cells in culture giving rise to small colonies.

Corticoids or corticosteroids or glucocorticoids: Hormones secreted by the cortex (external part) of adrenal glands. Synthetic corticoids and their derivatives are used frequently as anti-inflammatory drugs.

Corticotropin releasing factor (CRF): A peptidergic hormone synthesized in many regions of the brain. Within the hypothalamus, CRF is found in a number of regions; the CRF that is released in the median eminence to act on the pituitary appears to originate in parvicellular neurons in the paraventricular nucleus, and some of these same neurons also contain vasopressin. CRF causes the release of adrenocorticotropin hormone (ACTH) from the anterior pituitary.

Cyclin: Proteins that increase or decrease at different stages of the cell cycle. They regulate cyclin-dependent kinases which in turn phosphorylate critical targets at decision points in the cell cycle.

Dementia: Severe disease characterized by a gradual and irreversible impairment of intellectual functions. Dementia is a general term including several kinds of neurological disorders originating from diseases like cardiovascular disease, depression, senility, or other diseases. Alzheimer's disease is one of the many forms of dementia.

Demethylation: Removal of methyl groups, for example, from DNA cytosine residues.

DNA or Deoxyribonucleic acid: Macromolecules that constitute the genetic substance found in every cell. Genetic information in DNA is transmitted through the sequence of the four nucleotides, adenine (A), thymine (T), guanine (G) and cytosine (C).

Diabetes: A disease characterized by an absolute or relative lack of insulin. Individuals suffering from diabetes cannot metabolize glucose.

Diacylglycerol (DAG): Compound derived from membrane phospholipids and implicated in the transmission of receptor signalling, in particular in intracellular calcium mobilization.

Diastolic pressure: The pressure in arteries during the relaxation phase of the left ventricle.

Dizygotic: See twins and zygote.

Dopamine: A neurotransmitter and intermediate compound in the synthesis of other catecholamines.

Doubling: The process by which a cell divides to generate two daughter cells.

Elastin: Springy material that maintains normal skin tension; it stretches during movements of the underlying muscles and joints.

Elongation factor: A factor involved in the elongation of the nascent polypeptide chain during the translation process.

Endocytosis: Uptake by any cell (even the axon of nerve cells) of exogenous materials.

Endoplasmic reticulum: A system of membranes within the cytoplasm of the cell, including the rough and smooth endoplasmic reticulum. The rough endoplasmic reticulum is associated with polyribosomes and participates in protein synthesis.

β-Endorphin: A peptidergic hormone with analgesic properties, synthesized in the anterior pituitary.

Epigenetic factors: Cellular events that play a role in the functional differentiation of cells and in regulation of gene expression but do not involve the genetic code per se.

Essential fatty acid or essential nutrient: A fat or nutrient that cannot be manufactured by the body (e.g., linoleic acid found in safflower oil) or by cells.

Estrogens: Hormones secreted by ovaries. Estrogens have many physiological and behavioral effects that are manifested before and during ovulation in women.

Food and Drug Administration or FDA: An agency of the United States government responsible for monitoring the safety and effectiveness of drugs sold in the United States.

Functional decline: Progressive decline or loss of body functions.

Functional reserve: Capacity of organs (heart, lungs, etc.) to respond to unusual demands on the organism.

Genetic linkage studies: see linkage analysis.

Germ line cells: The germ line provides continuity from generation to generation, i.e., nonsexual germ cells and gametes.

Glaucoma: Increased pressure within the eye, leading to secondary degeneration of the optic nerve.

Glucocorticoids: see corticoids.

Glycogen: A polysaccharide. A long chain polymer of glucose formed in and largely stored in the liver. Glycogen is the chief carbohydrate storage material in animals.

G-protein: Intracellular proteins that bind guanine nucleotides, guanosine triphosphates (GTP), and guanosine diphosphate (GDP). Ion channels, adenylyl cyclase, phospholipase C, phopholipase A2 and phosphodiesterase are known to be influenced by G proteins.

Haemolymph: Equivalent of blood in insects.

Haploid: A single complement of chromosomes.

HDL-Cholesterol: Cholesterol packaged in high-density lipoproteins. HDL is comprised of fats and proteins and serves as a transport for fats in the blood. HDL is considered as the good cholesterol.

Hormone: Substance secreted by endocrine glands or synthesized by a given tissue, released directly in blood where it is transported to target organs. Hormones play important roles in the proper functioning of organs or in specific biochemical processes.

Hybrid cells or hybridomas: Fusion of differentiated cells from various origins and/or animal species.

Hypertension: Elevated blood pressure.

Hypophysis: See pituitary.

Hypothalamus: Region located at the base of the brain comprising many regulatory centers. These produce factors or hormones that act directly on the secretion of hormones by the pituitary.

Immune surveillance: The monitoring by the immune system for the presence of foreign tissues or cells (cancer cells, for example). This is done mainly by T killer cells of the T-cell series.

Immune tolerance: The tolerance of the immune system for foreign tissues or substances. This can lead to the failure to reject foreign grafts or the failure to produce antibodies directed against these antigens.

Insulin: A polypeptide hormone produced by the islets of Langerhans in the pancreas. Insulin transports glucose from the blood into cells, thus lowering plasma glucose levels.

Insulin-like growth factor (IGF): A growth factor implicated in the regulation of cellular proliferation.

Interferon: A class of lymphokines produced by immune cells and fibroblasts which have antiviral actions. A component of cell-mediated immune response. Interferons may also help regulate cell development.

Interleukin: A class of lymphokines with immune and other functions.

Interspersed repetitive sequences: Stretches of DNA sequences that display a repetitive pattern of nucleotide sequence.

Ischemia: Acute or chronic decrease in tissue oxygenation.

Ischemic disease: Disease caused by a limited blood supply and therefore, decreased oxygenation.

Karyotypic analysis: Analysis of the banding pattern and shape of the complete set of chromosomes in cells.

Killer cells: T cells involved in immune surveillance that kill cancer cells and foreign organisms.

LDL-Cholesterol: Cholesterol packaged in low-density lipoprotein. A high level of LDL is associated with an increased risk of developing cardiovascular diseases.

Leukemia: Cancer of the white blood cells.

Leukotrienes: A class of polyunsaturated fats with three double bonds. Leukotrienes oxidation products have potent bronchoconstrictor effects.

Linkage analysis: Studies designed to determine the location of the gene responsible for disorders. The traditional approach for this type of analysis is based on two principles: (A) the primary sequence of the human genome is not fixed, but varies in different individuals (polymorphism). Since such polymorphisms often either generate a new restriction site or abolish an existing site, they can be identified by digestion with appropriate restriction enzymes and analysis of the size of the generated fragment. (B) During meiosis, corresponding pairs of chromosomes line up together and exchange genetic material (crossover), and generate a new pair of chromosomes. Two regions of the genome that are on the same chromosome tend to stay together, while those on different chromosomes distribute to the germ cells, totally independently from each other. When a crossover occurs between two regions on the same chromosome, they become separated on two chromosomes. It follows that the closer the two regions are on the same chromosome, the more likely they are to stay together. A successful linkage analysis requires DNA materials from family members of affected patients. DNA is prepared from each individual and analyzed after restriction enzyme digestion. The strategy is to find a polymorphic marker (see below) that is closely linked to affected individuals and thus to the disease-causing gene in the family. Any markers that do not segregate with or against the disease state are discarded.

Linoleic acid: An essential polyunsaturated fatty acid, not synthesized in the body.

Lipofuscin: Pigmented granules considered to be slowly accumulated metabolic byproducts, particularly oxidized lipids. Lipofuscins develop with aging in cells of many tissues, especially the brain, muscle, liver, and kidney.

Luteinizing hormone (LH): A peptidergic hormone synthesized in the pituitary. LH stimulates the synthesis of sexual hormones from the ovary in women and from the testis in men.

Lymphokines: Lymphokines belong to the family of cytokines. Cytokines is a general name given to a large group of polypeptide regulators such as the interleukins or the tumor necrosis factor. Many of them are synthesized by cells of the immune system or in the central nervous system. Cytokines have effects on the immune system, on cell differentiation, on cell survival after lesion, on the response of the pituitary gland to stress, etc. They are organized in highly complex cascades of mutual facilitatory or inhibitory actions, and serve specific functions in the body and the central nervous system.

Lysosomes: A membrane-bound organelle in the cytoplasm. Site of intracellular digestion by enzymes.

Macrophages: A phagocytic white blood cell found in blood and tissues which processes and presents antigen to stimulate B-cells to produce antibodies.

Markers: Well-characterized genes or DNA sequences with specific localization on chromosomes. Genetic markers help map genes not yet characterized (see linkage analysis).

Menopause: The cessation of ovulation and menstruation in women.

Messenger RNA: The transcription product of DNA which is transported into the cytoplasm and serves as a template for protein synthesis.

Metabolism: Generic term that includes the sum of chemical modifications occurring in the body to meet energy requirements, to synthesize and maintain the level of many substances such as hormones, enzymes, antibodies etc, and to repair tissue damage.

Missense mutation: A mutation that changes a codon so that it codes for a different amino acid.

Mitosis: Process of cellular division regulating the maintenance of a constant number of chromosomes in daughter cells.

Monozygotic: See twins and zygote.

Mosaicism: Assembly of cells or tissues that contain different chromosomal content.

Mutation: Any change in DNA sequence. Mutations can be base changes (missense or nonsense mutations (see definitions)), additions (insertions) or deletions of one or many DNA bases. A single base change is referred to as a point mutation.

Neurofibrillary tangles and neuritic plaques: Two major structures seen by microscopic analysis of cerebral cortex tissue. The biochemistry and structure of these lesions are complex. Neuritic or senile plaques are essentially accumulations of amyloid proteins, while neurofibrillary tangles are intracellular accumulations of a (modified) protein tau involved in microtubules.

Nitrosamine: A carcinogenic substance found in foods and cigarette smoke and formed in the stomach from nitrites in foods.

Nonsense mutations: Base mutations in the DNA sequence encoding a protein, which result in a stop codon (UAG, UAA, UGA). As a result of nonsense mutations, synthesis of protein is terminated prematurely. The localization of the nonsense mutations on DNA determines the final length of the protein.

Noradrenaline: Also known as norepinephrine. In the brain, noradrenaline systems derive mainly from the locus ceruleus of the brain stem and spread through the central nervous system, including the spinal cord. Noradrenaline is also produced by the adrenal medulla and is released in response to various stress situations.

Nucleosome: Combination of DNA and histones that confer a beadlike structure to chromatin. Histones are relatively small proteins, positively charged and rich in arginine and lysine, that aggregate to form small ellipsoïds around which DNA molecules wind.

Organelles: Intracellular structures that have specific functions: mitochondria, Golgi apparatus, etc.

Oligonucleotide: Short sequence of nucleic acids.

Osteoporosis: Fragility of bone due to thinning of bone structure.

Oxytocin: A peptidergic hormone produced by hypothalamic neurons. Oxytocin is released into the general circulation via the posterior pituitary. Oxytocin causes uterine contraction and milk ejection.

Parthenogenesis: Asexual reproduction in multicellular organisms.

Phenotypic: The phenotype represents that part of the information in the genotype that is expressed as measurable characteristics of an individual.

Pituitary: Endocrine gland releasing stimulating factors or hormones such as thyroid-stimulating hormone, folliculo-stimulating hormone, growth hormone, adrenocorticotropic hormone etc., directly in blood. These hormones act on other endocrine glands in the organism (thyroid, adrenal gland, ovaries, etc.).

Placebo: An inert substance that has no pharmacological action.

Pleiotropic effect: A single event that triggers a series of different events (for example, a single mutation may lead to the partial or complete loss of several functions).

Point mutation: See mutation.

Polymorphism: Genetic variability between individuals.

Postmitotic cell: A cell that has stopped dividing and is in phase G_0 of the cellular mitotic cycle.

Presynaptic: See synapse.

Progeria: Diseases characterized by the early onset of degenerative processes observed during senescence.

Prostaglandins: A group of hormone-like substances formed from polyunsaturated fatty acids that have numerous effects on the body, including contraction of smooth muscle and dilation or contraction of blood vessels.

Provitamin: A substance in food that can be converted to a vitamin once it enters the body. Beta carotene is the provitamin for vitamin A.

Renin: An enzyme produced by the kidney that converts angiotensinogen in the plasma to angiotensin. Angiotensin causes a rise in blood pressure and an increased secretion of aldosterone from the adrenal cortex.

Retina: The layer of light-sensitive cells (rods and cones) lining the inside of the eye.

Retrovirus: An RNA-containing virus that uses the host cell machinery to synthesize the enzyme reverse transcriptase to make a DNA copy of itself. It may integrate into the host genome and become a carcinogenic virus.

Ribosomal RNA: RNA produced by ribosomal DNA which forms the nucleic acid portion of the ribosomes.

Ribosome: A cytoplasmic element made of proteins and ribosomal RNA, which is the site for the translation of messenger RNA into protein.

Self-tolerance: Immunological unresponsiveness to autoantigens (self-antigens) acquired during fetal life by a process of self-recognition. The mechanisms of induction of tolerance involve deletion of antigen-responsive clones of B cells, antigen-induced inactivation of B and T cells, and induction of antigen-specific T suppressor cells.

Senile plaques: Microscopic lesions composed of fragmented axon terminals and dendrites surrounding a core of amyloid seen in the cerebral cortex in Alzheimer's disease.

Somatic: Refers to the soma. Somatic cells include all cells except germ cells.

Somatomedin C: A 70-amino acid protein found in human serum, believed to mediate many of the effects of growth hormone.

Splicing: Process by which exons (sequences of DNA that specify amino acid sequences) of the same gene are combined to generate messenger RNA molecules. The combination of different exons of the same gene yields different messenger RNA molecules.

Superantigens: Endogenous or exogenous molecules that are the products of viruses, bacteria, or mycoplasmas. Certain endogenous superantigens encoded by viruses (already present at the time of birth) can lead to the deletion of some T-cell clones expressing T-cell receptors bearing specific variable regions.

Superoxides: Unstable forms of oxygen atom that affects the integrity of surrounding molecules.

Synapse: Structure composed of an enlarged axon terminal, a cleft, and a postsynaptic cell or dendrite. Neurotransmitters are released from presynaptic vesicles into the synaptic cleft and influence receptors on postsynaptic cell membranes.

Systolic blood pressure: The maximum pressure in major arteries when the heart contracts.

T lymphocyte (T cell): White cell responsible for cellular immunity and derived from the thymus during development.

Telomeres: Free ends of chromosomes.

Testosterone: A male sex hormone produced by the Leydig cells of the testis.

Thymus: A gland that atrophies at adult age, and is the source of T cells. The thymus produces hormones that stimulate T cells after they leave the thymus.

Totipotent cells: Cells that retain the capacity to differentiate into any specialized cell.

Transcription: The synthesis of RNA from a DNA template.

Transfection: The process by which foreign genes are introduced into a cell.

Transfer RNA or tRNA: RNA that transports amino acids to the specific site on the polyribosome messenger RNA. The anticodon sequence on tRNA molecules specifies the type of amino acid transported.

Translation: The process by which proteins are synthesized using the information contained on the messenger RNA.

Triglycerides: One of the three classes of lipids. Triglycerides are composed of three fatty acids and one glycerol molecule. They are either saturated or unsaturated.

Triplet repeats: Repeats of a sequence of three given nucleotides in DNA.

Trisomic: Chromosomal anomaly during meiosis or mitosis, in which a chromosome (or part thereof) does not migrate correctly at anaphase. This erroneous partitioning results in a daughter cell with the two chromosomes of a pair or a chromosome and part of the second one. This cell will therefore be trisomic for a particular chromosome.

Twins: Monozygotic twins originate from the same fertilized egg, and have an identical genetic composition. Dizygotic twins originate from two independent fertilized eggs and have a different genetic composition.

Vasopressin: A peptidergic hormone synthesized by hypothalamic neuroendocrine cells, transported to the nerve terminals in the neural lobe of the pituitary and then released into the general circulation. Vasopressin has vasoconstrictive and antidiuretic effects.

Vitellogenin: A precursor protein in the egg yolk.

X Chromosome: The sex chromosome. The presence of two X chromosomes in cells determine the female genotype.

Y Chromosome: The heteromorphic male sex chromosome. The chromosome that determines the male genotype.

Zygote: The cell formed by the fusion of two gametes.

Relevant Books on Aging

There are many books written on various aspects of aging and several of them are excellent. Here is a list of some of the most useful. Books intended for readers who are scientifically knowledgeable are indicated by *; books intended for lay readers are marked by **. However, all these books contain information that all readers may find useful.

Ageing. A Biomedical Perspective. Denis Bellamy. John Wiley & Sons, West Sussex, 1995; 410 pages.*
The book concerns aging research ranging from the level of molecules to that of populations. The information covers different knowledge frames, concepts, theories, and models. It integrates data into a perspective (as stated in the title) that is innovative and reviews previous assumptions about aging and models.

Aging and the Nervous System. Salvatore Giaquinto. John Wiley and Sons, New York, 1988; 224 pages.*
Societal, demographic, financial and practical issues are discussed in a book on neurochemistry, neuroanatomy, or neuropsychological changes in the elderly. This is an eclectic approach.

Aging in Good Health. Mark H. Beers and Stephan K. Urice. Pocket Books, a division of Simon and Schuster Inc., New York, 1992; 351 pages.**
This is a well-written book on changes and symptoms that occur in elderly persons and on what to do in problem situations.

The Anti-Aging Plan. Strategies and Recipes for Extending Your Healthy Years. Roy L.Walford and Lisa Walford. Four Walls Eight Windows, New York, 1994; 309 pages.**
The author is a physician who instigated the Biosphere-2 experiment, in which eight persons lived for two years in an artificial and sealed environment in the Arizona desert. During this time, they ate a 1800-calories-per-day diet, analogous to the experiments on food restriction in nonhuman species. They lost weight and had low cholesterol values. The results of the experiment itself receive a very limited description, and two thirds of the book are devoted to recipes for persons who would like to start a food restriction diet, or, more precisely, a caloric restriction diet involving rearrangement of fat, carbohydrate, and protein proportions. Whether this anti-aging plan for humans is really efficacious is uncertain.

Biology of Aging: Observations and Principles. Robert Arking. Prentice-Hall, New Jersey, 1991; 420 pages.*

This is a very good textbook on the many facets of aging, from epidemiological to molecular findings. It includes a review of theories of aging and research on life prolongation.

Brain Boosters. Beverly A. Potter and Sebastian Orfali. Ronin Publishing, CA, 1993; 257 pages.**

This book is about foods and drugs that make you smarter, even if you do not suffer from brain drain. It includes a list of physicians in the United Staes and a few other countries who prescribe life-extending drugs, with the acknowledgement that some of the information provided is not accepted by mainstream medicine as being scientifically valid. The Life Extension Foundation considers its information to be five-to-ten years ahead of conventional medicine. The cautionary note in the book contains the statment that the authors are "not responsible for specific application of this material". The book covers the war over vitamins, i.e., the opposing views of nutritional supplement supporters and the FDA.

Bypassing Bypass: the New Technique of Chelation Therapy. Elmer Cranton. Hampton Roads publishing company, New York, 1992; 266 pages.**

Chelation therapy has not been much considered or evaluated in official medical journals, but lay readers can have access to most of the information that probably exists on this alternative to cardiac surgery for coronary bypass. This book describes an example of therapeutic techniques that have evolved in parallel to those that were developed in major university hospitals and by the pharmaceutical industry.

Complete Guide to Aging and Health. Mark E. Williams. M.D. The American Geriatrics Society. Harmony books, New York, 1995; 494 pages.**

This is an encyclopedia of changes that occur with aging. Ailments and diseases that affect elderly people are described extensively with information on diagnostic procedures, differential diagnosis, and therapeutic measures. The rationale for the book is that the availability of a practical handbook enables elderly people to handle better the positive and negative aspects of aging. The material is of high quality.

The Essential Guide to Vitamins and Minerals. Health media of America and Elizabeth Somer, Harper Perennial, New York, 1992; 403 pages.**

This is an extensive and clearly written reference book on vitamins, which can serve as an alternative to lay people who do not wish to read the more technical medical textbooks.

Evolutionary Biology of Aging. Michael R. Rose. Oxford University Press, New York, 1991; 221 pages.*

The author demonstrates how a reformulation of the field of gerontology in terms of evolutionary biology would strengthen research and future applications for postponing aging.

Extending Life, Enhancing Life. A National Research Agenda on Aging. Edmond T. Lonergan, National Academy Press, Washington, 1991; 152 pages.*

In this monograph, a Committee from the United States Institute of Medicine outlines priorities for biological and nonbiological research about aging.

The Fountain of Age. Betty Friedan. Simon and Schuster, New York, 1993, 671 pages.**
This bestseller book contains a radical critic of the decline model of aging, based on research results indicating that the decline in intellectual functions is more a matter of pathology and diseases that of age itself. The book is based on the idea that one can grow and evolve throughout life. It is optimistic and well documented.

Handbook of the Biology of Aging. Third edition. Edward L. Schneider and John W. Rowe, Academic Press, New York, 1990; 489 pages.*
The foreword of this excellent book, one of a series of three books on biological, psychological (see the next book quoted below) and societal issues of aging, contains the following statement: "There is little doubt from the reading of these volumes that the subject matter of aging has become more sophisticated and also mainstream in many scientific disciplines. It is hoped that the handbooks' publication will motivate continued attention to research on aging and the well-being of the elderly in our society".

Handbook of the Psychology of Aging. James E. Birren and K. Warner Schaie. Academic Press, NewYork, 1990; 552 pages.*
The same comments as above apply to this handbook offering a complete description of psychological issues of aging.

Human Longevity. David W.E. Smith. Oxford University press, New York, 1993; 175 pages.*
This is a well-documented analysis of human longevity, covering the recent change in life span and the medical, psychological and societal factors that influence life span.

The Johns Hopkins Complete Guide for Preventing and Reversing Heart Disease. Peter O. Kwiterovich, Prima Publishing, Rocklin, 395 pages.**
As the author writes: "Nothing takes the place of experience and knowledge. Over the past seventeen years, I have worked with many adults and children who have problems with cholesterol and other fats in their bodies. I have learned firsthand what their concerns are and what needs to be done to help them increase their chances of living longer and better lives". Kwiterovich has written a precise and well-documented book on blood lipids, together with practical recommendations for everyday life. This is a reasonable and constructive book, that demonstrates how scientific information can be communicated to lay persons to achieve the goals of preventive medicine.

Life Extension. A Practical Scientific Approach. Durk Pearson and Sandy Shaw. Warner Books, New York, 1980; 858 pages.**
This 15-year-old best seller is now often found in garage sales. It stands as a precursor to more recent documents certifying that life prolongation is feasible for each of us, if we follow basic health rules and ingest vitamins and medications

marketed in the United States or elsewhere. The book was a revelation about free radicals to lay people. It contains a lot of documentation, but a few statements are disturbing, such as the recommendation to take L-DOPA to feel more energetic and build muscle mass, or to take a capsule of co-ergocrine with each cigarette (if you cannot stop smoking) to decrease the oxidative stress of smoking and to prevent coughing!

Lifespan, Who Lives Longer and Why. Thomas J. Moore. Simon and Schuster, New York, 1993; 318 pages.**
This book describes the role of infectious diseases in the evolution of human longevity, as well as the actual threat of infective disorders. It gives a description of disorders such as obesity and hypertension and covers longevity in nonhuman species, theories of aging, and life extension experiments. The author takes a journalistic approach that facilitates understanding and illustrates a successful presentation of complex issues for the lay reader.

Longevity, Senescence and the Genome. C.E. Finch. The University of Chicago Press, London, 1990; 922 pages.*
Certainly one of the best scientific book about aging in different species and about the biological mechanisms of senescence. This high quality compilation should be on the desk of any person working in the field of aging; it can also be fascinating reading for lay people who are somewhat acquainted with technical terms and concepts. One would have to wait a long time before another book is published that can be considered a competitor to Finch's.

Longevity, the Science of Staying Young. Kathy Keeton. Viking Penguin, New York, 1992; 332 pages.**
The author, president of Longevity Magazine, worked with a team and interviewed many scientists on the latest findings on aging and the prolongation of life. The book is written in a journalistic style, which greatly facilitates understanding of the technical issues. Recommendations for diet, stress reduction, and other potentially useful life-prolongation approaches are given. Material in this book is less sensational than that in Longevity Magazine.

The Melatonin Miracle. Nature's Age-Reversing, Disease-Fighting, Sex-Enhancing Hormone. Walter Pierpaoli, William Regelson, Carol Colman. Simon & Schuster, New York, 1995, 255 pages.**
This book is about melatonin and its age-reversing effects. The authors strongly support the use of melatonin to improve immune functions, prevent cancer and heart disease, normalize sleep, reduce cholesterol, prevent hypertension, and improve the enjoyment of sexuality. Admittedly, melatonin has interesting effects, but it seems premature to recommend it with such enthusiasm and to claim that people with AIDS, Alzheimer's disease, asthma, Parkinson's disease, and other diseases ought to take exogenous melatonin. The first two authors have published many scientific articles, but in this book, their recommendations to the public are far ahead of the scientific facts.

The Natural Health Guide to Antioxidants. Using Vitamins and Other Supplements to Fight Disease, Boost Immunity, and Maintain Optimal

Health. Nancy Bruning and the Editors of Natural Health Magazine, Bantam Books, New York, 1994; 198 pages.**
Despite its overly provocative title, critical conclusions and recommendations in this book are based on a comprehensive survey of medical knowledge.

Normal Human Aging: The Baltimore Longitudinal Study of Aging. Nathan W. Shock. NIH Publication No. 84–2450, Washington, 1984; 630 pages. *
This monography illustrates the importance of longitudinal studies of aging. It is a series of publications from the first 23 years of one important longitudinal study, The Baltimore Study, started in 1958.

Prescription for Longevity. Eating Right for a Long Life. James Scala. Dutton Books, New York, 1992; 299 pages.**
In this book, readers learn about vitamins and diet in a way that is sufficiently technical but also useful for everyday recommendations. Again, the efficacy of most of these recommendations is not formally proven.

Stress, the Aging Brain, and the Mechanisms of Neuron Death. Robert M. Sapolsky. A Bradford Book, The MIT Press, London, 1992; 428 pages.*
A fascinating book about the effects glucocorticoids and other hormones on the brain, by a recognized expert on behavioral and biochemical aspects of stress in higher mammals.

Time, Cells and Aging. Bernard L. Strehler. Academic Press, New York, 1977, 456 pages. *
This second edition of a 1962 book is among the many scientific syntheses that were written on aging during the last decades. Published less than 20 years ago, it is fascinating to read because the author gave his views on what type of research should be done, and listed 100 key biomedical research goals in aging. The wording of the questions would certainly be different now due to technical advances, but one realizes that many of these goals have not been reached, even yet.

Understanding Aging. Robin Holliday. Cambridge University Press, Melbourne, 1995; 207 pages.*
The author describes the causes and consequences of aging at the molecular and cellular level, with a review of the role of DNA repair, free radicals, erroneous synthesis of macromolecules, immune responses, and other factors. He compares aging patterns among species and illustrates the intricate relations between body systems as the organism ages. The reader will find answers to why aging occurs and why it may facilitate the evolution of species.

We Live Too Short and Die Too Long. How to Achieve and Enjoy Your Natural 100-year-plus Life span. Walter M. Borts. Bantam Books, New York, 1991; 351 pages.**
This book is written by an M.D. who specializes in geriatrics. It is well-documented and enjoyable. The advice is reasonable and presents as a down-to-earth list of healthy habits.

Work, Stress, Disease and Life Expectancy. Ben C. Fletcher. John Wiley and Sons, New York, 1991; 152 pages.*

This is one of a series of books entitled Studies in Occupational Stress. Stress models, as a function of environmental demands and life events are presented, with a review of the consequences of stress on health, and a reminder of the extent of the stress problem in the work place.

References

Abbott M.H., Abbey H., Bolling D.R., Murphy E.A. (1978) *The familial component in longevity. A study of the offsprings of nonagenerians. 3. Intrafamilial studies.* Amer. J. Med. Genet., 2:105–120.

Adami H.O., Persson I. (1995) *Hormone replacement and breast cancer, a remaining controversy?* J. Am. Med. Assoc., 274:178–179.

Adams W.L., Garry P.J., Rhyne R., Hunt W.C., Goodwin J.S. (1990) *Alcohol intake in the healthy elderly: changes with age in a cross-sectional and longitudinal study.* J. Am. Geriatr. Soc., 38:211–216.

Adelman R.C. (1980) *Hormone interaction during aging.* In R.T. Schimke (ed.), Biological Mechanisms in Aging, p. 686. Washington, DC: U.S. Department of Health and Human Services.

Akli S., Caillaud C., Vigne E., Stratford-Perricaudet L.D., Poenaru L., Perricaudet M., Kahn A., Peschanski M. (1993) *Transfer of foreign genes into the brain using adenoviral vectors.* Nature Genetics, 3:224–228.

Aloia J.F., Vaswani A., Yeh J.K., Ross P., Flaster E., Dilmanian D. (1994) *Calcium supplementation with and without hormone replacement therapy to prevent postmenopausal bone loss.* Ann. Intern. Med., 120:97–103.

The **Alpha**-tocopherol, beta carotene cancer prevention study group. (1994) *The effect of vitamin E and beta carotene on the incidence of lung cancer and other cancers in male smokers.* N. Engl. J. Med., 330:1029–1035.

Amenta F., Bongrani S., Cadel S., Ricci A., Valsecchi B., Zeng Y.C. (1994) *Neuroanatomy of aging brain: influence of treatment with L-Deprenyl.* Ann. N.Y. Acad. Sci., 717:33–44.

Ames B.N., Cathcart R., Schwiers E., Hochstein P. (1981) *Uric acid provides an antioxidant defence in humans against oxidant- and radical-caused aging and cancer: a hypothesis.* Proc. Natl. Acad. Sci. USA, 78:6858–6862.

Ames G.F., Mimura C.S., Shyamala V. (1990) *Bacterial periplasmic permeases belong to a family of transport proteins operating from Esherichia coli to human: Traffic ATPases.* FEMS Microbiol.Rev., 6:429–446.

Anderson A.M. (1982) *The great Japanese IQ increase.* Nature, 297:180–181.

Anderson W.F. (1992) *Human gene therapy.* Science, 256:808–813.

Antiplatelet Trialists' Collaboration (1994) *Collaborative overview of randomised trials of antiplatelet therapy. I: prevention of death, myocardial infarction, and stroke by prolonged antiplatelet therapy in various categories of patients.* Br. Med. J., 308:81–106.

Anton-Tay F., Diaz J.L., Fernandey-Guardiola A. (1971) *On the effect of melatonin upon human brain. Its possible therapeutic implications.* Life Sciences, 10:841–850.

Apostol S. and Clain L. (1975) *The effect of some radioactivity sources on the plankton of the Danube River.* Atomic Index, 9:354–368.

Arendt T., Allen Y., Marchbanks R.M., Schugens M.M., Sinden J., Lantos P.L., Garry J.A. (1989) *Cholinergic system and memory in the rat: effects of chronic ethanol, embryonic basal forebrain brain transplants and excitotoxic lesions of cholinergic basal forebrain projection system.* Neuroscience, 33:435–462.

Argyle M. (1987) *The Psychology of Happiness.* London: Methuen and Co.

Arking R. (1991) *Biology of Aging; Observations and Principles.* Prentice Hall, Englewood Cliffs, New Jersey.

Arlett C.F. (1986) *Human DNA repair defects.* J. Inher. Metab. Dis., 9(Suppl.1):69–84.

Ascadi G., Dickson G., Love D.R., Jani A., Walsh F.S., Gurusinghu A., Wolff J.A., Davies K.E. (1991) *Human dystrophin expression in mdx mice after intramuscular injection of DNA constructs.* Nature, 352:815–818.

Asencot M. and Lensky Y. (1984) *Juvenile hormone induction of "queenliness" on female honey bee (Apis mellifera L.) larvae reared on worker jelly and on stored royal jelly.* Comp. Biochem. Physiol., 78B:109–117.

Askari F.K. and McDonnell W.M. (1996) *Molecular medicine, antisense-oligonucleotide therapy.* N. Engl. J. Med., 334:316–318.

Aslanidis C., Jansen G., Amemiya C., Shutler G., Mahadevan M., Tsilfidis C., Chen C., Alleman J., Wormskamp N.G.M., Voois M., Buxton J., Johnson K., Smeets H.J.M., Lennon G.G., Carrano A.V., Korneluk R.G., Wieringa B., de Jong P.J. (1992) *Cloning of the essential myotonic dystrophy region and mapping of the putative defect.* Nature, 355:548–551.

Bachmann G.A. and Leiblum S.R. (1991) *Sexuality in sexagenarian women.* Maturitas, 13:43–50.

Bahnson C.B. (1979) *An historical family systems approach to coronary heart disease and cancer.* In K.E. Schaefer, U. Stave & W. Blankenburg (eds.). A New Image of Man in Medicine, Futura, Mount Kisco, NY.

Bahnson C.B. (1981) *Stress and cancer: the state of the art.* Part 2. Psychosomatics, 22(3):207–220.

Bammer K. (1982) *Stress, Spread and Cancer.* In K. Bammer & B.H. Newberry (eds.), Stress and Cancer, C.J. Hogrefe, Toronto.

Barni S., Lissoni P., Cazzaniga M., Ardizzoia A., Meregalli S., Fossati V., Fumigalli L., Brivio F., Tancini G. (1995) *A randomized study of low-dose subcutaneous interleukin-2 plus melatonin versus supportive care alone in metastatic colorectal cancer patients.* Oncology, 52(3):243–245.

Barrett-Conner E., Khaw K.T., Yen S.S.C. (1986) *A prospective study of dehydroepiandrosterone sulfate, mortality, and cardiovascular disease.* N. Engl. J. Med., 315:1519–1524.

Barros J., Silveira F., Coelho R. (1986) *Vitamin A in colon cancer.* Digest. Dis. Sci., 31:172S.

Beauvoir (de) S. (1972) *The Coming of Age.* G.P. Putnam's sons, New York.

Belchetz P.E. (1994) *Hormonal treatment of postmenopausal women.* N. Engl. J. Med., 339(15):1062–1071.

Bellen H.J. and Kiger J.A. Jr. (1987) *Sexual hyperactivity and reduced longevity of dunce females of Drosophila melanogaster.* Genetics, 115:153–160.

Benson H. (1983) *The relaxation response: Its subjective and objective historical precedents and physiology.* Trends Neurosci., 6:281–284.

Berkman L.F. and Syme L. (1979) *Social networks, host resistance, and mortality: a nine-year follow-up study of Alameda county residents.* Am. J. Epidemiol., 109(2):186–204.

Beutler E. (1992) *Gaucher disease: New molecular approaches to diagnosis and treatment.* Science, 256:794–799.

Bierman E.L. (1978) *The effect of donor age on the in vitro life span of cultured human arterial smooth muscle cells.* In Vitro, 14:951–955.

Bierman E.L. (1985) *Arteriosclerosis and aging.* In Handbook of the Biology of Aging, 2nd ed., C.E. Finch and E.L. Schneider (eds.), pp. 842–858. Van Nostrand, New York.

Birkenhager-Gillesse E.G., Derksen J., Lagaay A.M. (1994) *Dehydroepiandrosterone sulphate (DHEAS) in the oldest old, aged 85 and over.* Ann. N.Y. Acad. Sci., 719:543–552.

Blair S.N., Kohl H.W., Barlow C.E., Paffenbarger R.S., Gibbons L.W., Macera C.A. (1995) *Changes in physical fitness and all-cause mortality.* J. Am. Med. Assoc., 273:1093–1098.

Blichert-Toft M. (1975) *Secretion of corticotrophin and somatotrophin by the senescent adenohypophysis in man.* Acta endocrinol. (Copenhagen), 78:1–157.

Bliznakov E.G. (1981) *Coenzyme Q, the immune system and aging.* Biomedical and Clinical Aspects of Coenzyme Q. Vol. 3, pp. 311–323. Elsevier, North-Holland Biomedical Press.

Bliznakov E.G. and Hunt G.L. (1986) *The Miracle Nutrient Coenzyme Q10.* A Bantam Book, Bantam Doubleday Dell Publishing Group, Inc., USA.

Block G. (1991) *Vitamin C and Cancer Prevention: The Epidemiologic Evidence.* Supplement to the Am. J. Clinic. Nutr., 53:278–279S.

Block G., Patterson B., Subar A. (1992) *Fruit, vegetables, and cancer prevention: a review of the epidemiological evidence.* Nutrition and Cancer, 18(1):1–29.

Blot W.J., Li J.Y., Taylor P.R. et al (1993) *Nutrition intervention trials in Linxian, China: supplementation with specific vitamin/mineral combinations, cancer incidence, and disease-specific mortality in the general population.* J. Natl. Cancer Inst., 85:1483–1492.

Blum A. and Wolinsky H. (1995) *AMA rewrites tobacco history.* Lancet, 346:261.

Boat T.F., Welsh M.J., Beaudet A.L. (1989) *The Metabolic Basis of Inherited Disease,* pp. 2649–2680. McGraw-Hill, New York.

Bohus B. and Croiset G. (1990) *Neuropeptides, behaviour, and the autonomic nervous system. Neuropeptides: Basics and Perspectives.* Chapter VII, pp. 255–276. De Wied, Honorary Editor. Elsevier Science Publishers B.V.

Booth-Kewley S. and Friedman H.S. (1987) *Psychological predictors of heart disease: A quantitative review.* Psychol. Bull., 101:343–362.

Bortz W.M. (1991) *We Live Too Short and Die Too Long. How to Achieve and Enjoy Your Natural 100-Year-Plus Life Span.* Bantam books, New York.

Bossi S.R., Simpson J.R., Isacson O. (1993) *Age dependence of striatal neuronal death caused by mitochondrial dysfunction.* Clin. Neurosci. Neuropathol., 4(1):73–76.

Boyd J.W., Himmelstein D.U., Woolhandler S. (1995) *The Tobacco/health-insurance connection.* Lancet, 346:64.

Bradley A.J., McDonald I.R., Lee A.K. (1980) *Stress and mortality in a small marsupial (Antechinus stuartii, Macleay).* Gen. Comp. Endocrinol., 40:188–200.

Brecher E. (1984) *Love, Sex and Aging.* Consumers Union, New York.

Breslow L. and Enstrom J.E. (1980) *Persistance of health habits and their relationship to mortality.* Prev. Med., 9:469–483.

Breslow L.B. and Breslow N. (1993) *Health practices and disability: some evidence from Alameda County.* Prev. Med., 22:86–95.

Bretschneider J. and McCoy N. (1988) *Sexual interest and behavior in healthy 80- to 102 year-olds.* Arch. Sex. Behav., 17(2):109–129.

Brittberg M., Lindahl A., Nilsson A., Ohlsson C., Isaksson O., Peterson L. (1994) *Treatment of deep cartilage defects in the knee with autologous chondrocyte transplantation.* N. Engl. J. Med., 331:889–895.

Brody H. (1955) *Organization of cerebral cortex. III. A study of aging, in the human cerebral cortex.* J. Comp. Neurol., 102:511–556.

Brody H. (1978) *Cell counts in cerebral cortex and brainstem.* In Alzheimer's Disease: Senile Dementia and Related Disorders. Robert Katzman, Robert D. Terry and Katherine L. Bick (eds.), pp. 345–351, Raven Press, New York.

Brown M.S. and Goldstein J.L. (1986) *A receptor-mediated pathway for cholesterol homeostasis.* Science, 232:34–47.

Brown W., Ford J., Gershey E. (1980) *Variation of DNA repair capacity in progeria cells unrelated to growth conditions.* Biochem. Biophys. Res. Commun., 97:347–353.

Brown W.T., Zebrower M., Kieros F.J. (1984) *Progeria, a model disease for the study of accelerated aging.* In Molecular Biology of Aging, Basic Life Science. Vol. 35, pp. 375–386, A.V. Woodhead, A.D. Blackett, and A. Hollaender (eds.), Plenum, New York.

Bruning N. (1994) *The Natural Health, Guide to Antioxidants.* Nancy Bruning and the Editors of Natural Health Magazine, Bantam Books, New York.

Brunzell J.D. (1984) *Physiologic approach to hyperlipidemia.* In Yearbook of Endocrinology, T.B. Schwartz and W.G. Ryan (eds.), pp. 11–30, Year book Medical, Chicago.

Bucala R., Model P., Russel M., Cerami A. (1985) *Modification of DNA by glucose-6-phosphate induces DNA rearrangments in an Escherichia coli plasmid.* Proc. Natl. Acad. Sci. USA, 82:8439–8442.

Buiatti E., Palli D., Decarli A., Amadori D., Avellini C., Bianchi S., Bonaguri C., Cipriani F., Cocco P., Giacosa A., Marubini E., Minacci C., Puntoni R., Russo A., Vindigni C., Fraumeni J.F., Blot W.J. (1990) *A case-control study of gastric cancer and diet in Italy: II. Association with Nutrients.* Int. J. Cancer, 45:896–901.

Bulbrook R.D., Hayward J.L., Spicer C.C. (1971) *Relation between urinary androgen and corticoid excretion and subsequent breast cancer.* Lancet, 2:395–398.

Burnet F.M. (1974) *Intrinsic Mutagenesis: a Genetic Approach,* Wiley, New York.

Burnstein K. and Cidlowski J. (1989) *Regulation of gene expression by glucocorticoids.* Annu. Rev. Physiol., 51:683–699.

Buxton J., Shelbourne P., Davies J., Jones C., Van Tongeren R., Aslanidis C., de Jung P., Jansen G., Anvret M., Riley B., Williamson R., Johnson K. (1992) *Detection of an unstable fragment of DNA specific to individuals with myotonic dystrophy.* Nature, 355:547–548.

Byers T., Graham S., Haughey B.P., Marshall J.R., Swanson M.K. (1987) *Diet and lung cancer risk: Findings from the Western New York Diet Study.* Am. J. Epidem., 125:351–363.

Cade R., Mars D., Wagemaker H., Zauner C., Packer D., Privette M., Cade M., Peterson J., Hood-Lewis D. (1984) *Effect of aerobic exercise training on patients with systemic arterial hypertension.* Am. J. Medicine, 77:785–790.

Cadenas E. (1989) *Biochemistry of oxygen toxicity.* Annu. Rev. Biochem., 58:79–110.

Cairns J. (1985) *The history of mortality and the conquest of cancer.* In Accomplishments in Cancer Research, J.P. Lippincott, Philadelphia.

Carlberg C., Hooft van Huijsduijen R., Staple J., DeLamarter J.F., Becker-André M. (1994) *RZRs, a new family of retinoid-related orphan receptors that function as both monomers and homodimers.* Mol. Endocrinol., 8:757–770.

Carlson L.D., Scheyer W.J., Jackson B.H. (1957) *The combined effects of ionizing radiation and low temperature on the metabolism, longevity and soft tissue of the white rat.* Radiat. Res., 7:190–195.

Carney J.M., Starke-Reed P.E., Oliver C.N., Landum R.W., Cheng M.S., Wu J.F., Floyd R.A. (1991) *Reversal of age-related increase in brain protein oxidation, decrease in enzyme activity, and loss in temporal and spatial memory by chronic administration of the spin-trapping compound N-tert-butyl-α-phenylnitrone.* Proc. Natl. Acad. Sci. USA, 88:3633–3636.

Caroleo M.C., Frasca D., Nistico G., Doria G. (1992) *Melatonin as immunomodulator in immunodeficient mice.* Immunopharmacology, 23(2):81–89.

Carskadon M.A., Brown E.D., Dement W.C. (1982) *Sleep fragmentation in the elderly, relationship to daytime sleep tendency.* Neurobiol. Aging, 3:321–327.

Carson D.A. and Ribeiro J. (1993) *Apoptosis and diseases.* Lancet, 341:1251–1254.

Cartwright L.K., Wink P., Kmetz C. (1995) *What leads to good health in midlife women physicians? Some clues from a longitudinal study.* Psychosom. Med., 57:284–292.

Caskey C.T., Pizzuti A., Fu Y.H., Fenwick R.G., Nelson D.L. (1992) *Triplet repeat mutations in human disease.* Science, 256:784–789.

Casson P.R., Andersen R.N., Herrod H.G., et al (1993) *Oral dehydroepiandrosterone in physiologic doses modulates immune function in postmenopausal women.* Am. J. Obstet. Gynecol., 169:1536–1539.

Catania J. and Fairweather D.S. (1991) *DNA methylation and cellular aging.* Mutation Research, 256:283–293.

Cheng R.S.S. and Pomeranz B. (1987) *Electro-acupuncture analgesia could be mediated by at least two pain-relieving mechanisms, endorphin and non-endorphin systems.* Life science, 25:1957–1963.

Chiang H.-L. and Dice J.F. (1988) *Peptide sequences that target proteins for enhanced degradation during serum withdrawal.* J. Biol. Chem., 263:6797–6805.

Chiu H.C. (1989) *Dementia: A review emphasizing clinicopathological correlation and brain-behavior relationships.* Arch. Neurol., 46:806–814.

Chung S.S., Dim S.H., Yang W.I., Choi I.J., Lee W.Y., Moon J.G., Park H.S., Shin H.S., Dim D.S., Ahn Y.M. (1993) *Homogenous fetal dopaminergic cell transplantation in rat striatum by cell suspension methods.* Yonsei-Med-J., 34(2):145–151.

Clarke A.R., Purdie C.A., Harrison D.J., Morris R.G., Bird C.C., Hooper M.L., Wyllie A.H. (1993) *Thymocyte apoptosis induced by p53-dependent and independent pathways.* Nature, 362:849–852.

Cleaver J.E. (1982) *Normal reconstruction of DNA supercoiling and chromatin structure in Cockayne syndrome cells during repair of damage from ultraviolet light.* American J. Hum. Genetics, 34:566–575.

Cogen R. and Steinman W. (1990) *Sexual function and practice in elderly men of lower socioeconomic status.* J. Fam. Pract., 31:162–166.

Coleman D.L. (1978) *Obese and diabetes: Two mutant genes causing diabetes-obesity syndromes in mice.* Diabetologia, 14:41–48.

Coleman D.L., Schwizer R.W., Leiter E.H. (1984) *Effect of genetic background on the therapeutic effects of dehydroepiandrosterone (DHEA) in diabetes-obesity mutants and in aged normal mice.* Diabetes, 33:26–32.

Coleman P., Curcio C.A., Buell S.J. (1982) *Morphology of the aging central nervous system: not all downhill.* The Aging Motor System. Morhmer J.A., Pirozzolo F.J., Maletta S. (eds.), pp. 7–35, Praeger, New York.

Comfort A. (1979) *The Biology of Senescence.* Third edition, Churchill Livingstone, Edinburgh and London.

Comfort A. and Dial L.K. (1991) *Sexuality and aging. An overview.* Clin. Geriatr. Med., 7:1–7.

Comstock G.W., Bush T.L., Helzlsouer K.J. (1992) *Serum retinol, β-carotene, vitamin E, and selenium as related to subsequent cancer of specific sites.* Am. J. Epidemiol., 135:115–121.

Congdon C.C. (1987) *A review of certain low-level ionizing radiation studies in mice and guinea pig.* Health Physics, 52:595–597.

Corder E.H., Saunders A.M., Strittmatter W.J., Schmechel D.E., Gaskall P.C., Small G.W., Roses A.D., Haines J., Pericak-Vance M.A. (1993) *Gene dose of apolipoprotein E type 4 allele and the risk of Alzheimer's disease in late onset families.* Science, 261:921–923.

Corral-Debrinski M., Horton T., Lott M.T., Shoffner J.M., Beal F.M., Wallace D.C. (1992) *Mitochondrial DNA deletions in human brain: regional variability and increase with advanced age.* Nature genetics, 2:324–329.

Costa P.T., McCrae R.R., Arenberg D. (1983) *Recent longitudinal research on personality and aging.* In Schaie (ed.), Longitudinal Studies of Adult Psychological Development, Guilford, New York.

Costa A., Formelli F., Chiesa F., Decensi A., De Palo G., Veronesi U. (1994) *Prospects of chemoprevention of human cancers with the synthetic retinoid fenretinide.* Cancer Res., 54(suppl):2032s–2037s.

Couch J.R. (1993) *Antiplatelet therapy in the treatment of cerebrovascular disease.* Clin. Cardiol., 16:703–710.

Cox T. and MacKay C. (1982) *Psychosocial factors and psychophysiological mechanisms in the aetiology and develpment of cancers.* Soc. Sci. and Med., 16:381–396.

Cristofalo V.J. and Stanulis-Praeger B.M. (1982) *Cellular senescence in vitro.* In Advances in Tissue Culture. K. Maramorosch (ed.). Vol. 2, pp. 1–68, Academic Press, New York.

Cristofalo V.J., Phillips P.D., Sorger T., Gerhard G. (1989) *Alterations in the responsiveness of senescent cells to growth factors.* J. Gerontol., Biol. Sci., 44:55–62.

Cristofalo V.J. (1990) *Overview of biological mechanism of aging.* Annu. Rev. Gerontol. Geriatr., 10:1–22.

Culver K.W., Ram Z., Wallbridge S., Ishii H., Oldfield E.H., Blaese R.M. (1992) *In vivo gene transfer with retroviral vector-producer cells for treatment of experimental brain tumors.* Science, 256:1550–1552.

Cutler R.G. (1984) *Evolutionary biology of aging and longevity in mammalian species.* In Aging and Cell Function. J.E. Johnson Jr. (ed.), Plenum, New York.

Cutler R.G. (1985) *Antioxidants and longevity of mammalian species.* Basic Life Sci., 35:15–73.

Cyrulnick B. (1993) *Les Nourritures Affectives.* Odile Jacob (ed.), Paris.

Dalsky G.P., Stocke K.S., Ehsani A.A., Slatopolsky E., Lee W.C., Birge S.J. (1988) *Weight-bearing exercise training and lumbar bone mineral content in postmenopausal women.* Ann. Intern. Med., 108:824–828.

Debry G. and Pelletier X. (1991) *Physiological importance of w-3/w6 polyunsaturated fatty acids in man. An overview of still unresolved and controversial questions.* Experientia, 47:172–177.

De Leon M., McRae T., Tsai J., George A., Marcus D., Freedman M., Wold A., McEwen B. (1988) *Abnormal cortisol response in Alzheimer's disease linked to hippocampal atrophy.* Lancet, 2:391–392.

Dement W., Richardson G., Prinz P., Carskadon M., Kripke D., Czeisler C. (1985) *Changes of sleep and wakefulness with age.* In Handbook of the Biology of Aging, pp. 692–717, 2nd ed., C.E. Finch and E.L. Schneider (eds.), New York: Van Nostrand.

De Wied D. (1969) *Effects of peptide hormones on behavior.* In Frontiers in Neuroendocrinology. W.F. Ganong and L Martini (eds.), p. 97, Oxford University Press, New York.

Diamond M., Miner J., Yoshinaga S., Yamamoto K. (1990) *Transcription factor interactions: selectors of positive or negative regulation from a single DNA element.* Science, 249:1266–1272.

Dieber-Rotheneder M., Puhl H., Waeg G., Striegl G. (1991) *Effect of oral supplementation with D-Alpha-Tocopherol on the vitamin E content of human low density lipoproteins and resistance to oxidation.* J. Lip. Res., 32(8):1325–1332.

Dilman V.M. (1984) *Three models of medicine.* Med. Hypothesis, 15:185–208.

Doll R., Peto R., Wheatley K., Gray R., Sutherland I. (1994) *Mortality in relation to smoking: 40 years' observations on male British doctors.* Br. Med. J., 309:901–911.

Doria G. and Frasca D. (1987) *Thymic hormones as immunoregulatory molecules: an overview.* In Tumor Immunology and Immunoregulation by Thymic Hormones. Dammacco F. (ed.), pp. 109–121, Masson, Milan.

Dormandy T.L. (1989) *Free-radical pathology and medicine. A review.* J. Royal College of Physicians (London), 23:221–227.

Driscoll M. (1994) *Genes controlling programmed cell death: relation to mechanisms of cell senescence and aging?* In Molecular Aspects of Aging. K. Esser and G.M. Martin (eds.), pp. 45–58, John Wiley and sons, New York.

Drumm M.L., Pope H.A., Cliff W.H., Rommens J.M., Marvin S.A., Tsui L.-C., Collins F.S., Frizzell R.A., Wilson J.M. (1990) *Correction of the cystic fibrosis defect in vitro by retrovirus-mediated gene transfer.* Cell, 62:1227–1233.

Dubocovitch M.L. (1995) *Melatonin receptors: are there multiple subtypes?* Trends Pharmacol., 16:50–56

Ellis R.E. and Horvitz H.R. (1991) *Two C. Elegans genes control the programmed deaths of specific cells in the pharynx.* Development, 112(2):591–603.

Ellis R.E., Yuan J.Y., Horvitz H.R. (1991) *Mechanisms and functions of cell-death.* Annu. Rev. Cell. Biol., 7:663–698.

Emerich D.F., Black B.A., Kesslak J.P., Cotman C.W., Walsh T.J. (1992) *Transplantation of fetal cholinergic neurons into the hippocampus attenuates the cognitive and neurochemical deficits induced by AF64A.* Brain Res. Bull., 28(2):219–226.

Endicott J.A. and Ling V. (1989) *The biochemistry of P-glycoprotein-mediated multidrug resistance.* Annu. Rev. Biochem., 58:137–171.

Epstein C.J., Martin G.M., Schultz A.L., Motulsky A.G. (1966) *Werner's syndrome: A review of its symptomatology, natural history, pathological features, genetics, and relationship to the natural aging process.* Medicine, 45:177–221.

Epstein E.H. (1992) *Molecular genetics of epidermolysis bullosa.* Science, 256:799–804.

Esposito J.L. (1987) *The Obselete Self: Philosophical Dimensions of Aging*. University of California Press, Berkeley.

Esterbauer H., Dieber-Rotheneder M., Striegl G., Waeg G. (1991) *Role of vitamin E in preventing the oxidation of low-density lipoprotein*. Am. J. Clin. Nutr., 53:314S–321S.

Evans D.G.R., Fentiman I.S., McPherson K., Asbury D., Ponder B.A.J., Howell A. (1994) *Familial breast cancer*. Br. Med. J., 308:183–187.

Evans V. (1993) *Multiple pathways to apoptosis*. Cell Biol. Int., 17:461–475.

Eysenck H.J. (1984) *Personality, stress and lung cancer*. In Contributions to Medical Psychology. S. Rachman (ed.). Vol. 3, Pergamon Press, Oxford.

Eysenck H.J. (1988) *Personality, stress and cancer: prediction and prophylaxis*. Br. J. Med. Psychology, 61:57–75.

Eysenck H.J. and Grossarth-Maticek R. (1991) *Creative novation behaviour therapy as a prophylactic treatment for cancer and coronary heart disease: Part II- Effects of treatment*. Behav. Res. Ther., 29(1):17–31.

Fahn S. (1992) *A pilot trial of high-dose alpha-tocopherol and ascorbate in early Parkinson's disease*. Ann. Neurol., 32 Suppl:S128–132.

Fahrer H., Hoeflin F., Lauterburg B.H., Peheim E., Levy A., Vischer T.L. (1991) *Diet and fatty acids: can fish substitute for fish oil?* Clin. Exp. Rheumatol, 9:403–406.

Fairweather D.E., Fox M., Margison G.P. (1987) *The in vitro life span of MRL-5 cells is shortened by 5-azacytidine-induced demethylation*. Exp. Cell Res., 168:153–159.

Farmer G., Bargonetti J., Zhu H., Friedman P., Prywes R., Prives C. (1992) *Wild-type p53 activates transcription in vitro*. Nature, 358:83–86.

Fattoretti P., Viticchi C., Piantanelli L. (1982) *Age-dependent decrease of beta-adrenoceptor density in the submandibular glands of mice and its modulation by the thymus*. Arch. Gerontol. Geriatr., 1:229–240.

Felten D. (1991) *Neurotransmitter signaling of cells of the immune system: important progress, major gaps*. Brain Behav. Immun., 5:2–8.

Feng J., Funk W.D., Wang S.S., Wenrich S.L., Avilion A.A., et al (1995) *The RNA component of human telomerase*. Science, 269:1236–1241.

Fernandes G., Flescher E., Venkatraman J.T. (1990a) *Modulation of cellular immunity, fatty acid composition, fluidity and Ca 2+ influx by food restriction in aging rats*. In Aging: Immunology and Infectious Diseases. Vol. 2, No 3, pp. 117–125. Mary Ann Liebert, Inc., Publishers.

Fernandes G., Venkatraman J., Khare A., Horbach G.J.M.J., Friedrichs W. (1990b) *Modulation of gene expression in autoimmune disease and aging by food restriction and dietary lipids*. Proc. Soc. Exp. Biol. Med., 193:16–22.

Finch C.E. and Landfield P.W. (1985) *Neuroendocrine and autonomic functions in aging mammals*. In The Handbook of the Biology of Aging. C.E. Finch and E.L. Schneider (eds.), p. 567, Van Nostrand Reinhold, New York.

Finch C.E. (1990) *Longevity, Senescence and the Genome*. The University of Chicago Press, Ltd., London.

Fiore M.C., Smith S.S., Jorenby D.E., Baker T.B. (1994) *The effectiveness of the nicotine patch for smoking cessation: a meta-analysis*. J. Am. Med. Assoc., 271:1940–1947.

Fisher L.J. and Gage F.H. (1994) *Intracerebral transplantation: basic and clinical application to the neostriatum*. Faseb J., 8:489–496.

Fobair P. and Cordoba C.S. (1982) *Scope and magnitude of the cancer problem in psychosocial research*. In Psychosocial Aspects of Cancer. J. Cohen, J.W. Cullen and L.R. Martin (eds.), Raven Press, New York.

Foreman H.J. and Fischer A.B. (1981) *Antioxidant defenses.* In Oxygen and Living Processes. D.L. Gilbert (ed.), pp. 65–90, Springer-Verlag, New York.

Fox B.H. (1981) *Psychosocial factors in the immune system in cancer.* In Psychoneuroimmunology. R. Ader (ed.), pp. 103–158, Academic Press, New York.

Friedman D.B. and Johnson T.E. (1988a) *A mutation in the* ***age-1*** *gene in Caenorhabditis elegans lengthens life and reduces hermaphrodite fertility.* Genetics, 118:75–86.

Friedman D.B. and Johnson T.E. (1988b) *Three mutants that extend both mean and maximum life span of the nematode, Caenorhabditis elegans, define the* ***age-1*** *gene.* J. Gerontol., 43:B102–B109.

Fries F.G. and Crapo L.M. (1981) *Vitality and Aging.* Freeman, San Francisco.

Fryer J.H. (1962) *Studies of body composition on men aged 60 and over.* In Biological Aspects of Aging. N.W. Shock (ed.), pp. 59–78, Columbia University Press, New York.

Fu Y.H., Kuhl D.P.A, Pizzuti A., Pieretti M., Sutcliffe J.S., Richards S., Verkerk A.J.M.H., Holden J.J.A., Fenwick R.G., Warren S.T., Oostra B.A., Nelson D.L., Caskey C.T. (1991) *Variation of the CGG repeat at the fragile X site, results in genetic instability: resolution of the Sherman paradox.* Cell, 67:1047–1058.

Fuchs C.S., Stampfer M.J., Colditz G.A., Giovannucci E.L., Manson J.E., Kawachi I., Hunter D.J., Hankinson S.E., Hennekens C.H., Rosner B., Speizer F.E., Willett W.C. (1995) *Alcohol consumption and mortality among women.* N. Engl. J. Med., 332(19):1245–1250.

Fujiwara R., Tong Z.G., Matsuoka H., Shibata H., Iwamoto M., Yokoyama M.M. (1991) *Effects of acupuncture on immune response in mice.* Int. J. Neurosci., 57:141–150.

Gaby S.K. and Machlin L.J. (1991) *Vitamin E.* In Vitamin Intake and Health: A Scientific Review. S.K. Gaby, A Bendich, V Singh, and L. Machlin (eds.), New York.

Garfinkel D., Laudon M., Nof D., Zisapel N. (1995) *Improvement of sleep quality in elderly people by controlled-release melatonin.* Lancet, 346:541–543.

Garland M., Morris J.S., Stampfer M.J., Graham A., Colditz G.A., Spate V.L., Baskett C.K., Rosner B., Speizer F.E., Willett W.C., Hunter D.J. (1995) *Prospective study of toenail selenium levels and cancer among women.* J. Natl. Cancer Inst., 87(7):497–505.

Gattaz W.F., Kollisch M., Thuren T., Virtanen J.A., Kinnunen P.K.J. (1987) *Increased plasma phopholipase-A2 activity in schizophrenic patients: Reduction after neuroleptic therapy.* Biol. Psychiatry, 22:241–426.

Gey F.K., Moser U.K., Jordan P., Stähelin H.B., Eichholzer M., Lüdin E. (1993) *Increased risk of cardiovascular disease at suboptimal plasma concentrations of essential antioxidants: an epidemiological update with special attention to carotene and vitamin C.* Am. J. Clin. Nutr., 57(suppl):787S–797S.

Gillman M.W., Cupples A., Gagnon D., Millen Posner B., Ellison C., Castelli W.P. Wolf P.A. (1995) *Protective effects of fruits and vegetables on development of stroke in men.* J. Am. Med. Assoc., 273:1113–1117.

Glantz S.A., Barnes D.E., Bero L., Hanauer P., Slade J. (1995) *Looking through a keyhole at the tobacco industry.* J. Am. Med. Assoc., 274:219–224.

Goldberg A.P. and Schonfeld G. (1985) *Effects of diet on lipoprotein metabolism.* Annu. Rev. Nutr., 5:195–212.

Goldstein S. (1969) *Life span of cultured cells in progeria.* Lancet, 1:424 (abstract).

Goldstein S. and Moerman E.J. (1976) *Defective protein in normal and abnormal fibroblasts during aging in vitro.* In Interdisciplinary Topics of Gerontology. Vol. 10, pp. 24–43, Karger, Basel.

Gonzalez-Quintial R. and Theofilopoulos A.N. (1992) *Vb gene repertoires in aging mice.* J. Immunol., 149:230–236.

Gordon G.B., Bush D.E., Weisman H.F. (1988) *Reduction of atherosclerosis by administration of dehydroepiandrosterone. A study in the hypercholesterolemic New Zealand white rabbit with aortic intimal injury.* J. Clin. Invest., 82:712–720.

Goso C., Frasca D., Doria G. (1992) *Effects of synthetic thymic humoral factor (THF-γ2) on T cell activities in immunodeficient ageing mice.* Clin. Exper. Immunol, 87:346–351.

Goto M., Rubenstein M., Weber J., Woods K., Drayna D. (1991) *Genetic linkage of Werner's syndrome to five markers on chromosome 8.* Nature, 355:735–738.

Gowen M. and Mundy G.R. (1986) *Actions of recombinant interleukin 1, interleukin 2, and interferon-gamma on bone resorption in vitro.* J. Immunol., 136:2478–2482.

Grad B.R. and Rozencwaig R. (1993) *The role of melatonin and serotonin in aging: update.* Psychoneuroendocrinology, 18(4):283–295.

Greene W.A., Betts R.F., Ochitill H.N., Iker H.P., Douglas R.G. (1978) *Psychosocial factors and immunity: preliminary report.* Psychosom. Med., 40:87 (abstract).

Greenwald P., Sondik E., Lynch B.S. (1986) *Diet and Chemoprevention.* Annu. Rev. Public Health, 7:267–291.

Greer S. and Watson M. (1985) *Towards a psychobiological model of cancer: psychological considerations.* Soc. Sci. Med., 20:773–777.

Greider C.W. and Blackburn E.H. (1985) *Identification of a specific telomere terminal transferase activity in Tetrahymena extracts.* Cell, 43:405–413.

Grossman M., Raper S.E., Kozarsky K., Stein E.A., Engelhardt J.F., Muller D., Lupien P.J., Wilson J.M. (1994) *Successful ex vivo gene therapy directed to liver in a patient with familial hypercholesterolaemia.* Nature Genet., 6:335–341.

Gualde N. and Goodwin J. (1984) *Effect of irradiation on human T-cell proliferation: low dose irradiation stimulates mitogen-induced proliferation and functions of the suppressor/cytotoxic T-cell subset.* Cell. Immunol., 84:439–445.

Gudbjarnason S., Benediktsdóttir VE., Gudmundsdóttir E. (1991) *Balance between ω3 and ω6 fatty acids in heart muscle in relation to diet, stress and ageing.* In Health Effects of ω3 Polyunsaturated Fatty Acids in Seafoods. Simopoulos A.P., Kifer R.R., Martin R.E., Barlow S.M. (eds.). Vol. 66, pp. 292–305. World Rev. Nutr. Diet., Karger, Basel.

Guralnik J.M. and Kaplan G.A. (1989) *Predictors of healthy aging: prospective evidence from the Alameda County Study.* Am. J. Public Health, 79:703–708.

Gusella J.F., Wexler N.S., Conneally P.M., Naylor S.I., Anderson M.A.,Tanzi R.E., Watkins P.C., Ottina K., Wallace M.R., Sakaguchi A.Y., Young A.B., Shoulson I., Bonilla E., Martin J.B. (1983) *A polymorphic DNA marker genetically linked to Huntington's disease.* Nature, 306:234–238.

Gwatkin D.R. and Brandel S.K. (1982) *Life expectancy and population growth in the third world .* Sci. Am., 246(5):33–41.

Hall K.Y., Hart R.W., Benirschke A.K., Walford R.L. (1984) *Correlation between ultraviolet-induced DNA repair in primate lymphocytes and fibroblasts and species maximum achievable life span.* Mech. Ageing Dev., 24:163–174.

Halley E. (1693) *An estimate of the degree of mortality of mankind drawn from various tables of the births and funerals of the city of Breslau; with an attempt to ascertain the price of annuities upon lives.* Philosophical Transactions of the Royal Society, 17:596.

Halliwell B. (1981) *Free radicals, oxygen toxicity and aging.* In Age Pigments. R.S. Sohal (ed.), pp. 1–62, Elsevier North Holland, Amsterdam.

Halliwell B. and Gutteridge J.M.C. (1985) *Free Radicals in Biology and Medicine.* Clarendon Press, Oxford.

Hamilton J.B. and Mestler G.E. (1969) *Mortality and survival: A comparison of eunuchs with intact men and women in a mentally retarded population.* J. Gerontol., 24:395–411.

Hamilton J.B., Tereda H., Mestler G.E., Tirman W. (1969) *1. Coarse sternal hairs, a male secondary sex character that can be measured quantitatively: The influence of sex, age and genetic factors. 2. Other sex-differing characters: Relationship to age, to one another, and to values for coarse sternal hairs.* In Advances in Biology of Skin Hair Growth, Proceedings of a symposium held at the University of Oregon Medical School, 1967. Vol. 9, pp. 129–151. Pergamon Press, Oxford and New York.

Hanabuchi S., Koyanagi M., Kawasaki A., et al (1994) *Fas and its ligand in a general mechanism of T-cell mediated cytotoxicity.* Proc. Natl. Acad. Sci. USA, 91:4930–4934.

Hardeland R., Reiter R.J., Poeggeler B., Tan D.X. (1993) *The significance of the metabolism of the neurohormone melatonin: antioxidative protection and formation of bioactive substances.* Neurosc. Biobehav. Rev., 17:347–357.

Harding A.E., Thomas P.K., Baraitser M., Bradbury P.G., Morgan-Hughes J.A., Ponsford J.R. (1991) *X-linked recessive bulbospinal neuronopathy: a report of ten cases.* J. Neurol. Neurosurg. Psych., 45:1012–1019.

Harley C.B. and Villeponteau B. (1995) *Telomeres and telomerase in aging and cancer.* Curr. Opin. Genet. Dev., 5:249–255.

Harley H.G., Brook J.D., Rundle S.A., Crow S., Reardon W., Buckler A.J., Harper P.S., Housman D.E., Shaw D.J. (1992) *Expansion of an unstable DNA region and phenotypic variation in myotonic dystrophy.* Nature, 355:545–546.

Harman D. (1956) *Aging: A theory based on free radical and radiation biology.* J. Gerontol., 11:298–300.

Harman D. (1984) *Free radicals in aging.* Mol. Cell. Biol., 84:155–161.

Harrison D.E., Archer J.R., Astle C.M. (1984) *Effects of food restriction on aging: separation of food intake and adiposity.* Proc. Natl. Acad. Sci. USA, 81:1835–1838.

Hart R.W. and Turturro A. (1983) *Theories of aging.* In Review of Biological Research in Aging. M. Rothstein (ed.). Vol. 1, pp. 5–18, Alan Liss, New York.

Hayes D., Baylis S., Lee J., Halberg F. (1977) *Codling moth development and aging in different light regimens.* Chronobiologia, 4(2):118.

Hayflick L. and Moorhead P.S. (1961) *The serial cultivation of human diploid cell strains.* Exper. Cell Research, 25:585–621.

Hayflick L. (1985) *Theories of biological aging.* Exper. Gerontology, 20:145–159.

Heller T., Holt P.R., Richardson A. (1990) *Food restriction retards age-related histologic changes in rat small intestine.* Gastroenterology, 98:387–391.

Helzlsouer K.J., Block G., Blumberg J., Diplock A.T., Levine M., Marnett L.J., Schulplein R.J., Spence J.T., Simic M.G. (1994) *Summary of the round table discussion on strategies for cancer prevention: diet, food, additives, supplements, and drugs.* Cancer Res. (suppl), 54:2044s–2051s.

Henderson C. (1995a) *L'avenir thérapeutique des facteurs neurotrophiques dans les maladies neuro-dégénératives.* Médecine et Science, 11:1067–1069.

Henderson C. (1995b) *Neurotrophic factors as therapeutic agents in amyotrophic lateral sclerosis.* Adv. Neurol., 68:235–240.

Henderson V.W. and Finch C.E. (1989) *The neurobiology of Alzheimer's disease.* J. Neurosurg., 70:335–353.

Hengartner M.O., Ellis R.E., Horvitz H.R. (1992) *Caenorhabditis elegans gene ced-9 protects cells from programmed cell death.* Nature, 356(6369):494–499.

Hengartner M.O. and Horvitz H.R. (1994) *Activation of C. elegans cell death protein CED-9 by an amino-acid substitution in a domain conserved in Bcl-2.* Nature, 369:318–320.

Hervonen A., Jaatinen P., Sarviharju M., Kiianmaa K. (1992) *Interaction of aging and lifelong ethanol ingestion on ethanol-related behaviors and longevity.* Exp. Gerontol., 27:335–345.

Hockenbery D., Nuñes G., Milliman C., Schreiber R.D., Korsmeyer S.J. (1990) *Bcl-2 is an inner mitochondrial membrane protein that blocks programmed cell death.* Nature, 348:334–336.

Hodis H.N., Mack W.J., LaBree L., Cashin-Hemphill L., Sevanian A., Johnson R., Azen S.P. (1995) *Serial coronary angiographic evidence that antioxidant vitamin intake reduces progression of coronary artery atherosclerosis.* J. Am. Med. Assoc., 273:1849–1854.

Holloway L., Butterfield G., Hintz R.L., Gesundheit N., Marcus R. (1994) *Effects of recombinant human growth hormone on metabolic indices, body composition, and bone turnover in healthy elderly women.* J. Clin. Endocrinol. Metab., 79(2):470–479.

Holt P.R., Heller T.D., Richardson A.G. (1991) *Food restriction retards age-related biochemical changes in rat small intestine.* J. Gerontol., 46(3):B89–94.

Hong W.K., Lippman S.M., Itri L.M., et al (1990) *Prevention of second primary tumors with isotretinoin in squamous-cell carcinoma of the head and neck.* N. Engl. J. Med., 323:795–800.

Horowitz M.C. (1993) Cytokines and estrogen in Bone: anti-osteoporotic effects. Science, 260:626–627.

Hotta Y. and Benzer S. (1972) *Mapping of behaviour in Drosophila mosaics.* Nature, 240:527–535.

House J.S., Landis K.R., Umberson U. (1988) *Social relationships and health.* Science, 241:540–545.

Howard B.H., Fordis C.M., Sakamoto K., Holter W., Corsico C.D., Howard T. (1988) *Negative regulation of cell growth by interspersed repetitive DNA sequences.* In Vitro, 24(2):abstract 144, p. 47A.

Howe G.R., Hirohata T., Hislop T.G., Iscovich J.M., Yuan J.-M., et al (1990) *Dietary factors and risk of breast cancer: combined analysis of 12 case-control studies.* J. Natl. Cancer Inst., 82(7):561–569.

Humphries P., Kenna P., Farrar G.J. (1992) *On the molecular genetics of Retinitis Pigmentosa.* Science, 256:804–808.

The **Huntington**'s Disease Collaborative Research Group (1993) *A novel gene containing a trinucleotide repeat that is expanded and unstable on Huntington's disease chromosomes.* Cell, 72:971–983.

Hyde S.C., Gill D.R., Higgins C.F., Trezise A.E.O., MacVinish L.J., Cuthbert A.W., Ratcliff R., Evans M.J., Colledge W.H. (1993) *Correction of the ion transport defect in cystic fibrosis transgenic mice by gene therapy.* Nature, 362:250–255.

Ikeda Y., Ikeda K., Long D.M. (1989) *Protective effect of the iron chelator deferoxamine on cold-induced brain edema.* J. Neurosurg., 71:233–238.

Imaizumi T. (1989) *Intravascular release of platelet-activating factor-like lipid (PAF-LL) induced by cigarette smoking.* In Proceedings of the Third International Conference on Platelet-Activating Factor and Structurally Related Alkyl Ether Lipids, p. 78, Tokyo.

Ingram D.K., Wiener H.L., Chachich M.E., Long J.M., Hengemihle J., Gupta M. (1993) *Chronic treatment of aged mice with L-deprenyl produces marked striatal MAO-B inhibition but no beneficial effects on survival, motor performance, or nigral lipofuscin accumulation.* Neurobiol. Aging, 14:431–440.

Itoh N., Yonehara S., Ishii A., et al (1991) *The polypeptide encoded by the cDNA for human cell surface antigen Fas can mediate apoptosis.* Cell, 66:233–243.

Iwasaki K., Sunderland T., Kusiak J.W., Wolozin B. (1996) *Changes in gene transcription during aβ-mediated cell death.* Mol. Psychiatry, 1:65–71.

Iwashita Y., Kawaguchi S., Murata M. (1994) *Restoration of function by replacement of spinal cord segments in the rat.* Nature, 367:167–170.

Iwatsubo T, et al (1995) *Amyloid β protein (Aβ) deposition: Aβ42(43) precedes Aβ40 in Down syndrome.* Ann. Neurol., 37:294–299.

Jacques P.F., Chylack L.T., McGandy R.B., Hartz S.C. (1988) *Antioxidant status in persons with and without senile cataract.* Arch. Ophthalmol., 106:337–340.

Janiaud P. (1987) *Light, pineal, melatonin and chemical carcinogenesis.* Neuroendocrinol. Lett., 9(5):311.

Jarvik L.F., Falek A., Kallman F.J., Lorge I. (1960) *Survival trends in a senescent twin population.* Am. J. Hum. Genet., 12:170–179.

Jarvik L.F., Ruth V., Matsuyama S.S. (1980) *Organic brain syndrome and aging: A six-year follow-up of surviving twins.* Arch. Gen. Psychiatry, 37:280–286.

Jesberger J.A. and Richardson J.S. (1991) *Oxygen free radicals and brain dysfunction.* Intern. J. Neurosci., 57:1–17.

Jilka R.L., Hangoc G., Girasole G., Passeri G., Williams D.C., Abrams J.S., Boyce B., Broxmeyer H., Manolagas S.C. (1992) *Increased osteoclast development after estrogen loss: mediation by interleukin-6.* Science, 257:88–91.

Johnson T.E. (1990) *Increased life-span of **age-1** mutants in Caenorhabditis elegans and lower Gompertz rate of aging.* Science, 249:908–912.

Jouandet M.L., Tramp M.J., Herron D.M., Hermann A., Loftus W.C., Bazell, J., Gazzaniga M.S. (1989) *Brain prints: Computer-generated cerebral cortex in vivo.* J. Cog. Neurosci., 1:88–117.

Kahn A. (1993) *Thérapie Génique. L'ADN Médicament.* Médecine et Sciences. Editions John Libbey Eurotext, France.

Kajiwara K., Sandberg M.A., Berson E.L., Dryja T.P. (1993) *A null mutation in the human peripherin/RDS gene in a family with autosomal dominant retinitis punctata albescens.* Nature Genet., 3:208–212.

Kamei R., Hughes L., Miles L., Dement W. (1979) *Advanced sleep phase syndrome studied in a time isolation facility.* Chronobiologia, 6:115.

Kanter M.M. (1994) *Free radicals, exercise, and antioxidant supplementation.* Int. J. Sport Nutr., 4:205–220.

Kaplan G.A. and Haan M.N. (1989) *Is there a role for prevention among the elderly?* In Aging and Health Care: Social Science and Policy Perspectives. Ory M.G., Bond K. (eds.), Tavistock, London.

Kaplan H.S. (1990) *Sex, intimacy, and the aging process.* J. Am. Acad. Psychoanal., 18:185–205.

Kash F.W., Boyer J.I., Van Camp S.P., Verny L.S., Wallace J.P. (1993) *Effect of exercise on cardiovascular ageing.* Age Ageing, 22:5–10.

Kato A., Fujita H., Shimura N., Nakamura C., Hirayama Y., Kobayashi A., Ohi K. (1983) *Enhancement of PFC response by acupuncture β-endorphin.* Proc. Jap. Soc. Immmunol., 13:744–745.

Keast D. (1981) *Immune surveillance and cancer.* In Stress and Cancer. K. Bammer & B.H. Newberry (eds.), C.J. Hogrefe, Ontario.

Keeton K. (1992) *Longevity, the Science of Staying Young.* Viking, Penguin Books, USA.

Kelly M.J. (1988) *The pharmacology of vitamin E.* Progress in Medicinal Chemistry, 25:249–290.

Kemeny M.E., Cohen F., Zegans L.S., Conant M.A. (1989) *Psychological and immunological predictors of genital herpes recurrence.* Psychos. Med., 51:195–208.

Kenney R.A. (1985) *Physiology of aging.* Clin. Geriatr. Med., 1:37–59.

Kiehntopf M., Esquivel E.L., Brach M.A., Herrmann F. (1995) *Clinical applications of ribozymes.* Lancet, 345:1027–1028.

Kinlen L.J. and Rogot E. (1988) *Leukaemia and smoking habits among United States veterans.* Br. Med. J., 297:657–659.

Kitada T., Seki S., Kawakita N., Kuroki T., Monna T. (1995) *Telomere shortening in chronic liver diseases.* Biochem. Biophys. Res. Commun., 211:33–39.

Kitani K., Kanai S., Carrillo M.C., Go Y. (1994) *L-deprenyl increases the life span as well as activities of superoxide dismutase and catalase but not of glutathione peroxidase in selective brain regions in Fischer rats.* Ann. N.Y. Acad. Sci., 717:60–71.

Knoll J., Ecery Z., Keleman K., Nievel J., Knoll B. (1965) *Phenylisopropylmethylpropinylamine (E-250), a new spectrum psychic energizer.* Arch. Int. Pharmacodyn. Ther., 155:154.

Knoll J. (1988) *The striatal dopamine dependence of life span in male rats. Longevity study with (L-)deprenyl.* Mech. Ageing Dev., 46:237–262.

Kohn R.R. (1971) *Principals of mammalian aging.* Englewood Cliffs, Prentice Hall, NJ.

Kohn R.R. (1982) *Causes of death in very old people.* J. Am. Med. Assoc., 247:2793–2797.

Kohn R.R., Cerami A., Monnier V.M. (1984) *Collagen aging in vitro by nonenzymatic glycosylation and browning.* Diabetes, 33:57–59.

Koide R., Ikeuchi T., Onodera O., Tanaka H., Igarashi S., Endo K., Takahashi H., Kondo R., Ishikawa A., Hayashi T., Saito M., Tomoda A., Miike T., Naito H., Ikuta F., Tsuji S. (1994) *Unstable expansion of CAG repeat in hereditary dentatorubral-pallidoluysian atrophy (DRPLA).* Nature Genet., 6(1):9–13.

Korsmeyer S.J. (1992) *Bcl-2 initiates a new category of oncogenes: regulators of cell death.* Blood, 80:879–886.

Kozma A. and Stones M.J. (1983) *Predictors of happiness.* J. Gerontol., 38(5):626–628.

Knowles M.R., Hohneker K.W., Zhou Z., et al (1995) *A controlled study of adenoviral-vector-mediated gene transfer in the nasal epithelium of patients with cystic fibrosis.* N. Engl. J. Med., 333:823–831.

Krajewski S., Tanaka S., Takayama S., Schibler M.-J., Fenton W., Reed J.C. (1993) *Investigation of the subcellular distribution of the bcl-2 oncoprotein: residence in the nuclear envelope, endoplasmic reticulum, and outer mitochondrial membranes.* Cancer Res., 53(19):4701–4714.

Kremer J.M., Jubiz W., Michalek A., Rynes R.I., Bartholomew L.E., Bigouaette J., Timchalk M., Beeler D., Lininger L. (1987) *Fish oil fatty acid supplementation in active rheumatoid arthritis. A double-blinded, controlled, crossover study.* Ann. Intern. Med., 106:497–502.

Krieger D. (1982) *Cushing's Syndrome.* Monographs in Endocrinology. Vol. 22, Springer-Verlag, Berlin.

Kritz-Silverstein D. and Barrett-Connor E. (1996) *Long-term postmenopausal hormone use, obesity, and fat distribution in older women.* J. Am. Med. Assoc., 275:46–49.

Kruk P.A., Rampino N.J., Bohr V.A. (1995) *DNA damage and repair in telomeres: relation to aging.* Proc. Natl. Acad. Sci. USA, 92:258–262.

Kuczmarski R.J., Flegal K.M., Campbell S.M., Johnson C.L. (1995) *Increasing prevalence of overweight among US adults: the National Health and Nutrition Examinaion surveys, 1960–1991.* J. Am. Med. Assoc., 272:205–211.

Kvale G., Bjelke E., Gart J.J. (1975) *Dietary habits and lung cancer risks.* Int. J. Cancer, 31:397–405.

LaCroix A.Z., Guralnik J.M., Bergman L.F., Wallace R.B., Satterfield S. (1993) *Maintaining mobility in late life.* Am. J. Epidemiol., 137(8):858–869.

Lakowski B. and Hekimi S. (1996) *Determination of life-span in Caenorhabditis elegans by four clock genes.* Science, 272:1010–1013.

La Spada A.R., Wilson E.M., Lubahn D.B., Harding A.E., Fishbeck K.H. (1991) *Androgen receptor gene mutations in X-linked spinal and bulbar muscular atrophy* . Nature, 352:77–79.

Lee I.M., Hsieh C.C., Paffenbarger R.S. (1995) *Exercise intensity and longevity in men.* J. Am. Med. Assoc., 273:1179–1184.

Lee X., Si S.P., Tsou H.C., Peacocke M. (1995) *Cellular aging and transformation suppression: a role for retinoic acid receptor beta 2.* Exper. Cell. Res., 218(1):296–304.

Le Gal La Salle G., Robert J.J., Berrard S., Ridoux V., Stratford-Perricaudet L.D., Perricaudet M., Mallet J. (1993) *An adenovirus vector for gene transfer into neurons and glia in the brain.* Science, 259:986–988.

Lehrer S. (1979) *Possible pineal-suprachiasmatic regulation of development and life-span.* Arch. Ophtalmol., 97:359.

Lerner A.B. and Nordlund J.J. (1978) *Melatonin: Clinical pharmacology.* J. Neural Transm., suppl.13:339–347.

Leveille P.J., Weindruch R., Walford R.L., Bok D., Horwitz J. (1984) *Dietary restriction retards age-related loss of gamma crystallins in the mouse lens.* Science, 224:1247–1249.

Levy S., Herberman R., Lippman M., d'Angelo T. (1987) *Correlation of stress factors with sustained depression of natural killer cell activity and predicted prognosis in patients with breast cancer.* J. Clin. Oncol., 5:348–353.

Levy-Lahad E., Wijsman E.M., Nemens E., Anderson L., Goddard K.A.B., Weber J.L., Bird T., Schellenberg G.D. (1995a) *A familial Alzheimer's disease locus on Chromosome I.* Science, 269:970–973.

Levy-Lahad E., Wasco W., Poorkaj P., Romano D.M., Oshima J., Pettingell W.H., Yu C., et al (1995b) *Candidate gene for the chromosome 1 familial Alzheimer's disease locus.* Science, 269:973–977.

Liberman U.A., Weiss S.R., Bröll J., Minne H.W., et al (1995) *Effect of oral alendronate on bone mineral density and the incidence of fractures in postmenopausal osteoporosis.* N. Engl. J. Med., 30:1437–1443.

Lissoni P., Meregalli S., Fossati V., Paolorossi F., Barni S., Tancini G. (1994) *A randomized study of immunotherapy with low-dose subcutaneous interleukin-2 plus melatonin versus chemotherapy with cisplatin and etoposide as first-line therapy for advanced non-small cell lung cancer.* Tumori., 80(6):464–467.

Lissoni P., Barni S., Tancini G., Mainini E., Piglia F., Maestroni G.J., Lewinski A. (1995) *Immunoendocrine therapy with low-dose subcutaneous interleukin-2 plus melatonin of locally advanced or metastatic endocrine tumors.* Oncology, 52(2):163–166.

Lointier P., Levin B., Wargovich M., Boman B.M. (1986) *The effects of vitamin D on human colon carcinoma cells in vitro.* Gastroenterology, 90(3):1526.

Lonergan E.T. (1991) *Extending Life, Enhancing Life. A National Research Agenda on Aging.* Division of Health Promotion and Disease Prevention, Institute of Medicine. National Academy Press, Washington.

Lopes S.A. and Rene A. (1973) *Effect of 17-ketosteroids on glucose-6-phosphate dehydrogenase activity (G6PD) and on G6PD isoenzyme.* Proc. Soc. Exp. Biol. Med., 142:258–261.

Lowe S.W., Schmitt E.M., Smith S.W., Osborne B.A., Jacks T. (1993) *P53 is required for radiation-induced apoptosis in mouse thymocytes.* Nature, 362:847–849.

Luckinbill L.S., Arking R., Clare M.J., Cirocco W.C., Buck S.A. (1984) *Selection for delayed senescence in Drosophila Melanogaster.* Evolution, 38:996–1003.

Luke S., Birnbaum R., Verma R.S. (1994) *Centromeric and telomeric repeats are stable in nonagenarians as revealed by the double hybridization fluorescent in situ technique.* Genet. Anal. Tech. Appl., 11:77–80.

Lynch J.J. (1977) *The Broken Heart.* Basic Books, New-York.

Macieira-Coelho A. (1995) *The implications of the "Hayfick limit" for aging of the organism have been misunderstood by many gerontologists.* Gerontology, 41:94–97.

MacLennan D.H., Phillips M.S. (1992) *Malignant hyperthermia.* Science, 256:789–794.

Majumdar S., Shaw G., Offerman E.L., Thomson A., Bridges P. (1981) *Serum cortisol concentrations and the effect of chlormethiazole on them in chronic alcoholics.* Neuropharmacology, 20:1351–1352.

Makinodan T. and Kay M.M.B. (1980) *Age influence on the immune system.* Adv. Immunol., 29:287–295.

Malkin D., Jolly K.W., Barbier N., Look A.T., Friend S.H., Gebhardt M.C., Andersen T.I., Borresen A.-L., Li F.P., Garber J., Strong L.C. (1992) *Germline mutations of the p53 tumor-suppressor gene in children and young adults with second malignant neoplasms.* N. Engl. J. Med., 326(20):1309–1315.

Malluche H.H., Faugere M.-C., Ruch M., Friedler R. (1988) *Osteoblastic insufficiency is responsible for maintenance of osteopenia after loss of ovarian function in experimental Beagle dogs.* Endocrinology, 119:2649–2654.

Marantz-Henig M. (1987) *The Myth of Senility. Includes the Latest Information on Alzheimer's Disease.* An AARP book, Scott, Foresman and Company, Lifelong Learning Division, Glenview, Illinois.

Marczynski T.J., Artwohl J., Marczynska B. (1994) *Chronic administration of flumazenil increases life span and protects rats from age-related loss of cognitive functions: a benzodiazepine/GABAergic hypothesis of brain aging.* Neurobiol. of Aging, 15(1):69–84.

Marmot M. and Brunner E. (1991) *Alcohol and cardiovascular disease: the status of the U shaped curve.* Br. Med. J., 303:563–568.

Marsiglio W. and Donnelly D. (1991) *Sexual relations in later life: a national study of married persons.* J. Gerontol., 46:S338–S344.

Martin G., Sprague C., Epstein C. (1970) *Replicative life-span of cultivated human cells: effects of donor's age, tissue, and genotype.* Lab. Invest., 23:86–91.

Martin G.M. (1977) *Genetic syndromes in man with potential relevance to the pathobiology of aging.* In Genetic Effects of Aging, Birth Defects: Original Articles Series. Bergsma D., Harrison D.E. (eds.). Vol. 14D, The National Foundation-March of Dimes, New York.

Martin G.M. (1982) *Syndromes of Accelerated Aging.* National Cancer Institute Monograph (Research Frontiers in Cancer and Aging), 60:241–247.

Marwick C. (1995) Longevity requires policy revolution. J. Am. Med. Assoc., 273:1319, 1321.

Masoro E.J. (1988) *Minireview: food restriction in rodents: an evaluation of its role in the study of aging.* J. Gerontol., 43:B59–64.

Masoro E.J., Katz M.S., McMahan C.A. (1989) *Evidence for the glycation hypothesis of aging from the food-restricted rodent model.* J. Gerontol., 44:B20–B22.

Mayer P.J. and Baker G.T. (1985) *Genetic aspects of Drosophila as a model system of eukaryotic aging.* Int. Rev. Cytol., 95:61–102.

Mays-Hoopes L.L. (1989) *Development, aging, and DNA methylation.* Int. Rev. Cytol., 114:118–220.

McClung J.K., Danner D.B., Stewart D.A., Smith J.R., Schneider E.L., Lumpkin C.K., Dell'Orco R.T., Nuell M.J. (1989) *Isolation of a cDNA that hybrid selects antiproliferative mRNA from rat liver.* Biochem. Biophys. Res. Commun., 164(3):1316–1322.

McEwen B.S. and Brinton R.E. (1987) *Neuroendocrine aspects of adaptation.* Prog. Brain Res., 72:11–26.

McEwen B.S. and Stellar E. (1993) *Stress and the individual: Mechanisms leading to disease.* Arch. Intern. Med., 153(18):2093–2101.

McGinnis J.M., Lee P.R. (1995) *Healthy People 2000 at mid decade.* J. Am. Med. Assoc., 273:1123–1129.

McKay C.M. and Crowell M.F. (1934) *Prolonging the life span.* Science Monthly, 39:405–414.

McKay C., Crowell M., Maynard L. (1935) *The effects of retarded growth upon the length of life and upon ultimate size.* J. Nutr., 10:63–79.

McLaughlin M.E., Sandberg M.A., Berson E.L., Dryja T.P. (1993) *Recessive mutations in the gene encoding the β-subunit of rod phophodiesterase in patients with retinitis pigmentosa.* Nature Genet., 4:130–134.

McNamara S.G., Grunstein R.R., Sullivan C.E. (1993) *Obstructive sleep apnoea.* Thorax, 48:754–764.

Medawar P.B. (1952) An Unsolved Problem of Biology. H.K.Lewis, London.

Mendell J.R., Kissel J.T., Amato A.A., et al (1995) *Myoblast transfer in the treatment of Duchenne's muscular dystrophy.* N. Engl. J. Med., 333:832–838.

Meydani S.N., Barklund M.P., Liu S., Meydani M., Miller R.A., Cannon J.G., Morrow F.D., Rocklin R., Blumberg, J.B. (1990a) *Vitamin E supplementation enhances cell-mediated immunity in healthy elderly subjects.* Am. J. Clin. Nutr., 52:557–563.

Meydani S.N., Endres S., Woods M.M., Goldin B.R., Soo C., Morrill-Labrode A., Dinarello C.A., Gorbach S.L. (1990b) *Oral (n-3) fatty acid supplementation sup-*

presses cytokine production and lymphocyte proliferation: comparison between young and older women. J. Nutr., 121(4):547–555.

Miles L.E. and Dement W.C. (1980) *Sleep and aging.* Sleep, 3:119–120.

Miller R. (1994) *Aging and immune function: cellular and biochemical analyses.* Exp. Gerontol., 29:21–35.

Miyaura S., Eguchi H., Johnston J.M. (1992) *Effect of a cigarette smoke extract on the metabolism of the proinflammatory autacoid, platelet-activating factor.* Circ. Res., 70:341–347.

Mobbs C.V. (1990) *Neurotoxic effects of estrogen, glucose and glucocorticoids: Neurohumoral hysteresis and its pathological consequences during aging.* Rev. Biol. Res. Aging, 4:201–228.

Möller G. and Möller E. (1978) *Immunological surveillance against neoplasia.* In Immunological Aspects of Cancer. J.E. Castro (ed.), MTP Press, Lancaster.

Moore R.Y. and Lenn N.J. (1972) *A retinohypothalamic projection in the rat.* J. Comp. Neurol., 146:1–14.

Morales A.J., Nolan J.J., Nelson J.C., Yen S. (1994) *Effects of replacement dose of dehydroepiandrosterone in men and women of advancing age.* J. Clin. Endocrinol. Metab., 78:1360–1367.

Morrison N.A., Cheng Qi J., Tokita A., Kelly P.J., Crofts L., Nguyen T.V., Sambrook P.N., Eisman J.A. (1994) *Prediction of bone density from vitamin D receptor alleles.* Nature, 367:284–287.

Morley J.E. (1991) *Endocrine factors in geriatric sexuality.* Clin. Geriatr. Med., 7:85–93.

Mulligan T. and Moss C.R. (1991) *Sexuality and aging in male veterans: a cross-sectional study of interest, ability, and activity.* Arch. Sex. Behav., 20:17–25.

Munck A., Guyre P., Holbrook N. (1984) *Physiological functions of glucocorticoids in stress and their relation to pharmacological actions.* Endocr. Rev., 5:25–44.

Murasko D.M. and Goonewardene I.M. (1990) *T-cell function in aging: mechanisms of decline.* Annu. Rev. Gerontol. and Geriatr., 10:71–96.

Nair N.P.V., Haiharasubramanian N., Pilapil C., Isaac B., Thavundayil J.X. (1986) *Plasma melatonin -an index of brain aging in humans?* Biol. Psychiatry, 21:141–150.

Nakagawa S., Watanabe H., Ohe H., Nakao M. (1990) *Sexual behavior in Japanese males relating to area occupation, smoking, drinking and eating habits.* Andrologia, 22:21–28.

National Research Council (1989) *Diet and Health: Implications For Reducing Chronic Disease Risk.* National Academy Press, Washington DC.

National Research on Aging Planning Panel. (1982) Toward an independent old age: A national plan for research on aging. US Department of health and human services. Public Health Services (Ed.) NIH publication no 82-2453. Washington DC.

Naurath H.J., Joosten E., Riezler R., Stabler S.P., Allen R.H., Lindenbaum J. (1995) *Effects of vitamin B12, folate, and vitamin B6 supplements in elderly people with normal serum vitamin concentrations.* Lancet, 346:85–89.

Nelson M.E., Meredith C.N., Dawson-Hughes B., Evans W.J. (1988) *Hormone and bone mineral status in endurance-trained and sedentary postmenopausal women.* J. Clin. Endocrinol. Metab., 66:927–935.

Nichols B.E., Sheffield V.C., Vandenburgh K., Drack A.V., Kimura A.E., Stone E.M. (1993) *Butterfly-shaped pigment dystrophy of the fovea caused by a point mutation in codon 167 of the RDS gene.* Nature Genet., 3:202–206.

Nicoletti C., Borghesi-Nicoletti C., Yang X., Schulze D.H., Cerny J. (1991) *Repertoire diversity of antibody response to bacterial antigens in aged mice. II: Phosphorylcholine-antibody in young and aged mice differ in both $V_{H/}V_L$ gene repertoire and in specificity.* J. Immunol., 147:2750–2755.

Nicoletti C., Yang X., Cerny J. (1993) *Repertoire diversity of antibody response to bacterial antigens in aged mice. III: Phosphorylcholine antibody from young and aged mice differ in structure and protective activity against infection with Streptococcus pneumoniae.* J. Immunol., 150:543–549.

Nitenberg A., Antony I., Foult J.M. (1993) *Acetylcholine-induced coronary vasoconstriction in young, heavy smokers with normal coronary arteriographic findings.* Amer. J. Med., 95:71–77.

Noodén L.D. (1988) *Whole plant senescence.* In Senescence and Aging in Plants. L.D. Noodén and A.C. Leopold (eds.), pp. 391–442, Academic Press, San Diego.

Nuell M.J., Stewart D.A., Walker L., Friedman V., Wood C.M., Owens G.A., Smith J.R., Schneider E.L., Dell'Orco R., Lumpkin C.K., Danner D.B., McClung J.K. (1991) *Prohibitin, and evolutionarily conserved intracellular protein that blocks DNA synthesis in normal fibroblasts and Hela Cells.* Mol. Cell. Biol., 11(3):1372–1381.

Null G. and Feldman M. (1993) *Reverse the Aging Process Naturally.* Villard Books, New York.

Nuñez G., Hockenbery D., McDonnell T.J., Sorensen C.M., Korsmeyer S.J. (1991) *Bcl-2 maintains B-cell memory.* Nature, 353:71–73.

O'Connor K., Stravynski A. (1982) *Evaluation of a smoking typology by use of a specific behavioural substitution method of self-control.* Behav. Res. Ther., 20:279–288.

O'Hoy K.L., Tsilfidis C., Mahadevan M.S., Neville C.E., Barceló J., Hunter A.G.W., Korneluk R.G. (1993) *Reduction in size of the myotonic dystrophy trinucleotide repeat mutation during transmission.* Science, 259:809–812.

Omenn G.S., Goodman G., Thornquist M., Grizzle J., Rosenstock L., Barnhart S., et al (1994) *The β-carotene and retinol efficacy trial (CARET) for chemoprevention of lung cancer in high risk population, smokers and asbestos-exposed workers.* Cancer Res., 54(suppl):2038s–2043s.

Onsrud M. and Thorsby E. (1981) *Long-term changes in natural killer activity after external pelvic radiatherapy.* Int. J. Radiation Oncol., Biol. and Physics, 7:609–614.

Orentreich N., Brind J.L., Rizer R.L., Vogelman J.H. (1984) *Age changes and sex differences in serum dehydroepiandrosterone sulfate concentrations throughout adulthood.* J. Clin. Endocrinol. Metab., 59:551–555.

Orr H.T., Chung My, Banfi S., Kwiatkowski T.J., Servadio A., Beaudet A.L., McCall A.E., Duvick L.A., Ranum L.P.W., Zoghbi H.Y. (1993) *Expansion of an unstable trinucleotide CAG repeat in spinocerebellar ataxia type 1.* Nature Genet., 4:221–226.

Pacifici R.E. and Davies K.J.A. (1991) *Protein, lipid and DNA repair systems in oxidative stress: the free-radical theory of aging revisited.* Gerontology, 37:166–180.

Pagliusi S.R., Gerrard P., Abdallah M., Talabot D., Catsicas S. (1994) *Age-related changes in expression of AMPA-selective glutamate receptor subunits: is calcium-permeability altered in hippocampal neurons?* Neuroscience, 61(3):429–433.

Palmieri G., Pitcock J., Brown P., et al (1988) *Effect of calcitonin and vitamin D on osteoporosis.* Clin. Res., 36:884A (abstract).

Pastan I. and Gottesman M.M. (1991) *Multidrug resistance.* Annu. Rev. Med., 42:277–286.

Peacocke M. and Campisi J. (1991) *Cellular Senescence: A reflection of normal growth control, differentiation, or aging?* J. Cell. Biochem., 45:147–155.

Peltomäki P., Aaltonen L.A., Sistonen P., Pylkkänen L., Mecklin J.-P., Järvinen H., Green J.S., Jass J.R., Weber J.L., Leach F.S., Petersen G.M., Hamilton S.R., de la Chapelle A., Vogelstein B. (1993) *Genetic mapping of a locus predisposing to human colorectal cancer.* Science, 260:810–812.

Pereira-Smith O.M. and Smith J.R. (1983) *Evidence for the recessive nature of cellular immortality.* Science, 221:964–966.

Pereira-Smith O.M. and Smith J.R. (1988) *Genetic analysis of indefinite division in human cells. Identification of four complementation groups.* Proc. Natl. Acad. Sci. USA, 85:6042–6046.

Peterson R.C., Smith G.E., Ivnik R.J., Tangalos E.G., Schaid D.J., Thibodeau S.N., Kokmen E., Waring S.C., Kurland L.T. (1995) *Apolipoprotein E status as a predictor of the development of Alzheimer's disease in memory-impaired individuals.* J. Am. Med. Assoc., 273:1274–1278.

Peto R. (1994) *Smoking and death: the past 40 years and the next 40.* Br. Med. J., 309:937–939.

Phillips J.P., Campbell S.D., Michaud D., Charbonneau M., Hilliker A.J. (1989) *Null mutation of copper/zinc superoxide dismutase in Drosophila confers hypersensitivity to paraquat and reduced longevity.* Proc. Natl. Acad. Sci. USA, 86:2761–2765.

Piantanelli L., Gentile S., Fattoretti P., Viticchi C. (1985) *Thymic regulation of brain cortex beta-adrenoceptors during development and aging.* Arch. Gerontol. Geriatr., 4:179–185.

Piccinin G.L., Finali G., Piccirilli M. (1990) *Neuropsychological effects of L-deprenyl in Alzheimer's type dementia* . Clin. Neuropharmacol., 13:147–163.

Pierpaoli W., Dall'Ara, Pedrino E., Regelson W. (1991) *The pineal gland and aging: The effects of melatonin and pineal grafting on the survival of older mice.* Ann. N.Y. Acad. Sci., 621:291–313.

Pierpaoli W. and Regelson W. (1994) *Pineal control of aging: Effect of melatonin and pineal grafting on aging mice.* Proc. Natl. Acad. Sci. USA, 91:787–791.

Pierpaoli W., Regelson W., Colman C. (1995) *The Melatonin Miracle. Nature's Age-Reversing, Disease Fighting, Sex-Enhancing Hormone.* Simon & Schuster, New York.

Pitsikas N. and Algeri S. (1992) *Deterioration of spatial and nonspatial reference and working memory in aged rats: protective effect of life-long calorie restriction.* Neurobiol. Aging, 13:369–373.

Pittler S.J. and Baehr W. (1991) *Identification of a non-sense mutation in the rod photoreceptor cGMP phosphodiesterase beta subunit gene of the rd mouse.* Proc. Natl. Acad. Sci. USA, 88:8322–8326.

Planel H., Soleilhavoup J.P., Tixador R., Richoilley G., Conter A., Croute F., Caratero C., Gaubin Y. (1987) *Influence on cell proliferation of background radiation or exposure to very low chronic gamma radiation.* Health Physics, 52:571–578.

Pollock M.L., Foster C., Knapp D., Rod J.L., Schmidt D.H. (1987) *Effect of age and training on aerobic capacity and body composition of master athletes.* J. Applied Physiol., 62:725–731.

Pomeranz B. and Chiu D. (1976) *Naloxone blockade of acupuncture analgesia: endorphin implicated* . Life Science, 19:1757–1762.

Popp D.M. (1982) *An analysis of genetic factors regulating life span in congenic mice.* Mech. Ageing Dev., 18:125–134.

Potter B. and Orfali S. (1993) *Brain Boosters: Food and Drugs That Make You Smarter.* Ronin Publishing, Inc., USA.

Prinz P.N. and Halter J.B. (1983) *Sleep disturbances in the elderly, neurohormonal correlates.* In Sleep Disorders: Basic and Clinical Research. M. Chase (ed.), pp. 463–488, Spectrum, New York.

Proust J., Moulias R., Fumeron F., Bekkhoucha F., Busson M., Schmid M., Hors J. (1982) *HLA and longevity.* Tissue Antigens, 19:168–173.

Prowse K.R. and Greider C.W. (1995) *Developmental and tissue-specific regulation of mouse telomerase and telomere length.* Proc. Natl. Acad. Sci. USA, 92:4818–4822.

Reiter R.J. (1984) *The Pineal Gland.* Reiter (ed.), Raven press, New York.

Reiter R.J. (1992) *The ageing pineal gland and its physiological consequences.* Bioessays, 14:169–175.

Reiter R.J., Poeggeler B., Tan DX., Chen L., Manchester L.C., Guerrero J.M. (1993) *Antioxidant capacity of melatonin: a novel action not requiring a receptor.* Neuroendocrinol. Lett., 15:103–116.

Reiter R.J. (1995) *The pineal gland and melatonin in relation to aging: a summary of the theories and of the data.* Exper. Gerontol., 30:199–212.

Reiter R.J. and Robinson J. (1995) Melatonin. Bantam books, New York.

Richardson G. (1990) *Circadian rhythms and aging.* In Handbook of the Biology of Aging. E.L. Schneider and J.W. Rowe (eds.), 3rd ed., pp. 275–305, Academic Press, San Diego.

Rich-Edwards J.W., Manson J.E., Hennekens C.H., Buring J.E. (1995) *The primary prevention of coronary heart disease in women.* N. Engl. J. Med., 332:1758–1766.

Richter C.P. (1965) *Biological Clocks in Medicine and Psychiatry.* Thomas Springfield, IL.

Riley V. (1981) *Psychoneuroendocrine influences on immuno-competence and neoplasia.* Science, 217:1100–1109.

Rimm E.B., Stämpfer M.J., Ascherio A., Giovannucci E., Colditz G.A., Willett W.C. (1993) *Vitamin E consumption and the risk of coronary heart disease in men.* N. Engl. J. Med., 328(20):1450–1456.

Robertson O.H. and Wexler B.C. (1960) *Histological changes in the organs and tissues of senile castrated kokanee salmon: Oncorhynchus nerka kennerlyi.* Gen. Comp. Endocrinol., 2:458–472.

Robertson O.H. (1961) *Prolongation of the life span of kokanee salmon (Oncorhynchus nerka kennerlyi) by castration before beginning of gonad development.* Proc. Natl. Acad. Sci. USA, 47:609–621.

Robertson J.M., Donner A.P., Trevithick J.R. (1989) *Vitamin E intake and risk of cataracts in humans.* Ann. N.Y. Acad. Sci., 570:372–382.

Robinson G.E. (1987) *Regulation of honey bee age polyethism by juvenile hormone.* Behav. Ecol. Sociobiol., 20:329–338.

Rodin J. (1986) *Health, control and aging.* In Aging and the Psychology of Control. Baltes M. and Baltes P.B. (eds.), pp. 83–92, Lawrence Erlbaum, Hillsdale, N.J.

Rogot E., et al (1988) *A Mortality Study of One Million Persons.* National Institutes of Health, Bethesda, Md.

Röhme D. (1981) *Evidence for a relationship between longevity of mammalian species and lifespans of normal fibroblasts in vitro and erythrocytes in vivo.* Proc. Natl. Acad. Sci. USA, 78:5009–5013.

Rommens J.M., Iannuzzi M.C., Kerem B.S., Drumm M.L., Melmer G., Dean M., Rozmahel R., Cole J.L., Kennedy D., Hidaka N., Zsiga M., Buchwald M., Riordan J.R., Tsui L.-C., Collins F. (1989) *Identification of the cystic fibrosis gene: chromosome walking and jumping*. Science, 245:1059–1065.

Rose M.R. (1984) *Laboratory evolution of postponed senescence in Drosophila melanogaster*. Evolution: Int. J. Org. Evol., 38:1004–1010.

Rose M.R. and Graves J.L. (1990) *Evolution of aging*. In Review of Biological Research in Aging. M. Rothstein (ed.). Vol. 4, pp. 3–14, Liss, New York.

Rose R. (1985) *Psychoendocrinology*. In Textbook of Endocrinology (7th ed.). Wilson J., Foster D. (eds.), Saunders, Philadelphia.

Rosen D.R., Siddique T.,Patterson D., Figlewicz D.A., Sapp P., Hentati A., Donaldson D., Goto J., et al (1993) *Mutations in Cu/Zn superoxide dismutase gene are associated with familial amyotrophic lateral sclerosis*. Nature, 362:59–62.

Rosenberg S.A., Packard B.S., Aebersold P.M., et al (1988) *Use of tumor-infiltrating lymphocytes and interleukin 2 in the immunotherapy of patients with metastatic melanoma: a preliminary report*. N. Engl. J. Med., 319:1676–1680.

Rosenberg S.A., Aebersold P., Cornetta K., et al (1990) *Gene transfer into humans-immunotherapy of patients with advanced melanoma, using tumor-infiltrating lymphocytes modified by retroviral gene transduction*. N. Engl. J. Med., 323:570–578.

Rossolini G., Viticchi C., Basso A., Zaia A., Piantanelli L. (1991) *Thymus-induced recovery of age-related decrease of brain cortex α and β-adrenoceptors*. Int. J. Neurosci., 59:143–150.

Roth G.S. and Hess G.D. (1982) *Changes in the mechanisms of hormone and neurotransmitter action during aging: current status of the role of receptor and post receptor alterations*. Mech. Ageing Dev., 20:175–194.

Roth G.S. (1989) *Changes in hormone action with age: altered calcium mobilization and/or responsiveness impairs signal transduction*. In Endocrine Function and Aging. H.J. Armbrecht (ed.), Springer Verlag, New York.

Rothstein M. (1987) *Evidence for and against the error catastrophe hypothesis*. In Modern Biological Theories of Aging. H.R. Warner et al (eds.), Raven Press, New York.

Roughan P.A., Kaiser F.E., Morley J.E. (1993) *Sexuality and the older women*. Clin. Geriatr. Med., 9(1):87–106.

Rowe J.W. and Kahn R.L. (1987) *Human aging: Usual and successful*. Science, 237:143–149.

Rozencwaig R., Grad B.R., Ochoa J. (1987) *The role of melatonin and serotonin in aging*. Med. Hypothesis, 23:337–352.

Rudman D., Kutner M.H., Rogers C.M., Lubin M.F., Fleming G.A., Bain R.P. (1981) *Impaired growth hormone secretion in the adult population: relation to age and adiposity*. J. Clin. Invest., 67:1361–1369.

Rudman D., Feller A.G., Nagraj H.S., Gergans G.A., Lalitha P Y., Goldberg A.F., Schlenker R.A., Cohn L., Rudman I.W., Mattson D.E. (1990) *Effects of human growth hormone in men over 60 years old*. N. Engl. J. of Med., 323(1):1–6.

Russel R.M. and Suter P.M. (1993) *Vitamin requirements of elderly people: an update*. Am. J. Clin. Nutr., 58:4–14.

Ryan J.M., Ostrow D.G., Breakefield X.O., Gershon E.S., Upchurch L. (1981) *A comparison of the proliferative and replicative life span kinetics of cell cultures derived from monozygotic twins*. In Vitro, 17:20–27.

Ryder L.P., Svejaard A., Dausset J. (1981) *Genetics of HLA Disease Association.* Annu. Rev. Genetics, 15:169–188.

Sack R.L., Lewy A.J., Erb D.L.,Vollmer W.M., Singer C.M. (1986) *Human melatonin production decreases with age.* J. Pineal Res., 3:379–388.

Salk D. (1982) *Werner's syndrome: A review of recent research with an analysis of connective tissue metabolism, growth control of cultured cells and chromosomal aberrations.* Hum. Genet., 62:1–15.

Salk D, Fujiwara Y, Martin GM, eds. (1985) *Werner's Syndrome and Human Aging.* New York: Plenum.

Sandbrink R., Hartmann T., Masters C.L., Beyreuther K. (1996) *Genes contributing to Alzheimer's disease.* Mol. Psychiatry, 1:27–40.

Sandyk R., Anastasiadis P.G., Anninos P.A., Tsagas N. (1992) *Is postmenopausal osteoporosis related to pineal gland functions?* Int. J. Neurosci., 62(3–4):215–225.

Sapolsky R., Krey L., McEwen B. (1983) *Corticosterone receptors decline in a site-specific manner in the aged rat brain.* Brain Res., 289:235–240.

Sapolsky R., Krey L., McEwen B. (1986a) *The adrenocortical axis in the aged rat: impaired sensitivity to both fast and delayed feedback inhibition.* Neurobiol. Aging, 7:331–335.

Sapolsky R., Krey L., McEwen B. (1986b) *The neuroendocrinology of stress and aging: The glucocorticoid cascade hypothesis.* Endocrine Revs., 7:284–306.

Sapolsky R, Uno H., Rebert C., Finch C. (1990) *Hippocampal damage associated with prolonged glucocorticoid exposure in primates.* J. Neurosci., 10:2897–2902.

Sapolsky R.M. (1992) *Stress, the Aging Brain, and the Mechanisms of Neuron Death.* A Bradford Book. The MIT Press, Cambridge (Mass), London.

Sato K. and Sato F. (1984) *Defective Beta adrenergic response of cystic fibrosis sweat glands in vivo and in vitro.* J. Clin.Invest., 73:1763–1771.

Sato T., Saito H., Swensen J., Olifant A., Wood C., Danner D., Sakamoto T., Takita K., Kasumi F., Miki Y., Skolnick M., Nakamura Y. (1992) *The human prohibitin gene located on chromosome 17q21 is mutated in sporadic breast cancer.* Cancer Res., 52(No 6):1643–1646.

Saunders A.M., Schmader K., Breitner J.C.S., Benson D., Brown W.T., Goldfarb L., Goldgaber D., Manwaring M.G., Szymanski M.H., McCohn N., Dole K.C., Schmechel D.E., Strittmatter W.J., Pericak-Vance M.A., Roses A.D. (1993) *Apolipoprotein E ε4 allele distributions in late-onset Alzheimer's disease and in other amyloid-forming diseases.* Lancet, 342:710–711.

Savitsky K., Bar-Shira A., Gilad S., Rotman G., Ziv Y., Vanagaite L., Tagle D.A., Smith A., et al (1995) *A single ataxia telangiectasia gene with a product similar to PI-3 kinase.* Science, 268:1749–1753.

Scala J. (1992) *Prescription for Longevity; Eating Right for a Long Life.* Dutton, Penguin books, USA.

Schächter F., Faure-Delanef L., Guénot F., Rouger H., Froguel P., Lesueur-Ginot L., Cohen D. (1994) *Genetic associations with human longevity at the APOE and ACE loci.* Nature Genet., 6:29–32.

Schiavi R.C., Schreiner Engel P., Mandeli J., Schanzer H., Cohen E. (1990) *Healthy aging and male sexual function.* Am. J. Psychiatry, 147:766–771.

Schiavi R.C., White D., Mandeli J. (1992) *Pituitary-gonadal function during sleep in healthy aging men.* Psychoneuroendocrinology, 17:599–609.

Schieken R.M., Mosteller M., Goble M.M., Moskowitz W.B., Hewitt J.K., Eaves L.J., Nance W.E. (1992) *Multivariate genetic analysis of blood pressure and body size.* Circulation, 86(6):1780–1788.

Schmickel R.D., Chu E.H.Y., Trosko J.E., Chang C.C. (1977) *Cockayne syndrome: a cellular sensitivity to ultraviolet light.* Pediatrics, 60:135–139.

Schneider E.L. and Mitsui Y. (1976) *The relationship between in vitro cellular aging and in vivo human age.* Proc. Natl. Acad. Sci. USA, 73:3584–3588.

Schneider S.H., Amorosa L.F., Kachadurian A.K., Ruderman N.B. (1984) *Studies of the mechanisms of improved glucose control during regular exercise in Type-2 (non-insulin dependent) diabetics.* Diabetologia, 26:355–360.

Schnell L. and Schwab M.E. (1990) *Axonal regeneration in the rat spinal cord produced by an antibody against myelin-associated neurite growth inhibitors.* Nature, 343:269–272.

Schnell L., Schneider R., Kolbeck R., Barde Y.A., Schwab M.E. (1994) *Neurotrophin-3 enhances sprouting of corticospinal tract during development and after adult spinal cord lesion.* Nature 367:170–173.

Schnider S.L. and Kohn R.R. (1981) *Effects of age and diabetes mellitus on the solubility and nonenzymatic glucosylation of human skin collagen.* J. Clin. Invest., 67:1630–1635.

Schroder J.A., Messer R., Bachmann M., Bernd A., Muller W.E.G. (1987) *Superoxide radical-induced loss of nuclear restriction of immature mRNA: a possible cause of aging.* Mech. Ageing Dev., 41:251–266.

Schwartz A.G. (1979) *Inhibition of spontaneous breast cancer formation in female C3H (Avy/a) mice by long-term treatment with dehydroepiandrosterone.* Cancer, 39:1129–1132.

Schwartz A.G. and Tannen R.H. (1981) *Inhibition of 7–12-dimethylbenz(a)-anthracene and urethan-induced lung tumor formation in A/J mice by long-term treatment with dehydroepiandrosterone.* Carcinogenesis (London), 2:1335–1337.

Schwegler G., Schwab M.E., Kapfhammer J.P. (1995) *Increased collateral sprouting of primary afferents in the myelin-free spinal cord.* J. Neurosci., 15:2756–2767.

Seals D.R., Hagberg J.M., Hurley B.F., Ehsani A.A., Holloszy J.O. (1984a) *Effects of endurance training on glucose tolerance and plasma lipid levels in older men and women.* J. Am. Med. Assoc., 252:645–649.

Seals D.R., Allen W.K., Hurley B.F., Dalsky G.P., Ehsani A.A., Hagberg J.M. (1984b) *Elevated high-density lipoprotein cholesterol levels in older endurance athletes.* Am. J. Cardiol., 54:390–393.

Seligman M.E.P. and Elder G.H. (1986) *Learned helplessness and life-span development.* In Human Development and the Life Course: Multidisciplinary Perspectives. A.B. Sørenson, F.E. Weinert, L.R. Sherrod (eds.), Lawrence Erlbaum Associates, Publishers.

Selye H. (1936) *A syndrome produced by diverse nocuous agents.* Nature, 138:32.

Selye H. and Tuchweber B. (1976) *Stress in relation to aging and disease.* In Hypothalamus, Pituitary and Aging. Everitt A., Burgess J. (eds.), CC Thomas, Springlfield.

Shamberger R.J., Frost D.V. (1969) *Possible protective effect of selenium against human cancer.* Canad. Med. Assoc. J., 100:682.

Shaper A.G., Wannamethee G., Walker M. (1988) *Alcohol and mortality in British men: Explaining the U-shaped curve.* Lancet, December 3:1271–1273.

Shepherd J.C., Walldorf, U., Hug P. and Gehring W.J. (1989) *Fruit flies with additional expression of the elongation factor EF-1α live longer.* Proc. Natl. Acad. Sci. USA, 86:7520–7521.

Sherrington R., Rogaev E.I., Liang Y., Rogaeva E.A., Levesque G., Ikeda M., et al (1995) *Cloning of a gene bearing missense mutations in early-onset familial Alzheimer's disease.* Nature, 375:754–760.

Shmookler-Reis R.J. and Goldstein S. (1980) *Loss of reiterated DNA sequences during serial passage of human diploid fibroblasts.* Cell, 21:739–749.

Shock N.W. (1962) *The physiology of aging.* Sci. Am., 206:100–110.

Shock N. (1977) *System integration.* In Handbook of the Biology of Aging. Finch C., Hayflick L. (eds.), Van Nostrand, New York.

Shock N.W. (1983) *Aging of physiological systems.* J. Chronic Diseases, 36:137–142.

Shock N.W., Greulich R.C., Andres R., Arenberg D., Costa P.T., Lakatta E.G., Tobin J.D. (1984) *Normal Human Aging: The Baltimore Longitudinal Study on Aging.* US Department of Health and Human Services. NIH publication No 84–2450.

Shock N.W. (1985) *Longitudinal studies of aging in humans.* In Handbook of the Biology of Aging, 2nd ed. C.E. Finch and E.L. Schneider (eds.), pp. 721–743. Van Nostrand Reinhold, New York.

Simms H.S. and Berg B.N. (1957) *Longevity and the onset of lesions in male rats.* J. Gerontol., 12:244–252.

Slagboom P.E., Droog S., Boomsma D.I. (1994) *Genetic determination of telomere size in humans: a twin study of three age groups.* Am. J. Hum. Genet., 55:876–882.

Small G.W., Mazziotta J.C., Collins M.T., Baxter L.R., Phelps M.E., Mandelkern M.A., et al (1995) *Apolipoprotein E Type 4 allele and cerebral glucose metabolism in relatives at risk for familial Alzheimer's disease.* J. Am. Med. Assoc., 273:942–947.

Smith J.R., and Lumpkin C.K. (1980) *Loss of gene repression activity: a theory of cellular senescence.* Mech. Ageing Dev., 13:387–392.

Sohal R.S. and Allen R.G. (1990) *Oxidative stress as a causal factor in differentiation and aging: A unifying hypothesis.* Exp. Gerontol., 25(6):499–522.

Solomon D.H. and Hart R.G. (1994) *Antithrombotic therapies for stroke prevention.* Curr. Opin. Neurol., 7:48–53.

Solomon G.F. and Amkraut A.A. (1979) *Neuroendocrine aspects of the immune response and their implications for stress effects on tumor immunity.* Cancer Detect. Prev., 2:197–223.

Solomon G. (1987) *Psychoneuroimmunology: Interaction between control nervous system and immune system.* J. Neurosci. Res., 18:1–9.

Somer E. (1992) *The Essential Guide to Vitamins and Minerals.* Health Media of America, Harper Collins Publishers, New York.

Soong N.W., Hinton D.R., Cortopassi G., Arnheim N. (1992) *Mosaicism for a specific somatic mitochondrial DNA mutation in adult human brain.* Nature Genet., 2:318–323.

Sørensen A.B., Weinert F.E., Sherrod L.R. (1986) *Human Development and the Life Course: Multidisciplinary Perspectives.* Hillsdale, Lawrence Erlbaum Associates, NJ.

Sorrentino B.P., Brandt S.J., Bodine D., Gottesman M., Pastan I., Cline A., Nienhuis A.W. (1992) *Selection of drug-resistant bone marrow cells in vivo after retroviral transfer of human MDR1.* Science, 257:99–103.

Srivastava S., Zou Z., Pirollo K., Blattner W., Chang E.H. (1990) *Germ-line transmission of a mutated p53 gene in a cancer-prone family with Li-Fraumeni syndrome.* Nature, 348:747–749.

Stähelin H.B., Gey K.F., Eichholzer M., Lüdin E., Bernasconi F., Thurneysen J., Brubacher G. (1991) *Plasma antioxidant vitamins and subsequent cancer mortality in the 12-year follow-up of the Prospective Basel Study.* Am. J. Epidemiol., 133(8):766–775.

Stämpfer M.J., Hennekens C.H., Manson J.A., Colditz G.A., Rosner B., Willett W.C. (1993) *Vitamin E comsumption and the risk of coronary disease in women.* N. Engl. J. Med., 328(20):1487–1489.

Stein C.A. and Cheng Y.C. (1995) *Antisense oligonucleotides as therapeutic agents-is the bullet really magical?* Science, 261:1004–1012.

Stein M., Miller A., Testman R. (1991) *Depression, the immune system, and health and illness.* Arch. Gen. Psychiatry, 48:171.

Steinberg D., Parthasarathy S., Carew T.E., Khoo J.C., Witztum J.L. (1989) *Beyond cholesterol. Modifications of low-density lipoprotein that increase its atherogenicity.* N. Engl. J. Med., 320:915–924.

Stewart M.L., McDonald J.T., Levy A.S., et al (1985) *Vitamin/mineral supplement use: A telephone survey of adults in the United States.* J. Am. Diet Assoc., 85:1585–1590.

Strittmatter W.J., Saunders A.M., Schmechel D., Pericak-Vance M., Enghild J., Salvesen G.S., Roses A.D. (1993) *Apolipoprotein E: high avidity binding to β-amyloid and increased frequency of type 4 allele in late-onset familial Alzheimer disease.* Proc. Natl. Acad. Sci. USA, 90:1977–1981.

Stuart-Hamilton S. (1991) *The Psychology of Ageing.* Jessica Kingsley Publishers, London.

Suadicani P., Hein H.O., Gyntelberg F. (1992) *Serum selenium concentration and risk of ischaemic heart disease in a prospective cohort study of 3000 males.* Atherosclerosis, 96:33–42.

Suber M.L., Pittlers S.J., Qin N, Wright G.C., Holcombe V., Lee T.H., Craft C.M., Lolley R.N., Baehr W., Hurwitz R.L. (1993) *Irish setter dogs affected with rod/cone dysplasia contain a nonsense mutation in the rod cGMP phophodiesterase β-subunit gene.* Proc. Natl. Acad. Sci. USA, 90:3968–3972.

Suda Y., Suzuki M., Ikawa Y., Aizawa S. (1987) *Mouse embryonic stem cells exhibit indefinite proliferative potential.* J. Cell Physiol., 133:197–201.

Sugawara O, Oshimura M., Koi M., Annab L.A., Barett J.C. (1990) *Induction of cellular senescence in immortalized cells by human chromosome I.* Science, 247:707–710.

Sugita T., Ikenaga M., Suehara N., Kozuka T., Furuyama J., Yabuchi H. (1982) *Prenatal diagnosis of Cockayne syndrome using assay colony-forming ability in ultraviolet light irradiated cells.* Clin. Genet., 22:137–142.

Suzuki H., Hayakawa S., Tamura S., Wada S., Wada O. (1985) *Effect of age on the modification of rat plasma lipids by fish and soybean oil diets.* Biochem. Biophys. Acta, 836:390–393.

Swisshelm K., Disteche C.M., Thorvaldsen J., Nelson A. and Salk D. (1990) *Age-related increase in methylation of ribosomal genes and inactivation of chrosome-specific rRNA gene clusters in mouse.* Mutation Res., 237(3–4):131–146.

Swisshelm K., Ryan K., Tsuchiya K., Sager R. (1995) *Enhanced expression of an insulin growth factor-like binding protein (mac25) in senescent human mammary epithelial cells and induced expression with retinoic acid.* Proc. Natl. Acad. Sci. USA, 92:4472–4476.

Szarka C.E., Grana G., Engstrom P.F. (1994) *Chemoprevention of cancer.* Curr. Probl. Cancer, January/February: 6–79.

Taaffe D.R., Pruitt L., Reim J., Hintz R.L., Butterfield G., Hoffman A.R., Marcus R. (1994) *Effect of recombinant human growth hormone on the muscle strength response to resistance exercise in elderly men.* J. Clin. Endocrinol. Metab., 79(5):1361–1366.

Tacconi M.T., Lligona L., Salmona M., Pitsikas N., Algeri S. (1991) *Aging and food restriction: effect on lipids of cerebral cortex.* Neurobiol. Aging, 12:55–59.

Takata H., Suzuki M., Ishii T., Sekiguchi S., Iri H. (1987) *Influence of major histocompatibility complex region genes on human longevity among Okinawan-Japanese centenarians and nonagenarians.* Lancet, 2(8563):824–826.

Tan D.-X., Chen L.-D., Pöeggeler B., Manchester L.C., Reiter R.J. (1993a) *Melatonin: a potent, endogenous hydroxyl radical scavenger.* Endocrine J., 1:57–60.

Tan D.-X., Pöeggeler B., Reiter R.J., Chen L.-D., Chen S., Manchester L.C., Barlow-Walden L.R. (1993b) *The pineal hormone melatonin inhibits DNA-adduct formation induced by the chemical carcinogen safrole in vivo.* Cancer Letters, 70:65–71.

Tapp W.N. and Natelson B.H. (1986) *Life extension in heart disease: an animal model.* Lancet, 1:238–240.

Tassin J., Malaise F., Courtois Y. (1979) *Human lens cells have an in vitro proliferation capacity inversely proportional to the donor age.* Exp. Cell Res., 96:1–6.

The Writing Group for the PEPI Trial (1995) *Effects of estrogen or estrogen/progestin regiments on heart disease risk factors in postmenopausal women.* The postmenopausal estrogen/progestin interventions (PEPI) trial. J. Am. Med. Assoc., 273:199–207.

The Writing Group for the PEPI Trial (1996) *Effects of hormone replacement therapy on endometrial histology in postmenopausal women.* The postmenopausal estrogen/progestin interventions (PEPI) trial. J. Am. Med. Assoc., 275:370–375.

Thoman M.L. and Weigle W.O. (1989) *The cellular and subcellular bases of immunosenescence.* Adv. Immunol., 46:221–261.

Thomas T. (1995) *Dix ans de recherche sur les prédispositions génétiques au développement des tumeurs.* Médecine et Sciences, 11:336–348.

Tobin J.D. (1984) *Physiological indices of aging.* Normal Human Aging: the Baltimore Longitudinal Study of Aging, US Department of Health and Human Services, pp. 387–395. NIH publication No. 84-2450.

Toguchida J., Yamaguchi T., Dayton S.H., Beauchamp R.L., Herrera G.E., Ishizaki K., Yamamuro T., Meyers P.A., Little J.B., Sasaki M.S., Weichselbaum R.R., Yandell D.W. (1992) *Prevalence and spectrum of germline mutations of the p53 gene among patients with sarcoma.* N. Engl. J. Med., 336(20):1301–1308.

Tsujimoto Y. and Croce C.M. (1986) *Analysis of the structure, transcripts and protein products of Bcl-2, the gene involved in human follicular lymphoma.* Proc. Nat. Acad. Sci. USA, 83:5214–5218.

Tuszynski M.H., Peterson D.A., Ray J., Baird A., Nakahara Y., Gage F.H. (1994) *FIbroblasts genetically modified to produce nerve growth factor induce robust neuritic ingrowth after grafting to the spinal cord.* Exp. Neurol., 126:1–14.

Tuszynski M.H. and Gage F.H. (1995) *Bridging grafts and transient nerve growth factor infusions promote long-term central nervous system neuronal rescue and partial functional recovery.* Proc. Nat. Acad. Sci. USA, 92:4621–4625.

Uno H., Tarara R., Else J., Suleman M., Sapolsky R. (1989) *Hippocampal damage associated with prolonged and fatal stress in primates.* J. Neurosci., 9:1705.

Upton A.C. (1992) *The biological effects of low-level ionizing radiation.* Sci. Am., 246(2):29–37.

Vaillant G.E. (1977) *Adaptation to Life.* Boston, Little Brown, MA.
vanBlockxmeer F.M. and Mamotte C.D.S. (1992) *Apolipoprotein ε4 homozygosity in young men with coronary heart disease.* Lancet, 340:879–880.
Van Eenwyk J., Davis F., Bowen P. (1991) *Dietary and serum carotenoids and cervical intraepithelial neoplasia.* Int. J. Cancer, 48:34–48.
Vanyushin B., Nemirovsky L.E., Klimenko V.V., Vasiliev V.K., Belozersky A.N. (1973) *The 5-methylcytosine in DNA of rats.* Gerontologia (Basel), 19:138–152.
Vischer T.L. (1992) *Régime et polyarthrite rhumatoïde (PR).* Médecine et Hygiène, 50:838–840.
Vlassara H., Brownlee M., Cerami A. (1984) *Accumulation of diabetic rat peripheral nerve myelin by macrophages increases with the presence of advanced glycosylation endproducts.* J. Exp. Med., 160:197–207.
Voelker R. (1995) *Recommendations for antioxidants: how much evidence is enough?* J. Am. Med. Assoc., 271:1148–1149.
Vogel R. (1986) *Biology, human genetics and the life course.* In Human Development and the Life Course: Multidisciplinary Perspectives. A. Sørenson, F.E. Weinert, L.R. Sherrod (eds.), Lawrence Erlbaum Associates, Publishers.
Volpicelli J.R. (1995) *Naltrexone in alcohol dependence.* Lancet, 326:456.
Voronoff S. and Alexandrescu G. (1930) *La Greffe Testiculaire du Singe à l'Homme. Technique Opératoire, Manifestations Physiologiques, Evolution Histologique, Statistique.* Gaston Doin and Cie (eds.), Paris.
Walford R.L. (1979) *Multigene families, histocompatibility system, transformation, meiosis, stem cells and DNA repair.* Mech. Ageing Dev., 9:19–26.
Walford R.L. (1981) *Immunopathology of aging.* Annual Rev. of Gerontology and Geriatrics. C. Eisdorfer (ed.). Vol. 2, p. 3, Springer Publishing Co, New York.
Walford R.L. (1987) *MHC regulation of aging: An extension of the immunologic theory of ageing.* In Modern Biological Theories of Aging. H.R. Warner et al (eds.), pp. 243–260, Raven Press, New York.
Walford R.L. and Walford L. (1994) *The Anti-Aging Plan.* A Jalet-Miller book. Four Walls Eight Windows Ist Edition, New York.
Walker L.G. and Eremin O. (1995) *A new fad or the fifth cancer treatment modality?* Am. J. Surg., 170:2–4.
Wallace D.C. (1992) *Mitochondrial genetics: a paradigm for aging and degenerative diseases?* Science, 356:628–632.
Weaver D.R., Rivkees S.A., Carlson L.L., Reppert S.M. (1991) *Localization of melatonin receptors in mammalian brain.* In Suprachiasmatic Nucleus: the Mind's Clock. D.C. Klein, R.Y. Moore and S.M. Reppert (eds.), pp. 289–308, Oxford Press, New York.
Webster G.C. and Webster S.L. (1983) *Decline in synthesis of elongation factor one (EF-1) precedes the decreased synthesis of total protein in aging Drosophila melanogaster.* Mech. Ageing Dev., 22:121–128.
Weihe E., Nohr D., Michel S., Müller S., Zentel H.-J., Fink T., Krekel J. (1991) *Molecular anatomy of the neuro-immune connection.* Int. J. Neurosci., 59:1–23.
Weindruch R., Walford R.L., Fligiel S., Guthrie D. (1986) *The retardation of aging in mice by dietary restriction: Longevity, cancer, immunity and lifetime energy intake.* J. Nutr., 116:641–654.
Weismann A. (1891) *Essays Upon Heredity and Kindred Biological Problems.* Oxford Univ. Press (Clarendon), London, New York.

Weiss J.N. and Mellinger B.C. (1990) *Sexual dysfunction in elderly men.* Clin. Geriatr. Med., 6:185–196.

Weksler M. (1983) *Senescence of Immune System.* In The Biology of Immunologic Disease. F.J. Dixon and D.W. Fisher, Sinauer Associates Inc. (eds.), p. 295, Sunderland, Massachusetts.

Wells J., Wroblewski J., Keen J., Inglehearn C., Jubb C., Eckstein A., Jay M., Arden G., Bhattacherya S., Fitzke F., Bird A. (1993) *Mutations in the human retinal degeneration slow (RDS) gene can cause either retinitis pigmentosa or macular dystrophy.* Nature Genet., 3:213–218.

Widner H. (1993) *Immature neural tissue grafts in Parkinson's disease.* Acta Neurol. Scand. Suppl., 146:43–45.

Willard W.K. (1971) *Dynamics of Conotrachelus nenuphar (Coleoptera cureulioninae) populations following exposure to ionizing radiation.* Nuclear Science Abstracts, 28:30235.

Willett W.C. (1994) *Diet and health: What should we eat?* Science, 264:532–537.

Williams G.C. (1957) *Pleiotropy, natural selection and the evolution of senescence.* Evolution, 11:398.

Wilson V.L. and Jones P.A. (1983) *DNA methylation decreases in aging but not in immortal cells.* Science, 220:1055–1057.

Wilson M., Shirapurkar N., Poiner L. (1984) *Hypomethylation of hepatic nuclear DNA in rats fed with a carcinogenic methyl-deficient diet.* Biochem. J., 218:987–994.

Winston M.L. (1987) *The Biology of the Honey Bee.* Cambridge, Harvard University Press, MA.

Wispe L. (1991) *The Psychology of Sympathy.* L. Wispe (ed.), Plenum Press, NY.

Women's Health Study Research Group (1992) *The women's health study: rationale and background.* J. Myocardial Ischemia, 4:30–40.

Woods S.E. (1994) *Primary prevention of coronary heart disease in women.* Arch. Fam. Med., 3:361–364.

Wooster R., Neuhausen S.L., Mangion J., et al (1994) *Localization of a breast cancer susceptibility gene(2) to chromosome 13q12–13.* Science, 94:2088–2090.

Wrye H. (1979) *The crisis of cancer: intervention perspectives. Journal Writing With Women With Breast Cancer.* Read before the 87th American Psychiatric Association Convention, September, New York.

Yamanaka S., Katsuyuki M., Yukimura T., Okumura M., Yamamoto K. (1992) *Putative mechanism of hypotensive action of platelet-activating factor in dogs.* Circulation Res., 70:893–901.

Yang X., Stedra J., Cerny J. (1994) *Repertoire diversity of antibody response to bacterial antigens in aged mice. IV: Study of V_H and V_L gene utilization in splenic antibody foci by in situ hybridization.* J. Immunol., 152:2214–2221.

Yin X.M., Oltval Z.N., Korsmeyer S.J. (1994) *BH1 and BH2 domains of Bcl-2 are required for inhibition of apoptosis and heterodimerization with Bax.* Nature, 369:321–323.

Ylikorkala O., Orpana A., Puolakka J., Pyorala T., Viinikka L. (1995) *Postmenopausal hormonal replacement decreases plasma levels of endothelin-1.* J. Clin. Endocrinol. Metab., 80:3384–3386.

Young E., Haskett R., Murphy-Weinberg V., Watson S., Akil H. (1991) *Loss of glucocorticoid fast feedback in depression.* Arch. Gen. Psychiatry, 48:693–699.

Yu B.P., Lee D.W., Marler C.G., Choi J.H. (1990) *Mechanism of food restriction: protection of cellular homeostasis.* Proc. Soc. Exp. Biol Med., 193:13–15.

Yu C.E., Oshima J., Fu Y.H., Wijsman E.M., Hisama F., Alisch R., Matthews S., Nakura J., Miki T., Ouais S., Martin G.M., Mulligan J., Schellenberg G.D. (1996) *Positional cloning of the Werner's syndrome gene.* Science, 272:258–262.

Yuan J. and Horvitz H.R. (1992) *The Caenorhabditis elegans cell death gene ced-4 encodes a novel protein and is expressed during the period of extensive programmed cell death.* Development, 116(2):309–320.

Yuan J., Shaham S., Ledoux S., Ellis H.M., Horvitz H.R. (1993) *The C. elegans cell death gene ced-3 encodes a protein similar to mammalian interleukin-1β-converting enzyme.* Cell, 75:641–652.

Zachariae R., Bjerring P., Arendt-Nielsen L. (1989) *Modulation of Type I immediate and Type IV delayed immunoreactivity using direct suggestion and guided imagery during hypnosis.* Allergy, 44:537–542.

Zec R.F. (1995) *The neuropsychology of aging.* Exper. Gerontology, 30:431–442.

Ziegler R. (1989) *A review of epidemiologic evidence that carotenoids reduce the risk of cancer.* J. Nutr., 119:116–122.

Subject Index

Acamprosate (see also alcohol) 185, 186
Acetylcarnitine (see also alternative medicine) 201
Acetylsalycilic acid (see aspirin)
Acupuncture 205
Adenosine deaminase (ADA) 211–214
Adenovirus 213, 221
Adenylate cyclase 58
Adrenal gland 35, 38–39, 61, 196, 197, 215
Adrenocorticotropic hormone (ACTH) 35, 37, 65, 75
Advanced Glycation End products (AGE) 111
Aflatoxin 10, 101
Aging and
  biological rhythms 64–65
  chromosomes 93–94
  cognition 57
  happiness 166–167
  hearing 58
  immune system 66–70
  memory 57
  physical exercise 130, 172–177
  sexuality 30–32, 171–172
  sleep 65
  taste and smell 58
  vision 58
Alcohol 130, 144, 183–187
Allele 96
Alliance for Aging Research 137
Altered proteins theory (see theories of aging)
Alternative medicine 167, 204–206
Alzheimer's disease (see senile dementia of the Alzheimer's type, SDAT)
Amplification of triplet nucleotides (see also fragile X syndrome, spinal and bulbar muscular atrophy, muscular dystrophy, Huntington's disease, spinocerebellar ataxia type 1) 102–106
Amyloid
  deposition 40
  amyloid precursor protein (APP) 43
  β-peptide amyloid protein 43, 49
  βA4 peptide fragment 44
Amyotrophic Lateral Sclerosis (ALS) 118, 159, 219, 221, 224
  Familial ALS (FALS) 118
Androgen receptor (AR) 102
Angiotensin-converting enzyme (ACE) 91
Ankylosing Spondylitis (AS) 46
Antisense oligonucleotide 225
Antibiotics 11
Antioxidants (see also free radicals) 114, 130, 190
  Antioxidant capacity 140
Antiplatelet therapy 166
  Antiplatelet Trialists'Collaboration 166
Apnea (central, obstructive apnea) 138
ApoE proteins, Apo e alleles 49, 91–92, 139
Apoptosis (see also programmed cell death) 101, 121–123, 227
Arachidonic acid 118, 156
Arginine (see also human growth hormone) 200, 201
Ascorbic acid (see vitamin C)
Aspirin (acetylsalycilic acid) 166–167
Atherosclerosis 26, 96, 148, 166
Autonomic nervous system (ANS) 72
Average mortality rate (AMR) 8
Axonal regeneration 220

B-Lymphocytes (see also immune system) 66–70
Baltimore Longitudinal Study on Aging 51, 173
BCE-100 202
Bcl-2 protein (see also long life genes) 122
Benzodiazepins (see also stress) 188
The benzodiazepine/GABAergic hypothesis 204
Beriberi (see also vitamin B1) 146
Bioactive lipids 226–227
Bioceramic bone 208
Biological age 130
  measuring biological age 139–140
Biological rhythms (see aging)
Biotechnology companies 123
Biotin (see vitamin B7) 147
Biphosphonate 153–154
Blood-brain barrier 63
Blood pressure 157
Bone renewal 55
Brain-derived neurotrophic factor (BDNF) 219

Buformin (see also alternative medicine) 202

Caenorhabditis elegans genes (*ced 3,4, 9, age-1, fer 15, clock,1,2,3*) 87–88
Calcitonin gene-related peptide (CGRP) 75
Calcium 58, 151–152
Calcium channels 99
Cancer and
  diet 142
  prevention 163
  stress 33–34
  in adopted children 25
  with dominant inheritance 48
Cardiovascular functions 52
β-Carotene (see provitamin A)
The β-carotene and retinol efficacy trial (CARET) 160
Cartilage 208
Catalase (see also free radicals) 114
Cataracts 58, 111, 208
Catecholamines 35
Cellular replicative potential (see Proliferative Doubling Limit, PDL)
Centenarians 10, 11
Centrophenoxine (see also alternative medicine and BCE-100) 202
Ceramide 227
c-Fos (see also early genes) 71
Chelation therapy 205
Chinese medicine (see also acupuncture) 205
Cholesterol 34, 96, 144
Choline (see also alternative medicine) 201
Chromatin structure 110
Chromosome jumping, walking 98
Ciliary neurotrophic factor (CNTF) 219, 224
c-Jun (see also early genes) 71
Clock of aging theory (see theories of aging)
Clonal senescence 19
Clonidine 181, 200
Cobalamine (see vitamin B12)
Cockayne syndrome (see progerias)
Codergocrine (see also alternative medicine) 202
Coenzyme A 147
Coenzyme Q 137, 164
Collagen 54, 79, 99
Copper (see also minerals) 152
Coronary artery disease (see also heart disease, ischemic heart disease) 54, 142
Corticotropin-releasing factor (CRF) 35
Cortisol 34
Cross-linking 113
Cross-linking theory (see theories of aging)
Cyclic Adenosine Monophosphate (cAMP) 58, 70
cAMP-specific phosphodiesterase 89
Cyclosporin 206
Cystic Fibrosis (CF) 97–98
Cystic fibrosis Transmembrane Conductance Regulator (CFTR) protein 98
  CFTR gene 216, 218

Death genes 86–93
  in *Caenorhabditis Elegans* 87–88
  in *Drosophila Melanogaster* 88–90
  in humans 90–93
Death genes theory (see theories of aging)
Dementia 41
L-Deprenyl (selegiline; see also inhibitor of monoaminoxidase) 131, 201
Dexamethasone 39
Diabetes mellitus 108, 111
Diet and
  cancer, coronary heart disease, diabetes, longevity, osteoporosis 142–145
Diet restriction (see also food restriction) 111
Dietary guidelines 143–145
Diethylstilbestrol (DES) 50
Dihydroepiandrosterone (DHEA) 61, 137,196–198
Diseases and psychological factors 76
Dizygotic twins (see twins)
DNA damage-repair theory (see theories of aging)
DNA demethylation 21
DNA markers 41, 98
DNA methylation 110
DNA repair 41, 46
DNA replication 41
DNA responsive elements 36
L-Dopa 162
Dopamine-producing nerve cells 45
Down syndrome (see progerias)
Drosophila melanogaster 9, 16–17
  genes influencing life span (*drd; dunce, EF-1a*) 88–90
Dynemicin A 203–204
Dysdifferentiation theory (see theories of aging)
Dystrophin gene 215–216

Early genes, *c-Fos*, *c-Jun* 71
Elastin 54, 79
Endocrine stress response 38
Endocytosis 96
Endoplasmic reticulum 90
β-Endorphins 75

Enkephalin (see also acupuncture) 206
Epidermolysis Bullosa 99
Epigenetic factors 6
Epstein-Barr virus 50
Error-catastrophe theory (see theories of aging)
Essential fatty acids 156
Essential nutrients 19
Estrogens and 55, 61
breast cancer 194
17β-estradiol 55
Estrogen therapy (see also sex hormones) 193, 194
Evolutionary theory (see theories of aging)
Extrinsic causes of senescence 4

Face lifting (see plastic surgery) 206
Familial hypercholesterolemia 96
Fiber (see also diet) 144
Fibroblast growth factor (FGF) 219
Flumazenil 204
Fluoxetine (see Prozac)
FMR-1 gene (see Fragile X syndrome) 104
Folic acid 147, 149
Food and Drug Administration (FDA) 149, 150, 210
Food restriction 70, 130, 132–136, 140–141, 197
Forced expiratory capacity 53
Fragile X syndrome 102–103
(see also amplification of triplet nucleotides)
Framingham Heart Study 142
Free radicals 64, 112, 113–119
Free radical theory (see theories of aging)
Functional reserves 51, 73

Ganciclovir (see also gene therapy) 223
Garbage molecules 112
Gaucher disease 99
Gene therapy 136, 211–221
Genetic engineering techniques 210, 214–227
gene transfer techniques 136, 214–224
Genetic linkage studies 41
Genetic mutations 130
Genetic theories (see theories of aging)
Germline mutation (see mutations)
Gerontology 4
biological gerontology 4
Glaucoma 58
Glucocerebrosidase 99
Glucocorticoids 34, 35, 61, 189
glucocorticoid receptors type I, II 35
Glucose metabolism 52
glucose tolerance (see also physical exercise) 174–175
Glutathione and glutathione peroxidase (see also free radicals) 114, 158, 163–164
Glycation (DNA and protein) 111–112
Glycosylase 114
G-proteins (see GTP-binding proteins)
G-protein-coupled receptors 63
Grant Longitudinal Study 27
Growth hormone 75, 111, 137
human growth hormone (hGH) 198–200
growth hormone releasing hormone (GHRH) 199
GTP-binding proteins (G-proteins) 58, 70

Haemolymph 15
Happiness (see aging)
Hayflick's limit 19, 22–23
Heart disease (see also coronary artery disease; ischemic heart disease) 157
Helicase 41
Hemoglobin 118
Hepatitis B virus 50, 101
Hepatitis C virus 101
Herbal medicine 205
Herpes simplex virus 76, 213, 223
High-density lipoproteins (HDL) 91, 157, 172, 173, 185
HDL-cholesterol 175
Hippocampus 35–39
Homeostasis 44, 72–74, 116, 125, 126
Honeybees 13–16
Hormones 60
Human genetic diseases 95–110
Human leukocyte antigen (HLA) system (HLA-B27, HLA-DR1, HLA-DR 4) 45–46
Human papilloma virus 50
Huntington's disease (HD) (see also amplification of triplet nucleotides) 102, 105, 109
Hutchinson-Gilford syndrome (see progerias)
Hybrid cells 93
Hydrogen peroxide 113, 158
Hydroxyl radicals (see also free radicals) 113, 114, 190, 193
Hypercholesterolemia 216–217
Hypercortisolism 36
Hyperlipidemia (see also physical exercise) 175
Hypertension 174
Hypothalamo-pituitary-adrenal axis (HPA axis) 32, 34, 65, 80
Hypothalamus 60

Immune surveillance 63
Immune system 66–70, 148

humoral immunity 66
cellular immunity 66, 162
negative selection 67
positive selection 67
Immunocompetence 140
Immunological senescence 67
Immunological theory (see theories of aging)
Inhibitor of monoaminoxidase (MAO-I) 130, 201
Initial Mortality Rate (IMR) 8
Inositol triphosphates (IP_3) 48, 70
Insulin-like growth factor I (IGF-I or somatomedin C) 54, 176, 198, 219
Interferon-γ (Inf-γ) 66, 68
Interleukin 1 (IL-1) 66, 69, 156
Interleukin 2 (IL-2) 66, 68, 156, 191, 215
Interleukin 6 (IL-6) (see also bone renewal) 55, 156
Interleukin-1β Converting Enzyme (ICE) 122
Interspersed repeated sequences 21
Intrinsic causes of senescence 4
Iodine (see also minerals) 152
Iron (see also minerals) 117, 152–153
Ischemia 113
Ischemic heart disease (see also coronary artery disease, heart disease) 91, 163

Jogging 177
Juvenile arthritis 46
Juvenile hormone (JH) 15

Karyotypic analysis 94
Killer cells and stress (see also immune system) 34

Leber's hereditary optic neuropathy (LHON) 107
Leukotrienes 180
Li-Fraumeni syndrome (see also cancer) 101
Life extension
in non-human species 130–136
in humans 136–138
Life span (see also initial mortality rate, mortality
rate doubling time)
effect of castration on life span 60
(see also marsupial mouse, salmon)
life span of cells in culture 17–23
life span of humans 10–13
life span of invertebrates 13–17
maximum life span 7
mean life span 7, 12
potential life span 3
Life style 130
α-Linoleic acid 155
Lipid peroxidation 117
Lipofuscins or lipofuscin deposition 40, 112
Locus ceruleus 45
Long life genes 86–93
in *Caenorhabditis Elegans* 87–88
in *Drosophila Melanogaster* 88–90
in humans (see also *bcl-2* protein) 90–93
Lou Gehrig's disease (see amyotrophic lateral sclerosis) 118
Low-density lipoprotein (LDL) 49, 96
LDL-Cholesterol 116, 157, 158, 175
Low-density lipoprotein receptor (LDL Receptor) 96, 213
LDL-receptor gene 216
Luteinizing hormone (LH) 61
Luteinizing hormone releasing hormone (LHRH) 111
Lysosome 112

Macrophages 66
Macular degeneration 58, 145, 154
Magnesium (see also minerals)
Magnetic Resonance Imaging (MRI) 57
Major Histocompatibility Complex (MHC) 45
Malignant hyperthermia 99
Mammography (see also medical screening, preventive medicine) 139
Manganese (see also minerals)
Marriage 29–30
Marsupial mouse *Antechinus stuartii* 59
Masturbation 171
Maximum life span (see life span)
Mean life span (see life span)
Medical screening 138
Melatonin 62, 159, 164–165, 190–193
Melatonin receptors (ML–1, 2) 63
Menopause and 61, 122, 171
sexuality 32
sex hormones 193–196
Metabolic theories (see theories of aging)
Metformin (see also alternative medicine) 202
Methylation (5-methylcytosine) 110–111
Minerals and mineral supplements 130, 151–155, 182
calcium 151–152
copper 152
iodine 152
iron 152–153
magnesium 153
manganese 153
phosphorus (see also biphosphonate) 153–154
selenium 154, 163–164
zinc 154

Missense mutations (see mutations)
Mitochondrial DNA, genes 106–110
Mitochondrial encephalomyopathy, lactic acidosis, stroke-like symptoms (MELAS) 107
Mitochondrial monoaminoxidase 109
Molecular genetics 47
Monozygotic twins (see twins)
Mortality Rate Doubling Time (MRDT) 8
Mosaicism 43
Multidrug-resistance gene 214
Muscular dystrophy 116
Mutations 41
  germline mutations 47
  missense mutations 49
  nonsense mutations 99
  point mutations 101
  somatic mutations 47, 95–110
Myelin 111
Myoclonic epilepsy and ragged-red fiber disease (MERRF) 107
Myotonic dystrophy (DM) 102–105
  (see also amplification of triplet nucleotides)
Myotonic protein kinase 103
  MT-protein kinase gene (MT-PK gene) 104

Naltrexone (see also alcohol) 185
National Cancer Institute 143
National Research Council 143–144
Necrosis 121
Negative pleiotropy hypothesis 124
Nerve-growth factor 219, 222
Neuritic plaques 44
Neuro-endrocrine-immune system 72, 74–75
Neuroendocrine theory (see theories of aging)
Neurofibrillary tangles 44
Neurohormones 60
Neurological diseases (see also gene therapy) 218–225
Neuronal death 118
Neuronal graft 222–223
Neuropeptide Y (NPY) 75
Neurotransmitters 58
Neurotransmitter receptors 59
  β-adrenergic receptors 59
  glutamate receptors 59
Neurotrophic factors 219–221
Niacin (or nicotinamide) 146
Nicotine 178
  nicotine patch 181
3-Nitropropionic acid (3-NP) 109
Nitrosamine 162
Nonsense mutation (see mutations)
Nonsteroidal anti-inflammatory drug (NSAID) 166, 167

Nutritional status 140

Obesity 142
Oligodendrocyte 220
Omega-3, omega-6 fatty acids 155–157
Oncogene 100
Osteoblast 55
Osteoclast 55, 154
Osteoporosis (type I, II; see also physical exercise) 54, 156, 175–176
Oxidative phosphorylation (OXPHOS) 106–110
Oxygen peroxide (see also free radicals) 113, 114, 158

p53 gene (see also cancer, human genetic diseases) 100–102
Pantothenic acid (see vitamin B3)
Parkinson's disease (PD) 44, 201, 221–222
Parthenogenesis 14, 15
Pernicious anemia 148
Personality 25
  Type A 27–28
  Type B 27–28
  Type C 27–28
Phenformin (see also alternative medicine) 202
Phosphatase 154
Phosphatidylserine (see also alternative medicine) 201
Phosphodiesterase (PDE) 99
Phospholipase 114
  phospholipase A2 118
  phospholipase C 58
Phosphorus (see also minerals)
Physical exercise (see aging)
Pineal gland (see also melatonin) 64, 191
Plastic surgery 207
Platelet activating factor (PAF) 180
Polymorphism 17
Position Emission Tomography (PET) 57
Postmenopausal Estrogen/Progestin Intervention trial (PEPI trial) (see also sex hormones) 194
Postmitotic cells 22
Presenilin 1, 2 49
Preventive medicine 130
Progerias 6, 18, 80
  Cockayne syndrome 42
  Hutchinson-Gilford syndrome 41–42
  Down syndrome 43–44
  Werner syndrome 40–41
Progesterone (see also sex hormones) 194, 203
  progesterone receptors (see also RU486) 202
Programmed cell death (see also apoptosis) 101, 121–123

Prohibitin (and prohibitin gene) 92
Prolactin 75
Proliferation Doubling Limit (PDL) 17–23
Prostaglandin 180
Prostate cancer (see also testosterone, and thapsigargin) 196, 203
Prostate specific antigen testing 138
Prozac (fluoxetine) 188
Psychoneuroimmunology 75–76
Psychotherapy (see also stress) 187
Pulmonary functions 52
Pyridoxal (see vitamin B6)

Radiation 112, 119
Radical scavenger 159
Rate of senescence 7
RDS/peripherin (see retinitis pigmentosa)
Receptor-operated channels 59
Recombinant protein 212
Recommended Daily Intake (RDI) (see also minerals and vitamins) 149
Recommended Dietary Allowances (RDA) (see also minerals and vitamins) 143–144, 148, 149
Redwood conifers 23
Relaxation techniques (see also stress) 187–188
REM sleep 65
Renal functions 52
Replacement of body parts 207–209
Restriction enzyme 210, 212
Retina 63
Retinitis pigmentosa 99
Retinoblastoma 102
Retinoic Acid (retin-A) 146
Retinoid receptor 63
Retinoid (tretinoin, isotretinoin, fenretinide) 160
Retrovirus 213, 216
Rheumatoid arthritis 46
Riboflavin (see vitamin B2)
Ribozyme 226
RU 486 202–203

Salmon 60, 124
Saturated fats 144
Sea anemones 23
Selegiline (see L-deprenyl)
Selenium (see also antioxidants, minerals, smoking) 154, 163–164, 181–182
Selective death theory (see theories of aging)
Self-tolerance 67
Senile dementia of the Alzheimer's type (SDAT) 36, 43, 49–50, 73, 91–92, 201–202
Senile plaques 43
Sensitivity to UV 42
Sequoias 8
Serotoninergic antidepressants (see also smoking) 186
Sex hormones 193–196
Sex therapy 172
Sexuality (see aging)
Sleep (see aging, melatonin)
Smoking and 130, 177–183
  aspirin 167
  platelet activating factor (PAF) 180
  clonidine, vitamin A, C, E, β-carotene and selenium 181
Social integration 25
Social interactions 28–29,168–170
Somatic mutation (see mutations)
Somatic mutation theory (see theories of aging)
Somatomedin C (see insulin-like growth factor I)
Spinal and Bulbar Muscular Atrophy (SBMA) (see also amplification of triplet nucleotides) 102
Spinocerebellar Ataxia type 1 (see also amplification of triplet nucleotides) 105–106
Starch 144
Steroid-hormone-responsive elements 35
Stress 33–40, 187–189
Substantia nigra 45
Superantigens 67
Superoxide dismutase (SOD) (see also free radicals) 81, 113, 118, 131, 154, 158, 159, 201
Superoxide radicals (see also free radicals) 113, 114
Suprachiasmatic nucleus (SCN) 62
Systolic blood pressure 53

Tacrine (tetrahydroaminoacridine) 202
Telomerase 94–95
Telomeres (telomeric regions) 21, 94–95
Telomere theory (see theories of aging)
Testosterone 31, 137, 172, 195
Ticlopidine (see also antiplatelet therapy) 167
Thapsigargin 203
Theories of aging 77–84
  stochastic theories 77–80
    Altered proteins theory
    Cross-linking theory
    DNA damage-repair theory
    Error-catastrophe theory
    Somatic mutation theory
    Wear and tear theory
  programmed theories 80–82
    Clock of aging theory
    Death genes theory
    Dysdifferentiation theory
    Evolutionary theory

Free radical theory 80–82, 116
Genetic theories
Immunological theory
Metabolic theories
Neuroendocrine theory
Selective death theory
Telomere theory
Waste accumulation theory
Thiamine (see vitamin B1)
Thymidine kinase gene (see also gene therapy) 223
Thymus hormones 193
Thromboxane A2 180
Thymosins 75, 193
Thymus 66
thymus involution 67
T Lymphocytes 66–70
cytotoxic T cells 68
helper T cells 68
memory T cells 68
Tobacco (see smoking)
α-Tocopherol (see vitamin E; see also free radicals) 114, 115, 148–149
The α-tocopherol, β-carotene cancer prevention study group 160, 182
Tortoises, turtles 8, 23
Totipotent cells 5
Transplants of monkey testicles (see also replacement of body parts) 194
Triglyceride levels, metabolism 157, 173
Triplet repeats 102
Tumor necrosis factor (see also ceramide) 156, 215, 227
Tumor suppressor protein 100
Twins 8, 25–26

Uric acid (see also free radicals) 114
Unsaturated fats 155–158

Vasoactive intestinal peptide (VIP) 75
Vasopressin (VP) 35
Ventral tegmental ara (VTA; see also alcohol) 186
Vitaminotherapy 145
Vitamins 145–151
vitamin A, provitamin A (β-carotene; see also antioxidants, free radicals) 145–146, 159–161, 181–182
vitamin B1(thiamin) 146
vitamin B2 (riboflavin) 146
vitamin B3 (pantothenic acid) 147
vitamin B6 (pyridoxal) 147
vitamin B7 (biotin) 147
vitamin B12 (cobalamine) 147–148
vitamin C (ascorbic acid; see also antioxidants, free radicals) 114, 147–148, 161, 181–182
vitamin D 54, 148
1,25-dihydroxyvitamin D3 54, 176
vitamin E (α-tocopherol; see also antioxidants, free radicals) 114, 115, 148–149, 161, 181–182
vitamin K 149
Vitamin supplements 130, 150–151, 182
Vitellogenin 15
Voltage-gated channels 59

Waste accumulation theory (see theories of aging)
Wear and tear theory (see theories of aging)
Watanabe rabbit 96
Werner's syndrome (see progerias)
The Who/Monica Project 162

X-chromosome 26

Y-chromosome 26

Zinc (see minerals)